Magnetic, Dielectric, Electrical, Optical and Thermal Properties of Crystalline Materials

Magnetic, Dielectric, Electrical, Optical and Thermal Properties of Crystalline Materials

Editors

Khitouni Mohamed
Joan-Josep Suñol

Basel • Beijing • Wuhan • Barcelona • Belgrade • Novi Sad • Cluj • Manchester

Editors
Khitouni Mohamed
Qassim University
Buraidah
Saudi Arabia

Joan-Josep Suñol
University of Girona
Girona
Spain

Editorial Office
MDPI AG
Grosspeteranlage 5
4052 Basel, Switzerland

This is a reprint of articles from the Special Issue published online in the open access journal *Crystals* (ISSN 2073-4352) (available at: https://www.mdpi.com/journal/crystals/special_issues/516XCV9K09).

For citation purposes, cite each article independently as indicated on the article page online and as indicated below:

Lastname, A.A.; Lastname, B.B. Article Title. *Journal Name* **Year**, *Volume Number*, Page Range.

ISBN 978-3-7258-1673-6 (Hbk)
ISBN 978-3-7258-1674-3 (PDF)
doi.org/10.3390/books978-3-7258-1674-3

© 2024 by the authors. Articles in this book are Open Access and distributed under the Creative Commons Attribution (CC BY) license. The book as a whole is distributed by MDPI under the terms and conditions of the Creative Commons Attribution-NonCommercial-NoDerivs (CC BY-NC-ND) license.

Contents

About the Editors vii

Preface ix

Mohamed Khitouni and Joan-Josep Suñol
Magnetic, Dielectric, Electrical, Optical and Thermal Properties of Crystalline Materials
Reprinted from: *Crystals* **2024**, *14*, 252, doi:10.3390/cryst14030252 1

Hanen Rekik, Bechir Hammami, Mohamed Khitouni, Tarek Bachagha, Joan-Josep Suñol and Mahmoud Chemingui
Microstructure and Kinetics of Thermal Behavior of Martensitic Transformation in (Mn,Ni)Sn Heusler Alloy
Reprinted from: *Crystals* **2022**, *12*, 1644, doi:10.3390/cryst12111644 5

Abdulrahman Mallah, Mourad Debbichi, Mohamed Houcine Dhaou and Bilel Bellakhdhar
Structural, Mechanical, Electronic, Optical, and Thermodynamic Properties of New Oxychalcogenide $A_2O_2B_2Se_3$ (A = Sr, Ba; B = Bi, Sb) Compounds: A First-Principles Study
Reprinted from: *Crystals* **2023**, *13*, 122, doi:10.3390/cryst13010122 14

Abdulrahman Mallah, Mourad Debbichi, Mohamed Houcine Dhaou and Bilel Bellakhdhar
Structural, Mechanical, and Piezoelectric Properties of Janus Bidimensional Monolayers
Reprinted from: *Crystals* **2023**, *13*, 126, doi:10.3390/cryst13010126 24

Fanming Chen, Chengwen Liu, Lijie Zuo, Zhiyuan Wu, Yiqiang He, Kai Dong, et al.
Effect of Thermal Exposure on Mechanical Properties of Al-Si-Cu-Ni-Mg Aluminum Alloy
Reprinted from: *Crystals* **2023**, *13*, 236, doi:10.3390/cryst13020236 35

Soumaya Nasri, Mouhieddinne Guergueb, Jihed Brahmi, Youssef O. Al-Ghamdi, Frédérique Loiseau and Habib Nasri
Synthesis of a Novel Zinc(II) Porphyrin Complex, Halide Ion Reception, Catalytic Degradation of Dyes, and Optoelectronic Application
Reprinted from: *Crystals* **2023**, *13*, 238, doi:10.3390/cryst13020238 47

Shijian Tian, Libo Zhang, Yuan Liang, Ruikuan Xie, Li Han, Shiqi Lan, et al.
Room Temperature Ferromagnetic Properties of $Ga_{14}N_{16-n}Gd_2C_n$ Monolayers: A First Principle Study
Reprinted from: *Crystals* **2023**, *13*, 531, doi:10.3390/cryst13030531 64

Wael Ben Mbarek, Mohammed Al Harbi, Bechir Hammami, Mohamed Khitouni, Luisa Escoda and Joan-Josep Suñol
Nanostructured Mn–Ni Powders Produced by High-Energy Ball-Milling for Water Decontamination from RB5 Dye
Reprinted from: *Crystals* **2023**, *13*, 879, doi:10.3390/cryst13060879 74

Abdulrahman Mallah, Fatimah Al-Thuwayb, Mohamed Khitouni, Abdulrahman Alsawi, Joan-Josep Suñol, Jean-Marc Greneche and Maha M. Almoneef
Synthesis, Structural and Magnetic Characterization of Superparamagnetic $Ni_{0.3}Zn_{0.7}Cr_{2-x}Fe_xO_4$ Oxides Obtained by Sol-Gel Method
Reprinted from: *Crystals* **2023**, *13*, 894, doi:10.3390/cryst13060894 89

Muhammad Adil Mahmood, Rajwali Khan, Sattam Al Otaibi, Khaled Althubeiti,
Sherzod Shukhratovich Abdullaev, Nasir Rahman, et al.
The Effect of Transition Metals Co-Doped ZnO Nanotubes Based-Diluted Magnetic
Semiconductor for Spintronic Applications
Reprinted from: *Crystals* 2023, *13*, 984, doi:10.3390/cryst13070984 102

Reem Khalid Alharbi, Noura Kouki, Abdulrahman Mallah, Lotfi Beji, Haja Tar,
Azizah Algreiby, et al.
Processing and Investigation of $Cd_{0.5}Zn_{0.5}Fe_{2-x}Cr_xO_4$ ($0 \leq x \leq 2$) Spinel Nanoparticles
Reprinted from: *Crystals* 2023, *13*, 1121, doi:10.3390/cryst13071121 115

Maged S. Al-Fakeh, Munirah S. Alazmi and Yassine EL-Ghoul
Preparation and Characterization of Nano-Sized Co(II), Cu(II), Mn(II) and Ni(II) Coordination
PAA/Alginate Biopolymers and Study of Their Biological and Anticancer Performance
Reprinted from: *Crystals* 2023, *13*, 1148, doi:10.3390/cryst13071148 129

Dmitry A. Suslov, Petr M. Vetoshko, Alexei V. Mashirov, Sergei V. Taskaev,
Sergei N. Polulyakh, Vladimir N. Berzhansky and Vladimir G. Shavrov
Non-Collinear Phase in Rare-Earth Iron Garnet Films near the Compensation Temperature
Reprinted from: *Crystals* 2023, *13*, 1297, doi:10.3390/cryst13091297 149

Lingzhi Zhang, Yongkun Li, Rongfeng Zhou, Xiao Wang, Qiansi Wang, Lingzhi Xie, et al.
First-Principles Study of the Effect of Sn Content on the Structural, Elastic, and Electronic
Properties of Cu–Sn Alloys
Reprinted from: *Crystals* 2023, *13*, 1532, doi:10.3390/cryst13111532 160

Moufida Krimi, Mohammed H. Al-Harbi, Abdulelah H. Alsulami, Karim Karoui,
Mohamed Khitouni and Abdallah Ben Rhaiem
Optical, Dielectric, and Electrical Properties of Tungsten-Based Materials with the Formula
$Li_{(2-x)}Na_xWO_4$ (x = 0, 0.5, and 1.5)
Reprinted from: *Crystals* 2023, *13*, 1649, doi:10.3390/cryst13121649 181

Abdullah Hzzazi, Hind Alqurashi, Eesha Andharia, Bothina Hamad and M. O. Manasreh
Theoretical Investigations of the Structural, Dynamical, Electronic, Magnetic, and
Thermoelectric Properties of CoMRhSi (*M* = Cr, Mn) Quaternary Heusler Alloys
Reprinted from: *Crystals* 2024, *14*, 33, doi:10.3390/cryst14010033 195

About the Editors

Khitouni Mohamed

Prof. Khitouni Mohamed is a professor of Inorganic Chemistry at Qassim University (Saudi Arabia). He was previously a professor at the University of Sfax (Tunisia). 130 publications. More than 1550 citations. H index 22. Supervisor of 10 PhD theses. Research interest focused on mechanical alloying; soft magnetic materials, wastewater, magnetic shape memory, and high entropy alloys.

Joan-Josep Suñol

Prof. Joan-Josep Suñol is a professor of Applied Physics at the University of Girona Spain. More than 270 publications in scientific journals. More than 350 communications to congresses. Supervisor of 11 PhD theses. Research interest focused on mechanical alloying, nanocrystalline materials, soft magnetic materials, magnetic shape memory, and wastewater.

Preface

This reprint is a compilation of recent articles linked to the production and characterization (optical, electrical, dielectric, thermal, magnetic, thermodynamic, mechanical) of crystalline alloys and compounds.

As Guest editors of this Special issue, we are very happy with the final result and hope that the selected papers will be useful to researchers working with crystalline materials with improved functional properties. We would like to warmly thank the authors of the fifteen articles in this Special Issue for their contributions, and all of the reviewers for their efforts in ensuring high-quality publications. Finally, we will thank the mdpi editors for their continuous help.

Khitouni Mohamed and Joan-Josep Suñol
Editors

Editorial

Magnetic, Dielectric, Electrical, Optical and Thermal Properties of Crystalline Materials

Mohamed Khitouni [1,*] and Joan-Josep Suñol [2,*]

1 Department of Chemistry, College of Science, Qassim University, Buraidah 51452, Saudi Arabia
2 Department of Physics, University of Girona, Campus Montilivi s/n, 17003 Girona, Spain
* Correspondence: kh.mohamed@qu.edu.sa (M.K.); joanjosep.sunyol@udg.edu (J.-J.S.)

Citation: Khitouni, M.; Suñol, J.-J. Magnetic, Dielectric, Electrical, Optical and Thermal Properties of Crystalline Materials. *Crystals* **2024**, *14*, 252. https://doi.org/10.3390/cryst14030252

Received: 26 February 2024
Accepted: 1 March 2024
Published: 4 March 2024

Copyright: © 2024 by the authors. Licensee MDPI, Basel, Switzerland. This article is an open access article distributed under the terms and conditions of the Creative Commons Attribution (CC BY) license (https://creativecommons.org/licenses/by/4.0/).

This Special Issue entitled "Magnetic, Dielectric, Electrical, Optical and Thermal Properties of Crystalline Materials" is devoted to a general overview of the subject of crystalline materials and may extend to the nanocrystalline field. The articles published within it relate to the composition selection, the optimization of processing conditions and the structural and functional response. Thus, this Special Issue includes synthesis, characterization and applications studies. The properties, the structure and the mechanical, magnetic, dielectric, optical and thermal properties of these crystalline materials are the focus of interest [1–4].

As remarked in the Special Issue information, the keywords of the fifteen manuscripts published are a good indicator of the main fields of interest. The production techniques employed include rapid solidification, sol-gel, mechanical alloying, co-precipitation, liquid-phase epitaxy, click chemistry, and deposition techniques. The materials used are Heusler alloys, tungsten-based materials, copper–tin alloys, manganese–nickel alloys, gallium-based magnetic semiconductors, ferrimagnetic iron garnet films, cadmium base spinel, aluminum base alloys, oxychalcogenide compounds, porphyrin complexes, Janus bidimensional monolayers or cross-linked sodium alginate and polyacrylic acid biopolymers, and ferrite oxides. The structural information provided, based on the specific crystalline structures, includes crystallographic defects and solid-state structural transformations, such as the martensitic transformation and magnetic transformations, as well as ferromagnetic to paramagnetic transition. The mechanical properties analyzed include the elasticity, the strain, the creep mechanism, and the strengthening behavior. The studies of the electrical and dielectric properties are based in electrical conductivity or resistivity analysis, the electronic structure of the materials with ab initio theoretical studies, the study of the transport coefficients, orbital coupling, the semiconductor or piezoelectric behavior, and the study of the mechanisms of conduction. Magnetic analysis permits the determination of the ferromagnetic, ferrimagnetic, paramagnetic, or superparamagnetic behavior. Thermal analysis (characteristic temperatures, thermal treatments) is sometimes related to thermodynamics or kinetics, with parameters such as the activation energy. Optical analysis is linked to UV absorption, photoelectronic and photovoltaic analysis. Specific applications include dye degradation and antimicrobial, antioxidant, and anticancer activity.

Two articles are devoted to applications of Heusler alloys. Rekik et al. produce and analyze the structure and the thermal and kinetic behavior of the martensitic transformation of a Mn rich Mn–Ni–Sn compound. The Heusler alloys of the Mn–Ni–Sn family are candidates for implementation in magnetic refrigeration devices due to their magnetocaloric effect. Thus, the transition temperature control and the activation energy of the transformation are of interest. Hzzazi et al. produce quaternary Co–M–Rh–Si (M = Cr, Mn) Heusler alloys. These materials are candidates for spintronic and thermoelectric applications due to their high transport coefficients (electrical conductivity, Seebeck coefficient) and can provoke highly spin-polarized currents. The density functional theory (DFT) determines the total energy, phonon, and elastic behavior and complements the magnetic and functional response analysis.

One of the main fields of interest of crystalline materials is their environmental applications [5]. Two articles introduce an environmental application, namely the degradation of dyes. Nasri et al. synthesize porphyrin complexes. In this case, the degradation mechanism is a catalytic effect. The current–voltage response and the impedance spectroscopy measurements confirm that these complexes are candidates for optoelectronic applications such as in photovoltaic devices. Ben Mbarek et al. produce a binary alloy Mn–Al produced by mechanical alloying which provokes the degradation of azo dyes by means of a redox mechanism. UV absorption and infrared spectroscopy measurements confirm the discoloration of the dissolutions.

There are some studies based on the density functional theory [6]. This is an optimal approach for theoretical calculations of the electron density and energy distribution, the spin bad structures, the spin channels, and the spin-polarized density. It permits us to understand the interactions between electrons and ions. Zhang et al. apply DFT (considering the first-principles plane-wave pseudopotential method) to understand the effect of tin content in the structure, the elastic response, and the electronic properties of binary Cu–Sn alloys. The best content regarding structural stability is 3.125 at. % of Sn, whereas the best content for plasticity and elastic anisotropy is 6.25 at. % Sn. Tian et al. apply the DFT generalized gradient approximation and the Perdew–Nurke–Erzerhof formalism in diluted magnetic semiconductors (Ga–N-based ferromagnetic monolayers) to calculate the electronic and magnetic properties. Mallah et al. also produce, characterize, and analyze DFT monolayers. In this study, there are bidimensional Janus (two faces with two different local environments) monolayers. These monolayers consist of A_2XX', Si_2XX' and A_2PAs. These materials are candidates for application in piezoelectric nanodevices. The best results are observed for the materials containing chalcogenide atoms. In a second work, the same research team use the same methods to calculate the band structures and the density of states for several oxychalcogenides, including $A_2O_2B_2Se_3$ (A = Sr, Ba; B = Bi, Sb) compounds, and the optical and thermodynamic properties are analyzed. The reflectivity spectra confirm that these oxychalcogenides are optimal candidates for optical applications. Regarding mechanical behavior, some compounds are ductile, whereas $Ba_2O_2Sb_2Se_3$ is brittle.

Some articles characterize metallic alloys and compounds [7]. Chen et al. produce a cast heat-resistant aluminum alloy (Al–Si–Cu–Ni–Mg) by annealing the master alloy produced in an electric-smelling furnace. It is well known that the thermal treatment conditions (temperature, time, pressure, atmosphere) influence the microstructure and the mechanical response (tensile strength, elongation, creep). The authors analyze the main creep mechanism at low temperatures and low stress and detect that is a grain-boundary creep. Ferromagnetic rare-earth (gallium and gadolinium) iron garnet films were prepared via liquid-phase epitaxy. The authors analyze the structure and the magnetization near the compensation temperature, discovering the suppression effect of the non-colinear phase. The compensation temperature definition is as follows: the temperature at which the magnetizations of oppositely directed magnetic sub-lattices fully compensate one other. The authors used a magneto-optic approach to build the temperature–magnetic field (T-H) diagrams of the magnetic states in two- and three sub-lattice ferromagnetic structures.

Regarding properties, one article demonstrates the applicability of several cross-linked sodium alginate and polyacrylic acid biopolymers based on nanoscale natural polysacaccharides. This is a complex procedure, and nano-sized metallic polymers are prepared from the biopolymers. There are environmental and health applications due to the antimicrobial and anticancer activity (drug delivery) and the antioxidant and non-toxic (biocompatibility, biodegradability) behavior.

A large number of articles are devoted to oxides [8]. Mahmood et al. synthesized ZnO nanotube-based magnetic semiconductors and analyzed the effect of the addition of cobalt and gadolinium on the structure and the dielectric and magnetic properties. Both additions improve the electrical conductivity and the ferromagnetic behavior (exception

Gd doping >3 at. %) due to the magnetic impurities replaced in the ZnO oxide. These materials are candidates for spintronic electronic charge, quantum Hall effect, or resistive switching devices. Alharbi et al. synthetize complex ferrites with spinel structures as nanoparticles. The compositions are $Cd_{0.5}Zn_{0.5}Fe_{2-x}Cr_xO_4$ ($0 \leq x \leq 2$). The vibrations of the metal–oxygen bonds in the spinel structure provoke the detection of two absorption bands in infrared spectroscopy analysis. The nanocrystalline size decreases as the content of Cr increases. These materials are potential candidates for electronics (power transformers) and telecommunications applications. Krimi et al. produce lithium–sodium–tungsten-based oxides and analyze the optical, dielectric, and electric properties. The influence of the partial substitution of lithium by sodium is checked. The addition of sodium favors the formation of an orthorhombic crystallographic structure. Likewise, the increase in the disorder and the charge number provokes an increase in the conductivity. The last referred article in this Special Issue corresponds to nickel–zinc ferrites produced by means of the sol-gel method by Mallah and coworkers. Complex hyperfine structures are detected due to the different atomic neighbors having different magnetic moments. Concerning the magnetic behavior, the ferromagnetism is influenced by the presence or absence of Ni^{2+} or Zn^{2+} ions in tetrahedral locations.

Finally, we express our thanks to all of the contributors to this Special Issue, all of the reviewers for their help in improving the quality and soundness of the manuscripts, and especially the MDPI staff. The editors hope that the articles presented in this Special Issue are of interest to researchers producing crystalline alloys and compounds with magnetic, dielectric, electrical, optical, and thermal properties.

Author Contributions: Conceptualization, M.K. and J.-J.S.; writing—original draft preparation, M.K. and J.-J.S.; writing—review and editing, M.K. and J.-J.S. All authors have read and agreed to the published version of the manuscript.

Acknowledgments: The authors thank all of the researchers contributing to this Special Issue.

Conflicts of Interest: The authors declare no conflict of interest.

List of Contributions:

1. Rekik, H.; Hammami, B.; Khitouni, M.; Bachagha, T.; Suñol, J.-J.; Chemingui, M. Microstructure and Kinetics of Thermal Behavior of Martensitic Transformation in (Mn,Ni)Sn Heusler Alloy. Crystals 2022, 12, 1644. https://doi.org/10.3390/cryst12111644.
2. Hzzazi, A.; Alqurashi, H.; Andharia, E.; Hamad, B.; Manasreh, M.O. Theoretical Investigations of the Structural, Dynamical, Electronic, Magnetic, and Thermoelectric Properties of CoMRhSi (M = Cr, Mn) Quaternary Heusler Alloys. Crystals 2024, 14, 33. https://doi.org/10.3390/cryst14010033.
3. Nasri, S.; Guergueb, M.; Brahmi, J.; O. Al-Ghamdi, Y.; Loiseau, F.; Nasri, H. Synthesis of a Novel Zinc(II) Porphyrin Complex, Halide Ion Reception, Catalytic Degradation of Dyes, and Optoelectronic Application. Crystals 2023, 13, 238. https://doi.org/10.3390/cryst13020238.
4. Mbarek, W.B.; Al Harbi, M.; Hammami, B.; Khitouni, M.; Escoda, L.; Suñol, J.-J. Nanostructured Mn–Ni Powders Produced by High-Energy Ball-Milling for Water Decontamination from RB5 Dye. Crystals 2023, 13, 879. https://doi.org/10.3390/cryst13060879.
5. Zhang, L.; Li, Y.; Zhou, R.; Wang, X.; Wang, Q.; Xie, L.; Li, Z.; Xu, B. First-Principles Study of the Effect of Sn Content on the Structural, Elastic, and Electronic Properties of Cu–Sn Alloys. Crystals 2023, 13, 1532. https://doi.org/10.3390/cryst13111532.
6. Tian, S.; Zhang, L.; Liang, Y.; Xie, R.; Han, L.; Lan, S.; Lu, A.; Huang, Y.; Xing, H.; Chen, X. Room Temperature Ferromagnetic Properties of $Ga_{14}N_{16-n}Gd_2C_n$ Monolayers: A First Principle Study. Crystals 2023, 13, 531. https://doi.org/10.3390/cryst13030531.
7. Mallah, A.; Debbichi, M.; Dhaou, M.H.; Bellakhdhar, B. Structural, Mechanical, and Piezoelectric Properties of Janus Bidimensional Monolayers. Crystals 2023, 13, 126. https://doi.org/10.3390/cryst13010126.
8. Mallah, A.; Debbichi, M.; Dhaou, M.H.; Bellakhdhar, B. Structural, Mechanical, Electronic, Optical, and Thermodynamic Properties of New Oxychalcogenide $A_2O_2B_2Se_3$ (A = Sr, Ba; B = Bi, Sb) Compounds: A First-Principles Study. Crystals 2023, 13, 122. https://doi.org/10.3390/cryst13010122.

9. Chen, F.; Liu, C.; Zuo, L.; Wu, Z.; He, Y.; Dong, K.; Li, G.; He, W. Effect of Thermal Exposure on Mechanical Properties of Al-Si-Cu-Ni-Mg Aluminum Alloy. *Crystals* **2023**, *13*, 236. https://doi.org/10.3390/cryst13020236.
10. Suslov, D.A.; Vetoshko, P.M.; Mashirov, A.V.; Taskaev, S.V.; Polulyakh, S.N.; Berzhansky, V.N.; Shavrov, V.G. Non-Collinear Phase in Rare-Earth Iron Garnet Films near the Compensation Temperature. *Crystals* **2023**, *13*, 1297. https://doi.org/10.3390/cryst13091297.
11. Al-Fakeh, M.S.; Alazmi, M.S.; EL-Ghoul, Y. Preparation and Characterization of Nano-Sized Co(II), Cu(II), Mn(II) and Ni(II) Coordination PAA/Alginate Biopolymers and Study of Their Biological and Anticancer Performance. *Crystals* **2023**, *13*, 1148. https://doi.org/10.3390/cryst13071148.
12. Mahmood, M.A.; Khan, R.; Al Otaibi, S.; Althubeiti, K.; Abdullaev, S.S.; Rahman, N.; Sohail, M.; Iqbal, S. The Effect of Transition Metals Co-Doped ZnO Nanotubes Based-Diluted Magnetic Semiconductor for Spintronic Applications. *Crystals* **2023**, *13*, 984. https://doi.org/10.3390/cryst13070984.
13. Alharbi, R.K.; Kouki, N.; Mallah, A.; Beji, L.; Tar, H.; Algreiby, A.; Alnafisah, A.S.; Hcini, S. Processing and Investigation of $Cd_{0.5}Zn_{0.5}Fe_{2-x}Cr_xO_4$ ($0 \leq x \leq 2$) Spinel Nanoparticles. *Crystals* **2023**, *13*, 1121. https://doi.org/10.3390/cryst13071121.
14. Krimi, M.; Al-Harbi, M.H.; Alsulami, A.H.; Karoui, K.; Khitouni, M.; Ben Rhaiem, A. Optical, Dielectric, and Electrical Properties of Tungsten-Based Materials with the Formula $Li_{(2-x)}Na_xWO_4$ ($x = 0$, 0.5, and 1.5). *Crystals* **2023**, *13*, 1649. https://doi.org/10.3390/cryst13121649.
15. Mallah, A.; Al-Thuwayb, F.; Khitouni, M.; Alsawi, A.; Suñol, J.-J.; Greneche, J.-M.; Almoneef, M.M. Synthesis, Structural and Magnetic Characterization of Superparamagnetic $Ni_{0.3}Zn_{0.7}Cr_{2-x}Fe_xO_4$ Oxides Obtained by Sol-Gel Method. *Crystals* **2023**, *13*, 894. https://doi.org/10.3390/cryst13060894.

References

1. Radwan-Pragłowska, N.; Radwan-Pragłowska, J.; Łysiak, K.; Galek, T.; Janus, Ł.; Bogdał, D. Commercial-Scale Modification of NdFeB Magnets under Laser-Assisted Conditions. *Nanomaterials* **2024**, *14*, 431. [CrossRef]
2. Solozhenko, V.L.; Matar, S.F. High-Pressure Phases of Boron Pnictides BX (X = As, Sb, Bi) with Quartz Topology from First Principles. *Crystals* **2024**, *14*, 221. [CrossRef]
3. Hossen, M.F.; Shendokar, S.; Aravamudhan, S. Defects and Defect Engineering of Two-Dimensional Transition Metal Dichalcogenide (2D TMDC) Materials. *Nanomaterials* **2024**, *14*, 410. [CrossRef]
4. Puente-Córdova, J.G.; Luna-Martínez, J.F.; Mohamed-Noriega, N.; Miranda-Valdez, I.Y. Electrical Conduction Mechanisms in Ethyl Cellulose Films under DC and AC Electric Fields. *Polymers* **2024**, *16*, 628. [CrossRef]
5. Al-Kadhi, N.S.; Al-Senani, G.M.; Algethami, F.K.; Shah, R.K.; Saad, F.A.; Munshi, A.M.; Rehman, K.u.; Khezami, L.; Abdelrahman, E.A. Calcium Ferrite Nanoparticles: A Simple Synthesis Approach for the Effective Disposal of Congo Red Dye from Aqueous Environments. *Inorganics* **2024**, *12*, 69. [CrossRef]
6. Wang, N.; Wu, Y. First-Principles Investigation into the Interaction of H_2O with α-$CsPbI_3$ and the Intrinsic Defects within It. *Materials* **2024**, *17*, 1091. [CrossRef]
7. Hantoko, R.; Anggono, A.D.; Lubis, A.M.H.S.; Ngafwan. Microstructural Analysis of Friction Stir Welding Using CuZn and Zn Fillers on Aluminum 6061-T6: A Comparative Study. *Eng. Proc.* **2024**, *63*, 6. [CrossRef]
8. Chen, Y.; Li, A.; Jiang, S. Wettability and Mechanical Properties of Red Mud–Al_2O_3 Composites. *Materials* **2024**, *17*, 1095. [CrossRef]

Disclaimer/Publisher's Note: The statements, opinions and data contained in all publications are solely those of the individual author(s) and contributor(s) and not of MDPI and/or the editor(s). MDPI and/or the editor(s) disclaim responsibility for any injury to people or property resulting from any ideas, methods, instructions or products referred to in the content.

Article

Microstructure and Kinetics of Thermal Behavior of Martensitic Transformation in (Mn,Ni)Sn Heusler Alloy

Hanen Rekik [1], Bechir Hammami [2], Mohamed Khitouni [1,2,*], Tarek Bachagha [1], Joan-Josep Suñol [3] and Mahmoud Chemingui [1]

[1] Laboratory of Inorganic Chemistry (LR-17-ES-07), Faculty of Science, University of Sfax, Sfax 3018, Tunisia
[2] Department of Chemistry, College of Science, Qassim University, Buraidah 51452, Saudi Arabia
[3] Department of Physics, Campus Montilivi, University of Girona, 17071 Girona, Spain
* Correspondence: kh.mohamed@qu.edu.sa

Abstract: In this work, scanning electron microscopy, X-ray diffraction, and differential scanning calorimetry were used to investigate the solidification structure, thermal behavior, and kinetics of the martensitic transformations of the (Mn,Ni)Sn as-spun and annealed ribbons synthesized by melt-spinning. At room temperature, the as-spun and annealed (Mn,Ni)Sn ribbons exhibited a cubic single-phase Heusler $L2_1$ structure. The kinetics of the martensitic transformation (MT) was studied, together with their microstructure evolution and cooling rate dependence. The mechanism was also investigated. Additionally, a high dependence between the cooling rates and energy activation (Ea) was detected. A more detailed characterization of MT and account of thermodynamic parameters were examined after annealing.

Keywords: Heusler alloys; rapid solidification; martensitic transition; thermal analysis; energy activation; kinetics

Citation: Rekik, H.; Hammami, B.; Khitouni, M.; Bachagha, T.; Suñol, J.-J.; Chemingui, M. Microstructure and Kinetics of Thermal Behavior of Martensitic Transformation in (Mn,Ni)Sn Heusler Alloy. *Crystals* **2022**, *12*, 1644. https://doi.org/10.3390/cryst12111644

Academic Editor: Shouxun Ji

Received: 20 October 2022
Accepted: 13 November 2022
Published: 16 November 2022

Publisher's Note: MDPI stays neutral with regard to jurisdictional claims in published maps and institutional affiliations.

Copyright: © 2022 by the authors. Licensee MDPI, Basel, Switzerland. This article is an open access article distributed under the terms and conditions of the Creative Commons Attribution (CC BY) license (https://creativecommons.org/licenses/by/4.0/).

1. Introduction

The reversible first-order martensitic transition (MT) in Heusler Ni–Mn–X (X = Ga, Sn, In, and Sb) materials has garnered a lot of interest [1–4]. In these systems, MT occurs between a martensite phase with both a clearly reduced magnetic susceptibility and diversity of structural configurations and the ferromagnetic austenite phase, which has the cubic $L2_1$ structure. This latter structure, which can be 10M, 14M, 4O, or $L1_0$ structures, varies in composition and fabrication techniques [5]. Recent research demonstrated the efficiency of the melt-spinning process in producing highly textured, homogeneous, polycrystalline ribbons [6,7] with significantly improved magnetic characteristics [8]. As compared to those obtained through conventional casting, the rapid solidification of alloys via the melt-spinning technique can result in improved mechanical properties [9]. A nonequilibrium position of the atoms may also be caused by rapid solidification from the liquid phase. This makes it possible to change the atomic order, making research into melt-spun ribbon materials very important. Recently, some intriguing findings on the Ni–Mn–X alloy ribbons' physical characteristics were published. According to Hernando et al. [10], the martensite phase that forms in Ni–Mn–X ribbons has a different crystal structure than materials that are arc-melted in bulk. In comparison to the bulk alloy, the Heusler Ni–Mn–X ribbons' MT always starts at a lower temperature. For instance, Krenke et al. [11] reported that the martensitic transformation of $Ni_{50}Mn_{37}Sn_{13}$ ribbons occurred in the range of 300 K, but Santos et al. [12] observed that the change took place at around 212 K. The reduced degree of atomic order and internal tension introduced during rapid solidification, according to Feng et al. [13], explained why the transition temperature of the $Ni_{50}Mn_{28+x}Ga_{22-x}$ (x = 0, 1, 2, 3) ribbons is approximately 10 K lower than that of the corresponding bulk alloys. The alloy composition, preparation conditions, and external parameters can be used to control the MT temperatures in ternary

Ni–Mn–X (X = In, Sn, and Sb) alloys (magnetic field and hydrostatic pressure [14]). It has been suggested that a multitude of factors can affect the value of MT temperatures. By varying the composition or substituting $3d$ transition metals such as Cr, Fe, Co, and Cu, the valence electron concentration (e/a) and MT temperatures can be changed [15–18]. The variation in electron concentration and the Mn–Mn interatomic distance are also responsible for the compositional dependence of the phase-transition temperature [19,20]. However, superior physical properties with potential applications originating from first-order magnetic-induced martensite transformation, e.g., inverse magnetocaloric effect [21] and large magnetoresistance [22], have been reported to be comparable to bulk alloys. Additionally, it has been found that annealing for a short time [21] can significantly improve the physical properties of ribbons. Moreover, many fundamental aspects associated with the melt-spun rapid solidification process remain unclear, and research on Mn–Ni–X ribbons is still in its early stages; aspects under investigation include rapidly solidified phase competition and selection, microstructures, nonequilibrium thermodynamics, and kinetics of solid-phase transition. It is necessary to perform more studies to promote the potential applications of these materials.

At present, no studies on the effect of annealing on the MT of $Mn_{51}Ni_{39}Sn_{10}$ (at.%) alloy have been performed. Thus, the purpose of this study is to investigate the impact of annealing on the modification of the microstructure and the behavior of phase transformation temperatures of Heusler $Mn_{51}Ni_{39}Sn_{10}$ alloy.

2. Experimental Procedure

High-purity (99.99%) constituent metals were used to produce the as-cast ingots with the nominal composition $Mn_{51}Ni_{39}Sn_{10}$ utilizing the Bühler MAM-1 compact arc melting process. To ensure good initial homogeneity, these alloys experienced four melting cycles. With a circular nozzle of 0.5 mm and an argon overpressure, the samples were induction-melted in quartz crucibles before being ejected onto the polished surface of a copper wheel rotating at a linear speed of 48 ms^{-1} (Figure 1). The obtained as-quenched ribbons were flakes having dimensions of 1.2–2.0 mm in width and 4–12 mm in length. The ribbons were fixed in a quartz tube filled with argon gas, followed by annealing at 1273 K for 1 h, and then quenched in ice water. Following that, obtained samples were named as-spun and annealed. Scanning electron microscopy (SEM) was used on a ZEISS DSM-960A microscope fitted with an X-ray energy dispersive spectroscopy (EDS) microanalysis system to analyze the microstructure and elemental compositions. At room temperature (RT), the materials' structural characteristics were determined using X-ray diffractograms (XRD) on a Siemens D500 X-ray powder diffractometer with Cu–Kα radiation (λ = 1.5418 Å). Using the Maud Program, the sample structures were determined [23]. Calorimetry was used to verify the structural transformation of austenite to martensite. Under a nitrogen atmosphere, the cyclic tests (cooling–heating) were recorded at various rates of 10, 15, 20, 30, and 40 K/min. The Mettler-Toledo DSC30 device was used to perform DSC scans below RT while using a liquid nitrogen cooling system. The phase-transition activation energy was determined based on the DSC measurements after these measurements were used to analyze the typical temperatures of MT.

Figure 1. (a) Melt-spun chamber and (b) the obtained as-quenched ribbons.

3. Results and Discussion
3.1. SEM Analysis

The typical SEM images of the wheel surface of (Mn,Ni)Sn alloy are presented in Figure 2a. The austenite structure's granular microstructure is clearly visible on the wheel surface. This ribbon easily cleaves along this usual direction because it is mechanically weak and brittle. In Figure 2b, the alloy's free surface is also shown. These samples, which were obtained at high quenching rates, had a microstructure that was granular and completely crystalline. Around 1–2 μm was the value for the typical grain size. Furthermore, these typical grain size values are considerably lower than those seen in bulk alloys with coarse-grained microstructures and grain sizes ranging from 10 to 100 μm [1].

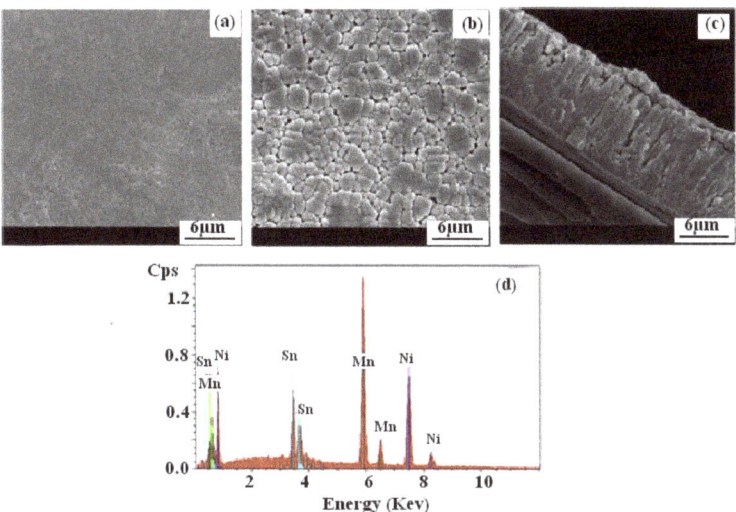

Figure 2. SEM images of wheel surface (**a**), free surface (**b**), and the cross-section microstructure (**c**) of the as-spun and annealed ribbons and the associated EDS analysis (**d**).

Figure 2c displays the cross-sections perpendicular to ribbon planes. The SEM images show that the samples were entirely crystalline. Additionally, a collinear granular columnar microstructure was visible. The ribbon was about 8 μm thick. With the longest axis aligned perpendicular to the ribbon plane, the thin layer of tiny equiaxed grains crystallized along the whole ribbon thickness. Figure 2d displays the results of the EDX analysis of the as-spun ribbon, confirming that mixed metallic elements were present. The nominal composition of the as-spun ribbon (51.2 at.% Ni; 39.3 at.% Mn; 9.5 at.% Sn) and the composition analysis results were in good accordance.

3.2. Structural Analysis

When choosing the parameters for a thermal study, it is frequently crucial to understand the crystal structure at RT [24]. The martensite–austenite transition must occur below RT in order to detect a cubic phase. On the other hand, if the phase is orthorhombic, monoclinic, or tetragonal, heating the alloy at normal temperature could produce the same transition. The XRD patterns of (Mn,Ni)Sn ribbons examined at RT are shown in Figure 3. Utilizing the Maud software program, miller indexes were assigned. After fitting, one can observe an austenite phase of the cubic $L2_1$ structure in both alloys. A crystalline structure cubic Heusler $L2_1$ structure, with lattice parameters of as-spun and annealed ribbons 5.995(1) and 5.990(1) Å, respectively, was confirmed by the reflections indexed as (311) and (331). On the basis of this XRD result, the martensite–austenite transition might be found using a DSC scan of the ribbon alloy that was cooled from RT. Recently, some interesting results on the physical characteristics of the Heusler Mn–Ni–Sn alloy were published. Coll et al. [24] reported that Mn–Ni–Sn alloys are completely single-phase at RT, with the cubic austenite phase thermally evolving into the structurally modulated orthorhombic martensite phase. However, the alloy's composition has a significant impact on the martensitic transition (MT). A single-phase $L2_1$ cubic austenite structure was seen at RT in a recent study using the as-spun $Ni_{50}Mn_{37}Sn_{6.5}In_{6.5}$ alloy [25], whereas the current phase in $Ni_{50}Mn_{42.5}Sn_{7.5}$ alloys is of the 14M monoclinic type [26].

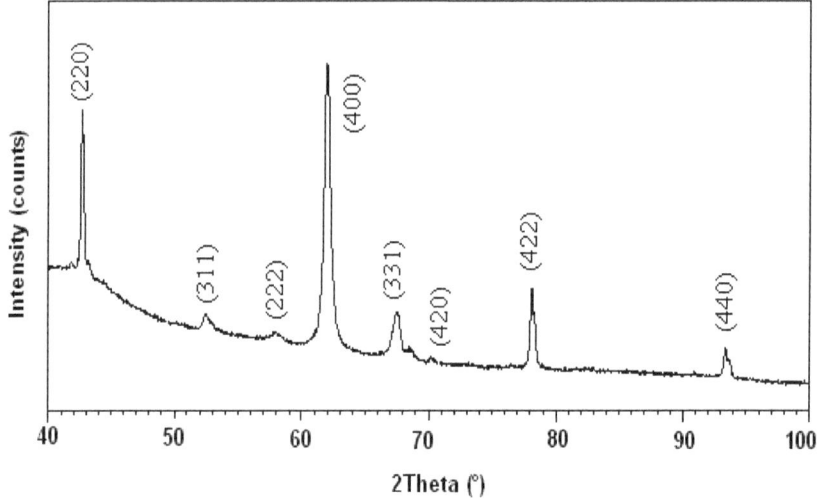

Figure 3. X-ray diffraction patterns for as-spun and annealed ribbons measured at RT.

3.3. Thermal Analysis

Figure 4 presents the thermal analysis curve of the as-spun and annealed ribbons using the DSC method. In the cooling and heating curves, it is observed that there are distinct exothermic and endothermic peaks that correlate to the martensitic transition. The temperatures for martensite start and finish and austenite start and finish are determined to be M_s = 300 K, M_f = 275 K, A_s = 293 K, and A_f = 310 K and M_s = K, M_f = K, A_s = K, and A_f = K. Sharmaa and Suresh [27] recently reported the characteristic martensitic transformation temperatures for ternary $Mn_{50}Ni_{40}Sn_{10}$ ribbons, which they determined to be M_s = 223 K, M_f = 182 K, A_s = 190 K, and A_f = 227 K. On the other hand, the transition temperatures for ternary $Mn_{50}Ni_{41}Sn_9$ ribbons were similarly established by Zhida Han et al. [28] as M_s = 300 K, M_f = 280 K, A_s = 290 K, and A_f = 320 K.

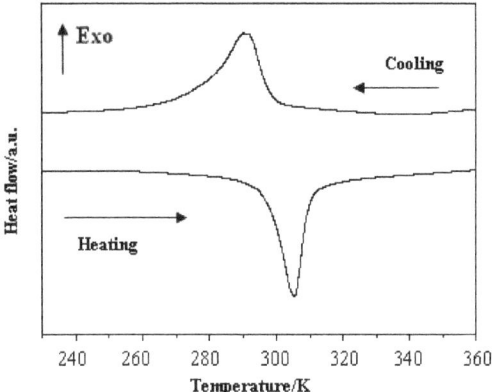

Figure 4. DSC cyclic scan for the as-spun and annealed ribbons obtained at a heating/cooling rate of 10 K min^{-1}. Arrows indicate cooling and heating.

The rise in elastic and surface energy during the development of martensite may be the cause of the measured hysteresis. Thus, supercooling is implied by the martensite's nucleation. The difference in temperatures at the peak sites, ΔT ($\Delta T = A_s - M_f$), is used to calculate the width of the hysteresis. For this ribbon, a value of about 14 K was obtained after cooling and heating. The intersection of a baseline and the tangents to each peak were used to identify the beginning and ending temperatures of the change. This made it very evident that the structural transition from the austenite to the martensite phase upon cooling and the opposite transition upon heating were both first-order processes. The martensite transformation temperature T_0 (the temperature at which the Gibbs energies of the martensitic and parent phases are related to the M_s and A_f parameters by the equation $T_0 = 1/2(M_s + A_f)$ [29]) can also be used to describe the transformation area. In Table 1, the calculated value of T_0 is displayed. As seen, the value of T_0 decreases as cooling rates rise. In general, the evolution of the electron to the atomic ratio (e/a), the Mn–Mn interatomic distance, and grain size can be linked to variations in transition temperatures. The electron concentration has a significant influence on the characteristic temperatures, including martensitic structural and transition temperatures. The number of 3d and 4s electrons in Mn, Ni, and Sn, as well as the sum of the 5s and 5p electrons in Mn, gives rise to the valence electron numbers of 7, 10, and 4, respectively. Additionally, it should be emphasized that an MT for Ni–Mn–Sn systems can only occur in the electron concentration range of 8.0–8.2 [30]. However, it should be highlighted that an MT can only happen for Mn–Ni–Sn in the electron concentration range of 7.9–8.2 [24]. The average valence electrons per atom (e/a) parameter was added to further define this alloy. For the alloy $Mn_{51}Ni_{39}Sn_{10}$, the calculated value of (e/a) is equal to 7.87. Other Ni–Mn–(In,Sn) Heusler alloys showed comparable results [8,11]. Indeed, Heusler alloys' structural transition temperatures can be changed by doping or modifying the composition [31]. According to Sanchez-Alorcos et al. [32], the valence electron concentration (e/a ratio) affects the martensitic transition temperature.

Table 1. Structural transition temperatures and the calculated values of T_0 recorded at different cooling rates.

Rates (K/min)	M_s (±1) (K)	M_f (±1) (K)	A_s (±1) (K)	A_f (±1) (K)	T_0 (±1) (K)
10	300	275	293	310	305
15	298.13	273.98	290.3	310	304
20	296.89	269.24	290.3	311	303.9
30	295.9	267.1	293.84	311	303.45
40	295.54	266.2	296.4	311	303.27

3.4. Kinetics

The dependence of the MT temperature interval was determined using calorimetric experiments with cooling rates ranging from 10 to 40 K min^{-1} (Figure 5). Prior to the measurements, careful calibrations with various rates were carried out. It is obvious that when cooling rates rise, the MT peak changes to lower temperatures. This effect is slightly more pronounced for greater cooling rates. It is clear that M_s and M_f both diminish as the cooling rate rises, but M_s exhibits a considerably larger reliance that may also result from the DSC sample's slow thermal conduction. Similar results were reported most recently by Zheng et al. and Bachaga et al. [33,34].

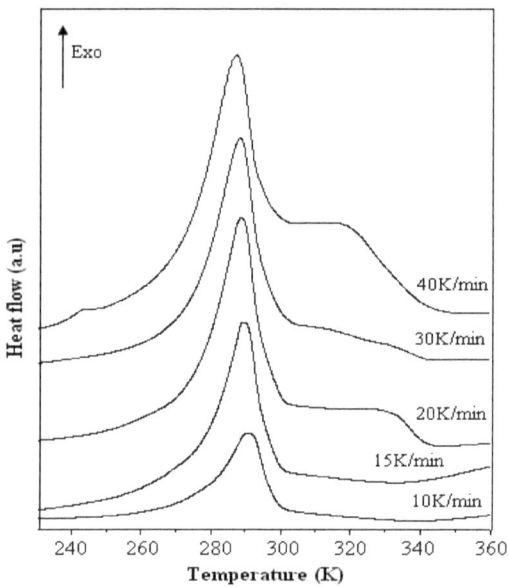

Figure 5. DSC charts of (Mn,Ni)Sn annealed ribbons recorded at different cooling rates of 10, 15, 20, 25, 30, and 40 K min^{-1}.

Based on the current result that the austenite phase's crystal structure is cubic L2$_1$, it is crucial to investigate the transformation's kinetics, specifically the activation energy, in order to gain a thorough understanding of the theoretical basis of MT. By investigating this, heat-treatment parameters may be adjusted. It is evident that when the cooling rate increases, the corresponding MT temperature decreases and the transition temperature range tends to widen. The Kissinger relation [35–37] can be used to obtain a transition parameter, as shown below:

$$\ln\left(\frac{B}{T^2}\right) = \frac{-Ea}{RT} + cte$$

where R is the gas constant, B is the cooling rate, T is the transition peak temperature, and c is the constant coefficient. In heating experiments, Ea is considered the average activation energy of the process. Nevertheless, in cooling experiments, this interpretation is doubtful. Tranchida et al. [38] consider Ea/R as a phenomenological parameter to obtain information about transformation tendencies. In this work, we applied Ea as a phenomenological parameter to check the cooling trend as well as to compare its values with the scientific literature. The slope of the plots of Ln (B/T^2) vs. $1/T$ shown in Figure 6 is used to calculate the value of Ea for annealed ribbons processed at various cooling rates using the approach described above. The relationship between Ln (B/T^2) and $1/T$ can be plotted linearly, and the calculated value of Ea is approximately 331.46 (0.02) kJ mol^{-1}. This value is comparable to that determined for the Ni$_{49}$Mn$_{39}$Sn$_{12}$ ribbons by Zheng et al. [33].

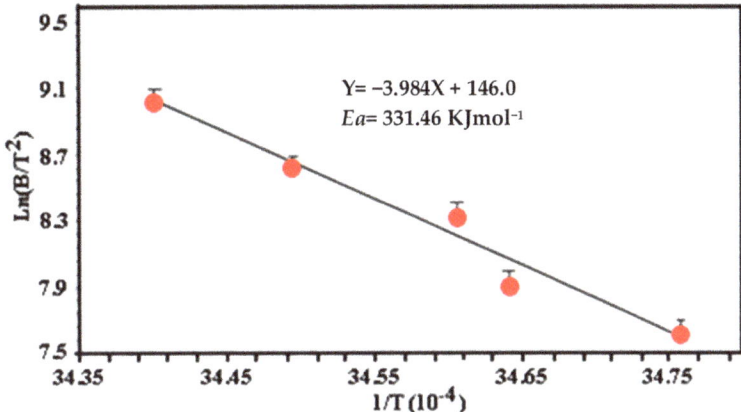

Figure 6. The Ln (B/T^2) vs. $1/T$ plot at different cooling rates of annealed ribbons.

Numerous studies have used Kissinger or Kissinger-like methods [39] to determine the MT and precipitation kinetics of various compositions with first-order transformation, such as Ti–Al–V [40], Cu–Ni–Al [41,42], and Fe–C(N) [43–45], as well as the ordering transition kinetics in Ni–Mn–Ga [36]. These studies were based on calorimetric results. The slope of linear curves of Ln (B/T^2) vs. $1/T$ appears to have a critical point, as reported by Fernandez et al. [46], who attributed this phenomenon to estimated error. For 50.8 at.% Ni–Ti SMA, Hsu et al. [46] discovered that martensitic substructure could change from coarse twins to fine twins or stacking faults with variations in cooling speeds varying from 0.5 to 25 K min^{-1}. Additionally, stacking faults in the plane (001) surface with a low cooling rate have an atomic displacement that is twice as large as stacking faults with a higher cooling rate. Due to the fact that twinning and atomic displacements in the MT process significantly increase at relatively low cooling rates, an increase in activation energy is required for the transition results, which is likely the cause of the rate dependency of activation energy in our work. The only factor affecting the phase interface velocity is supercooling, which causes the temperature rate to drop to extremely low levels [47]. Because slow cooling is responsible for a stronger chemical driving force and, thus, a lower activation energy, rapid cooling of the alloy would enable more supercooling in a shorter amount of time. The varied crystal structures related to varying cooling rates around the critical rate may also be a contributing factor, which leads to different activation energies being induced. It is likely that cooling the sample at a rate lower than the critical rate will make it easier for local or even long-term atomic diffusion, which could result in the appearance of a different variety of martensitic structure from that produced by cooling at a high rate or a non-martensite structure. However, it is not yet clear what exactly causes the rate of dependence. Researchers are still working to update some of these secrets, which demands more investigation.

4. Conclusions

In summary, the effect of annealing on the microstructure, structural, and MT of $Mn_{51}Ni_{39}Sn_{10}$ (at.%) shape memory alloy was studied. On the basis of the experimental results obtained, some conclusions can be cited.

- A cubic $L2_1$ structure was detected, at RT, for both alloys.
- The phase transformation temperatures increased remarkably after annealing.
- A high dependence between the cooling rates and the Ea was detected.

Author Contributions: Conceptualization, M.K. and J.-J.S.; formal analysis, H.R. and M.C.; data curation, B.H.; writing—original draft preparation, H.R., T.B. and B.H.; writing—review and editing; supervision, M.K. and J.-J.S. All authors have read and agreed to the published version of the manuscript.

Funding: This research received no external funding.

Institutional Review Board Statement: Not applicable.

Informed Consent Statement: Not applicable.

Data Availability Statement: Data can be requested from the authors.

Conflicts of Interest: The authors declare no conflict of interest.

References

1. Krenke, T.; Acet, M.; Wassermann, E.F.; Moya, X.; Mañosa, L.; Planes, A. Martensitic transitions and the nature of ferromagnetism in the austenitic and martensitic states of Ni-Mn-Sn alloys. *J. Phys. Rev. B* **2005**, *72*, 014412. [CrossRef]
2. Zhao, X.G.; Hsieh, C.C.; Lai, J.H.; Cheng, X.J.; Chang, W.C.; Cui, W.B. Effects of annealing on the magnetic entropy change and exchange bias behavior in melt-spun Ni-Mn-In ribbons. *J. Scr. Mater.* **2010**, *63*, 250–253. [CrossRef]
3. Zheng, H.X.; Wu, D.Z.; Xue, S.C.; Frenzel, J.; Eggeler, G.; Zhai, Q.J. Martensitic transformation in rapidly solidified Heusler Ni49Mn39Sn12 ribbons. *J. Acta Mater.* **2011**, *59*, 5692–5699. [CrossRef]
4. Raj Kumar, D.M.; Sridhara Rao, D.V.; Rama Rao, N.V.; Manivel Raja, M.; Singh, R.K.; Suresh, K.G. In-situ phase transformation studies of Ni48Mn39In13 melt-spun ribbons. *J. Intermet.* **2012**, *25*, 126–130. [CrossRef]
5. Krenke, T.; Moya, X.; Aksoy, S.; Acet, M.; Entel, P.; Manosa, L.; Planes, A.; Elerman, Y.; Yücel, A.; Wassermann, E.F. Electronic aspects of the martenistic transition in Ni-Mn based Heusler alloys. *J. Magn. Magn. Mater.* **2007**, *310*, 2788–2789. [CrossRef]
6. Sánchez Llamazares, J.L.; Sanchez, T.; Santos, J.D.; Pérez, M.J.; Sanchez, M.L.; Hernando, B. Martensitic phase transformation in rapidly solidified Mn50Ni40In10 alloy ribbons. *J. Appl. Phys. Lett.* **2008**, *92*, 012513. [CrossRef]
7. Xuan, H.; Xie, K.; Wang, D.; Han, Z.; Zhang, C.; Gu, B. Effect of annealing on the martensitic transformation and magnetocaloric effect in Ni44.1Mn44.2Sn11.7 ribbons. *J. Appl. Phys. Lett.* **2008**, *92*, 242506. [CrossRef]
8. Hernando, B.; Sánchez Llamazares, J.L.; Santos, J.D.; Escoda, L.; Suñol, J.J.; Varga, R.; Baldomir, D.; Serantes, D. Thermal and magnetic field-induced martensite-austenite transition in Ni50.3Mn35.3Sn14.4 ribbons. *J. Appl. Phys. Lett.* **2008**, *92*, 042504. [CrossRef]
9. Anantharman, T.R.; Suryanarayana, C. *Rapidly Solidified Metals: A Technological Overview*; Trans Tech Publications: Pfaffikon, Switzerland, 1987.
10. Hernando, B.; Sanchez Llamazares, J.L.; Prida, V.M.; Baldomir, D.; Serantes, D.; Ilyn, M. Magnetocaloric effect in preferentially textured Mn50Ni40In10 melt spun ribbons. *J. Appl. Phys. Lett.* **2009**, *94*, 222502. [CrossRef]
11. Krenke, T.; Duman, E.; Acet, M.; Wassermann, E.F.; Moya, X.; Mañosa, L. Inverse magnetocaloric effect in ferromagnetic Ni-Mn-Sn alloys. *J. Nat. Mater.* **2005**, *4*, 450–454. [CrossRef]
12. Santos, J.D.; Sanchez, T.; Alvarez, P.; Sanchez, M.L.; Sanchez Llamazares, J.L.; Hernando, B. Microstructure and magnetic properties of Ni50Mn37Sn13 Heusler alloy ribbons. *J. Appl. Phys.* **2008**, *103*, 07B326. [CrossRef]
13. Feng, Y.; Sui, J.H.; Chen, L.; Cai, W. Martensitic transformation behaviors and magnetic properties of Ni-Mn-Ga rapidly quenched ribbons. *J. Mater. Lett.* **2009**, *63*, 965–968. [CrossRef]
14. Manosa, L.; Moya, X.; Planes, A.; Gutfleisch, O.; Lyubina, J.; Barrio, M.; Tamarit, J.; Aksoy, S.; Krenke, T.; Acet, M. Effects of hydrostatic pressure on the magnetism and martensitic transition of Ni-Mn-In magnetic superelastic alloys. *J. Appl. Phys. Lett.* **2008**, *92*, 012515. [CrossRef]
15. Wang, D.; Zhang, C.; Han, Z.; Xuan, H.; Gu, B.; Du, Y. Large magnetic entropy changes and magnetoresistance in Ni45Mn42Cr2Sn11 alloy. *J. Appl. Phys.* **2008**, *103*, 033901. [CrossRef]
16. Krenke, T.; Duman, E.; Acet, M.; Moya, X.; Manosa, L.; Planes, A. Effect of Co and Fe on the inverse magnetocaloric properties of Ni-Mn-Sn. *J. Appl. Phys.* **2007**, *102*, 033903. [CrossRef]
17. Liu, J.; Scheerbaum, N.; Hinz, D.; Gutfleisch, O. Magnetostructural transformation in Ni-Mn-In-Co ribbons. *J. Appl. Phys. Lett.* **2008**, *92*, 162509. [CrossRef]
18. Wang, D.H.; Zhang, C.L.; Xuan, H.C.; Han, Z.D.; Zhang, J.R.; Tang, S.L.; Gu, B.X.; Du, Y.W. The study of low-field positive and negative magnetic entropy changes in Ni43Mn46−xCuxSn11 alloys. *J. Appl. Phys.* **2007**, *102*, 013909. [CrossRef]
19. Moya, X.; Manosa, L.; Planes, A.; Krenke, T.; Acet, M.; Wassermann, E.F. Lattice dynamics of Ni–Mn–Al Heusler alloys. *J. Mater. Sci. Eng. A* **2006**, *481–482*, 227–230. [CrossRef]
20. Yuhasz, W.M.; Schlagel, D.L.; Xing, Q.; McCallum, R.W.; Lograsso, T.A. Metastability of ferromagnetic Ni–Mn–Sn Heusler alloys. *J. Alloys Compd.* **2010**, *492*, 681–684. [CrossRef]
21. Xuan, H.C.; Deng, Y.; Wang, D.H.; Zhang, C.L.; Han, Z.D.; Du, Y.W. Effect of annealing on the martensitic transformation and magnetoresistance in Ni–Mn–Sn ribbons. *J. Phys. D Appl. Phys.* **2008**, *41*, 215002. [CrossRef]

22. Hernando, B.; Sanchez-Llamazares, J.L.; Santos, J.D.; Prida, V.M.; Baldomir, D.; Serantes, D.; Varga, R.; González, J. Magnetocaloric effect in melt spun Ni50.3Mn35.5Sn14.4 ribbons. *Appl. Phys. Lett.* **2008**, *92*, 132507. [CrossRef]
23. Lutterotti, L.; MAUD; CPD. (IUCr), No. 24. 2000. Available online: http://www.iucr.org/iucr-top/comm/cpd/Newsletters/Newsletter (accessed on 19 October 2022).
24. Coll, R.; Escoda, L.; Saurina, J.; Sanchez-Llamazares, J.L.; Hernando, B.; Sunol, J.J. Martensitic transformation in Mn–Ni–Sn Heusler alloys. *J. Therm. Anal. Calorim.* **2010**, *99*, 905–909. [CrossRef]
25. Rekik, H.; Chemingui, M.; Marzouki, A.; Bosh, E.; Escoda, L.; Sunol, J.J.; Khitouni, M. Structural and Magnetic Changes due to the Martensitic Transformation in Rapidly Solidified Ni50Mn37Sn6.5In6.5 Ribbons. *J. Supercond. Nov. Magn.* **2015**, *28*, 2165–2170. [CrossRef]
26. Bachaga, T.; Rekik, H.; Krifa, M.; Sunol, J.J.; Khitouni, M. Investigation of the enthalpy/entropy variation and structure of Ni–Mn–Sn (Co, In) melt-spun alloys. *J. Therm. Anal. Calorim.* **2016**, *126*, 1463–1468. [CrossRef]
27. Sharmaa, J.; Suresh, K.G. Investigation of multifunctional properties of Mn50Ni40−xCoxSn10 (x = 0–6) Heusler alloys. *J. Alloys Compd.* **2015**, *620*, 329–336. [CrossRef]
28. Han, Z.; Chen, X.; Zhang, Y.; Chen, J.; Qian, B.; Jiang, X.; Wang, D.; Du, Y. Martensitic transformation and magnetocaloric effect in Mn–Ni–Nb–Sn shape memory alloys: The effect of 4d transition-metal doping. *J. Alloys Compd.* **2012**, *515*, 114. [CrossRef]
29. Kaufman, L.; Hullert, M. Thermodynamics of martensite transformation. In *Martensite*; Olson, G.B., Owen, W.S., Eds.; ASM International: Cambridge, UK, 1992; pp. 41–58.
30. Schlagel, D.L.; Yuhasz, W.M.; Dennis, K.W.; McCallum, R.W.; Lograsso, T.A. Temperature dependence of the field-induced phase transformation in Ni50Mn37Sn13. *J. Scr. Mater.* **2008**, *59*, 1083. [CrossRef]
31. Planes, A.; Manosa, L.; Acet, M. Magnetocaloric effect and its relation to shape-memory properties in ferromagnetic Heusler alloys. *J. Phys. Condens. Matter.* **2009**, *21*, 233201. [CrossRef]
32. Sanchez-Alarcos, V.; Recarte, V.; Perez-Landazabal, J.I.; Gomez-Polo, C.; Rodriguez-Velamazan, J.A. Role of magnetism on the martensitic transformation in Ni–Mn-based magnetic shape memory alloys. *J. Acta. Mater.* **2012**, *60*, 459–468. [CrossRef]
33. Zheng, H.; Wang, W.; Wu, D.; Xue, S.; Zhai, Q.; Frenzel, J.; Luo, Z. Athermal nature of the martensitic transformation in Heusler alloy Ni–Mn–Sn. *J. Intermet.* **2013**, *36*, 90–95. [CrossRef]
34. Bachaga, T.; Zhang, J.; Ali, S.; Sunol, J.J.; Khitouni, M. Impact of annealing on martensitic transformation of Mn50Ni42.5Sn7.5 shape memory alloy. *Appl. Phys. A* **2019**, *125*, 146. [CrossRef]
35. Kissinger, H.E. Variation of Peak Temperature with Heating Rate in Differential Thermal Analysis. *J. Res. Natl. Bur. Stand.* **1956**, *57*, 217–221. [CrossRef]
36. Kostov, A.I.; Zivkovic, Z.D. Thermodilatometry investigation of the martensitic transformation in copper-based shape memory alloys. *J. Thermochim. Acta* **1997**, *291*, 51–57. [CrossRef]
37. Fernandez, J.; Benedetti, A.V.; Guilemany, J.M.; Zhang, X.M. Thermal stability of the martensitic transformation of Cu–Al–Ni–Mn–Ti. *J. Mater. Sci. Eng. A* **2006**, *723*, 438–440. [CrossRef]
38. Tranchida, D.; Gloger, D.; Gahleitner, M. A critical approach to the Kissinger analysis for studying non-isothermal crystallization of polymers. *J. Therm. Anal. Calorim.* **2017**, *129*, 1057–1064. [CrossRef]
39. Malinov, S.; Guo, Z.; Sha, W.; Wilson, A. Differential scanning calorimetry study and computer modeling of β ⇒ α phase transformation in a Ti-6Al-4V alloy. *J. Met. Mater. Trans. A* **2001**, *32*, 879. [CrossRef]
40. Lipe, T.; Morris, M.A. Effect of thermally activated mechanisms on the martensitic transformation of modified Cu-Al-Ni alloys. *J. Acta. Met. Mater.* **1995**, *43*, 1293–1303. [CrossRef]
41. Recarte, V.; Pérez-Landazabal, J.I.; Ibarra, A.; No, M.L.; Juan, J.S. High temperature β phase decomposition process in a Cu–Al–Ni shape memory alloy. *J. Mater. Sci. Eng. A* **2004**, *378*, 238–242. [CrossRef]
42. Liu, C.; Brakman, C.M.; Korevaar, B.M.; Mittemeijer, E.J. The tempering of iron- carbon martensite; dilatometric and calorimetric analysis. *J. Met. Trans. A* **1988**, *19*, 2415–2426.
43. Guo, Z.; Sha, W.; Li, D. Quantification of phase transformation kinetics of 18 wt.% Ni C250 maraging steel. *J. Mater Sci. Eng. A* **2004**, *373*, 10–20. [CrossRef]
44. Mittemeijer, E.J.; Van Gent, A.; Van Der Schaaf, P. Analysis of transformation kinetics by nonisothermal dilatometry. *J. Met. Mater. Trans. A* **1986**, *17*, 1441–1445. [CrossRef]
45. Vazquez, J.; Villares, P.; Jiménez-Garay, R. A theoretical method for deducing the evolution with time of the fraction crystallized and obtaining the kinetic parameters by DSC, using non-isothermal techniques. *J. Alloys Compd.* **1997**, *257*, 259–265. [CrossRef]
46. Hsu, T.Y. *Martensitic Transformation and Martensite*; Science Press: Beijing, China, 1999; Chapter 1.
47. Ostuka, K.; Ren, X.; Takeda, T. Experimental test for a possible isothermal martensitic transformation in a Ti-Ni alloy. *J. Scr. Mater.* **2001**, *45*, 145–152.

Article

Structural, Mechanical, Electronic, Optical, and Thermodynamic Properties of New Oxychalcogenide $A_2O_2B_2Se_3$ (A = Sr, Ba; B = Bi, Sb) Compounds: A First-Principles Study

Abdulrahman Mallah [1], Mourad Debbichi [2], Mohamed Houcine Dhaou [3,*] and Bilel Bellakhdhar [4]

1. Department of Chemistry, College of Science, Qassim University, Buraydah Almolaydah, Buraydah 51452, Saudi Arabia
2. Laboratoire de la matière condensée et nanosciences, Département de Physique, Faculté des Sciences de Monastir, Monastir 5019, Tunisia
3. Department of Physics, College of Science, Qassim University, Buraydah Almolaydah, Buraydah 51452, Saudi Arabia
4. Jeddah College of Technology, Jeddah 21361, Saudi Arabia
* Correspondence: m.dhaou@qu.edu.sa

Abstract: The structural, mechanical, electronic, and optical characteristics of Alkali chalcogenide and oxychalcogenides, i.e., $A_2O_2B_2Se_3$ (A = Sr, Ba; B = Bi, Sb), were investigated using density functional theory (DFT). After full relaxation, the obtained structural parameters are in good agreement with the experimental parameters. Furthermore, the calculated elastic stiffness C_{ij} shows that all of the studied compounds followed the mechanical stability criteria. Ductility for these compounds was analyzed by calculating Pugh's ratio; we classified the $Sr_2O_2Bi_2Se_3$, $Sr_2O_2Sb_2Se_3$, and $Ba_2O_2Bi_2Se_3$ as ductile, and the $Ba_2O_2Sb_2Se_3$ as brittle. The Debye temperature and acoustic velocity were estimated. In addition, electronic and chemical bonding properties were studied from the analysis of the band structure and density of state. The main features of the valence and conduction bands were analyzed from the partial density of states. Electronic band structures are mainly contributed to by Se-4p and Bi-6p/Sb-5p states. Direct band gaps are 0.90, 0.47, and 0.73 eV for $Sr_2O_2Bi_2Se_3$, $Sr_2O_2Sb_2Se_3$, and $Ba_2O_2Sb_2Se_3$, respectively. The $Ba_2O_2Bi_2Se_3$ compound has an indirect band gap of 1.12 eV. Furthermore, we interpreted and quantified the optical properties, including the dielectric function, absorption coefficient, optical reflectivity, and refractive index. From the reflectivity spectra, we can state that these compounds will be useful for optical applications.

Keywords: DFT; alkali chalcogenide; mechanical property; ab initio calculations; oxychalcogenide

Citation: Mallah, A.; Debbichi, M.; Dhaou, M.H.; Bellakhdhar, B. Structural, Mechanical, Electronic, Optical, and Thermodynamic Properties of New Oxychalcogenide $A_2O_2B_2Se_3$ (A = Sr, Ba; B = Bi, Sb) Compounds: A First-Principles Study. *Crystals* **2023**, *13*, 122. https://doi.org/10.3390/cryst13010122

Academic Editor: Sergio Brutti

Received: 12 December 2022
Revised: 24 December 2022
Accepted: 28 December 2022
Published: 10 January 2023

Copyright: © 2023 by the authors. Licensee MDPI, Basel, Switzerland. This article is an open access article distributed under the terms and conditions of the Creative Commons Attribution (CC BY) license (https://creativecommons.org/licenses/by/4.0/).

1. Introduction

Materials based on transition metal ions have attracted a lot of attention due to their unusual electrical, magnetic, and structural features [1,2]. Mixed-anion compounds containing oxide and chalcogenide anions have been extensively studied since the discovery of superconductivity in F-doped LaOFeAs at critical temperatures T∼26 K [3]. This result has accelerated studies on new layered materials. One of the most important compounds, layered oxychalcogenides, are mixed-anion compounds; chalcogenide and oxide anions are indirectly bounded via one or more cations, creating a stack of alternating oxide and chalcogenide layers [4,5].

The oxychalcogenides have attracted much interest owing to their rich and diverse chemistry, and are characterized by the coexistence of ionic oxide anions and more covalent chalcogenide anions. For example, oxide chalcogenides with the chemical formula $A_2MO_2X_2Ch_2$ (where A = Sr, Ba; X = Cu, Ag; Ch = S, Se and M = first-row transition metal) were first reported by Zhu et al. [6,7] and are isostructural with $Sr_2Mn_3Sb_2O_2$ structure. It was also found that it is possible to synthesize compounds with different thicknesses of the oxides and chalcogenide layers by varying the element ratios and heating conditions [8].

Currently, the most explored of such materials are bismuth chalcogenide compounds, particularly BiCh$_2$ (Ch = S, Se)-based compounds; many of them possess layered or tunneled structures. Bi$_2$O$_2$Ch (Ch = S, Se) have been exploited for their photocatalytic activities and thermoelectric properties [9,10]. The quaternary oxychalcogenides with 1111 stoichiometry BiMOCh (Ch = S, Se; M = Cu, Ag), due to their outstanding mechanical, chemical, electrical, and optical properties, have gained extensive attention as ionic and transparent conductors. The BiS$_2$-based superconductive compounds YO$_{1-x}$F$_x$BiS$_2$ (Y = La, Nd, Pr, and Ce), with the highest Tc of 10 K, have then been discovered and NdO$_{1-x}$F$_x$BiS$_2$ (x = 0.1–0.7) was found to have a maximum Tc of less than 5.6 K [11].

One such development is the discovery of new superconductivity compounds that exhibit zero resistance below a critical temperature (Tc). This progress requires the synthesis of new products with novel and fascinating properties. Recently, alkali chalcogenide and oxychalcogenides, i.e., A$_2$O$_2$B$_2$Se$_3$ (A = Sr, Ba; B = Bi, Sb) materials were synthesized by direct combination of SrO or BaO with Bi$_2$Se$_3$ or Sb$_2$Se$_3$ [12]. Insulating behavior has been revealed for all compounds. The crystal structure and chemical composition of these compounds were determined by X-ray powder diffraction (XRPD). Their structures consist of electronically active quasi-one-dimensional Sb-Se or Bi-Se ribbons isolated from one another by SrO units. Moreover, the structures of A$_2$O$_2$B$_2$Se$_3$ compounds share structural features with numerous bismuth and antimony chalcogenides, such as the ternary AB$_2$X$_4$ (A = Sr, Ba; B = Bi, Sb; X = S, Se). Since their structural and electronic similarity with the LnOBiX$_2$ (X = S, Se and Ln = La, Nd, Ce, Pr, Yb) superconductors [11,12], this new family provides a one-of-a-kind opportunity to consider the impacts of dimensionality on superconductivity.

The modeling of physical properties (by means of DFT techniques) has become a very useful tool for understanding the structural, electronic, mechanical, optical, and thermodynamic properties of various materials. In the present work, we would make a theoretical study of the electronic structure, elastic and optical properties of A$_2$O$_2$B$_2$Se$_3$ (A = Sr, Ba and B = Bi, Sb) materials using first-principle calculations. In Section 2, we provide the computational details of the calculations. Our results are presented in Section 3. A summary of the results is provided in Section 4.

2. Computational Methodology

The first-principles calculations were performed with the Quantum espresso code [13] using the generalized gradient approximation (GGA) in the form of Perdew–Burke–Ernzerhof (PBE) [14] to describe the exchange–correlation functional (XC). The plane waves of the electronic wave functions were expanded on the basis of a plane wave set with an energy cut-off of 40 Ry. the irreducible Brillouin zone was integrated with Monkhorst–Pack [15] $4 \times 4 \times 4$ k-point mesh. For the density of state (DOS) calculations, we used a mesh of $12 \times 12 \times 12$ with the tetrahedron method integration in order to obtain the high-quality charge density. An ultrasoft pseudo-potential [16] was adopted to describe the ionic cores and the valance electron interactions. In order to improve the convergence of the solution of the self-consistent Kohn–Sham equations, the energy levels were broadened by the Methfessel–Paxton [17] smearing with a Gaussian spreading $\sigma = 0.01$ Ry. The total energy convergence in the iterative solution of the Kohn–Sham equations [18] was set at 1.0×10^{-7} Ry to obtain well-converged ground state energy. All structures were fully relaxed by using the BFGS algorithm [19] with a threshold force of 10^{-3} Ry/Bohr.

3. Results and Discussion

3.1. Crystal Structure

A$_2$O$_2$B$_2$Se$_3$ (A = Sr, Ba; B = Bi, Sb) materials crystallized in monoclinic structures with the $P2_1C$ (No. 14) space group as displayed in Figure 1. The structures of these materials are described in detail by Jessica et al. [12], consisting of double-chain–quasi-one-dimensional ribbons of edge-linked BSe$_4$O square pyramids, connected to SrO fragments by the apical B-O bond.

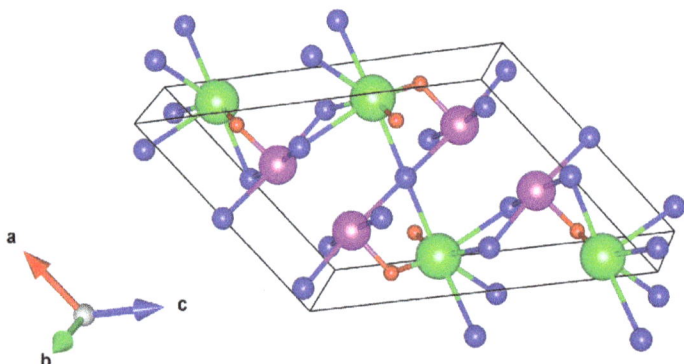

Figure 1. Crystal structure of the A$_2$O$_2$B$_2$Se$_3$ (A = Sr, Ba, and B = Bi, Sb) compounds, space group P2$_1$/c (No. 14). Green, violet, red, and blue represent A, B, O, and Se atoms.

Table 1 summarizes our calculated structural parameters (a, b, c, and angle β) and the formation energies for all compounds compared with the available experimental data. It appears from our results that optimized equilibrium parameters for Sr$_2$O$_2$Bi$_2$Se$_3$, Sr$_2$O$_2$Sb$_2$Se$_3$, and Ba$_2$O$_2$Bi$_2$Se$_3$ agree well with the experimental values reported in Reference [12]. For Ba$_2$O$_2$Sb$_2$Se$_3$, there are no corresponding experimental data for the lattice parameters. Moreover, by examining the structural parameters, the substitution of Sr by Ba and Sb by Bi tends to increase the lattice parameters, which may be interpreted in terms of the larger ionic radii of Ba^{2+} (0.134 Å) and Bi^{3+} (0.96 Å) compared to that of Sr^{2+} (1.12 Å) and Sb^{3+} (0.76 Å).

The relative stability of the A$_2$O$_2$B$_2$Se$_3$ (A = Sr, Ba; B = Bi, Sb) materials was examined by calculating their formation energies, E$_{form}$, defined as:

$$E_{form} = \frac{E(A_2O_2B_2Se_3) - 2E(A) - 2 \times \frac{1}{2}E(O_2) - 2E(B) - 3E(Se)}{9} \quad (1)$$

where E(A$_2$O$_2$B$_2$Se$_3$) is the total energy of one unit cell of the A$_2$O$_2$B$_2$Se$_3$ compound. E(A), E(B), and E(Se) are the energies of each atom for A, B, and Se in their stable bulk phases, respectively. E(O$_2$) is the total energy per O$_2$ molecule.

The calculated formation energies of the different compounds are regrouped in Table 1. As seen, the calculated values are all negatives, confirming the relative stability of all compounds.

Table 1. Calculated lattice parameters a, b, c (Å), β (°), formation energy, E$_{form}$ (eV) of the studied A$_2$O$_2$B$_2$Se$_3$ (A = Sr, Ba and B = Bi, Sb) materials compared with the available experimental data from Reference [12]. $\alpha = \gamma = 90°$.

	Sr$_2$O$_2$Bi$_2$Se$_3$	Sr$_2$O$_2$Sb$_2$Se$_3$	Ba$_2$O$_2$Bi$_2$Se$_3$	Ba$_2$O$_2$Sb$_2$Se$_3$
$a_{calc.}$	9.200	9.199	9.463	9.729
$a_{exp.}$	9.499	9.424	9.820	-
$b_{calc.}$	4.005	3.957	4.095	4.122
$b_{exp.}$	4.084	4.057	4.190	-
$c_{calc.}$	13.111	12.987	13.535	13.723
$c_{exp.}$	13.437	13.341	13.904	-
$\beta_{calc.}$	122.828	123.000	123.298	124.040
$\beta_{exp.}$	122.861	121.955	123.692	-
E$_{form}$	−2.145	−2.173	−2.139	−2.147

3.2. Mechanical Properties

The mechanical properties of these compounds are important for their potential technological and industrial applications. Elastic constants, needed to determine the mechanical stability of these compounds, were calculated. They give an overview of the mechanical and dynamical characteristics, especially the stability and stiffness of the present materials.

For monoclinic crystal [20,21], 13 independent elastic constants, i.e., C_{11}, $C_{12} = C_{21}$, C_{22}, $C_{13} = C_{31}$, $C_{23} = C_{32}$, C_{33}, C_{44}, $C_{15} = C_{51}$, $C_{25} = C_{52}$, $C_{35} = C_{53}$, C_{55}, $C_{46} = C_{64}$, and C_{66} remained with the explicit form of the tensor reduced to:

$$\begin{pmatrix} C_{11} & C_{12} & C_{13} & 0 & C_{15} & 0 \\ C_{12} & C_{22} & C_{23} & 0 & C_{25} & 0 \\ C_{13} & C_{23} & C_{33} & 0 & C_{35} & 0 \\ 0 & 0 & 0 & C_{44} & 0 & C_{46} \\ C_{15} & C_{25} & C_{35} & 0 & C_{55} & 0 \\ 0 & 0 & 0 & C_{46} & 0 & C_{66} \end{pmatrix}$$

In Table 2, we present the calculated thirteen independent elastic constants C_{ij} for the monoclinic lattice structures.

Table 2. The calculated elastic constants of the $A_2O_2B_2Se_3$ (A = Sr, Ba and B = Bi, Sb) compounds in the units of GPa.

	$Sr_2O_2Bi_2Se_3$	$Sr_2O_2Sb_2Se_3$	$Ba_2O_2Bi_2Se_3$	$Ba_2O_2Sb_2Se_3$
C_{11}	106.86	107.04	102.65	54.30
C_{22}	133.74	137.47	125.26	108.80
C_{33}	122.92	126.51	118.39	109.62
C_{44}	39.56	35.86	51.37	43.83
C_{55}	32.48	34.53	31.00	29.42
C_{66}	26.02	28.87	26.00	28.07
C_{12}	31.42	31.87	32.16	22.79
C_{13}	35.93	36.20	35.23	26.23
C_{23}	56.48	55.99	57.86	46.94
C_{15}	5.37	4.85	6.93	4.14
C_{25}	2.10	1.05	5.57	11.66
C_{35}	10.66	12.69	6.35	9.04
C_{46}	0.48	2.40	2.10	3.14

It is known that, for the monoclinic structure, mechanical stability requires elastic constants satisfying the following Born's criteria [22]:

$C_{ii} > 0$ $(i = 1, 6)$, $(C_{11} + C_{22} + C_{33} + 2(C_{12} + C_{13} + C_{23})) > 0$, $(C_{33}C_{55} - C_{35}^2) > 0$, $(C_{44}C_{66} - C_{46}^2) > 0$,
$(C_{22} + C_{33} - 2C_{23}) > 0$, $[C_{22}(C_{33}C_{55} - C_{35}^2) + 2C_{23}C_{25}C_{35} - C_{23}^2C_{55} - C_{25}^2C_{33}] > 0$,

and

$\{2[C_{15}C_{25}(C_{33}C_{12} - C_{13}C_{23}) + C_{15}C_{35}(C_{22}C_{13} + C_{12}C_{23}) + C_{25}C_{35}(C_{11}C_{23} - C_{12}C_{13})]$
$- [C_{15}^2(C_{22}C_{33} - C_{23}^2) + C_{25}^2(C_{11}C_{33} - C_{13}^2) + C_{35}^2(C_{11}C_{22} - C_{12}^2)] + gC_{55}\} > 0$,

where $g = C_{11}C_{22}C_{33} - C_{11}C_{23}^2 - C_{22}C_{13}^2 - C_{33}C_{12}^2 + 2C_{12}C_{13}C_{23}$. Therefore, these four compounds are structurally and mechanically stable.

The macroscopic mechanical properties of the different crystals, namely Young's modulus, bulk modulus, Poisson's ratio, and shear modulus, can be determined by the obtained elastic constants using the equations presented in Reference [23].

The bulk modulus (B) expresses the response of a material to a volume change of the hydrostatic pressure, whereas the shear modulus (G) reflects the resistance of a material to a shape change, and is deduced from elastic and compliance constants. Voigt [24] proposed to express the polycrystalline bulk modulus B_V and shear modulus G_V via the combinations of elastic constants C_{ij}. Similarly, Reuss and Angew [25] determined the bulk modulus B_R and shear modulus G_R expressions in terms of compliance constants S_{ij}. Moreover, Hill [26] calculated B and G from the average of the Voigt and Reuss bounds as:

$$B = \frac{B_V + B_R}{2}, G = \frac{G_V + G_R}{2}.$$

The polycrystalline Young's modulus (E) and Poisson's ratio (ν) are calculated using the relationships [26]

$$E = \frac{9BG}{3B + G}, \nu = \frac{3B - 2G}{6B + 2G}.$$

The values of these polycrystalline elastic constants are listed in Table 3. We can see from this table that all compounds exhibit good mechanical properties. The $Sr_2O_2Sb_2Se_3$ compound has larger bulk and shear modulus, indicating that it presents better mechanical properties compared to the others, which mainly attributes to the more stable Sr-O-Sb bond than Sr-O-Bi, Ba-O-Bi, and Ba-O-Sb. As also seen, the bulk modulus B is larger than the shear modulus G for all compounds, implying that G limits their stabilities.

Accordingly, we calculate "Pugh's criterion" ($D = B/G$), which is proposed to judge a metal's ductility and brittleness; the critical value that separates brittle and ductile materials is around 1.75 [27]. As shown in Table 3, Pugh's ratio is $D > 1.75$ for $Sr_2O_2Bi_2Se_3$, $Sr_2O_2Sb_2Se_3$, and $Ba_2O_2Bi_2Se_3$; thus, they behave in a ductile manner. The calculated D ratio for $Ba_2O_2Sb_2Se_3$ is less than 1.75, so it exhibits brittle behavior.

Table 3. The calculated polycrystalline elastic constants (B (GPa), G (GPa), and E (GPa)), and Poisson's ratio ν and Pugh's ratio (D) for $A_2O_2B_2Se_3$ (A = Sr, Ba and B = Bi, Sb) materials.

	B	G	E	ν	D	Type
$Sr_2O_2Bi_2Se_3$	66.523	34.755	88.801	0.277	1.914	ductile
$Sr_2O_2Sb_2Se_3$	67.333	35.566	90.723	0.275	1.893	ductile
$Ba_2O_2Bi_2Se_3$	64.817	35.201	89.417	0.270	1.841	ductile
$Ba_2O_2Sb_2Se_3$	46.913	30.764	75.737	0.230	1.525	brittle

3.3. Electronic Structure

Band structure calculations are required to provide more information for the fabrication and development of electronic and optoelectronic devices. To explore the electronic properties of the $A_2O_2B_2Se_3$ compounds, the energy band structures and the partial density of states (PDOS) of the four compounds are illustrated in Figure 2.

One can notice that band structures show similar shapes but with different gap values. It can be clearly seen that $Sr_2O_2Bi_2Se_3$, $Sr_2O_2Sb_2Se_3$, and $Ba_2O_2Sb_2Se_3$ show direct band gaps of 0.90, 0.47, and 0.73 eV, respectively, with the valence band maximum (VBM) and the conduction band minimum (CBM) lying at the Γ point. However, $Ba_2O_2Bi_2Se_3$ has an indirect band gap of 1.12 eV where the VBM is at the Γ point and the CBM is at the Y point. The difference between the experimental value and the theoretical one can be attributed to some aspects, such as the experimental environment and the exchange–correlation description considered in the present study.

From the partial density of states (PDOS), it can be observed that the VBMs of all compounds mainly consist of Se-$4p$ states, while the CBM is composed of Bi-$6p$/Sb-$5p$ states. We can see the contribution of the A states at the lowest-lying states. A strong hybridization between Sb and Se states was observed in the conduction band for the $Ba_2O_2Sb_2Se_3$ compound.

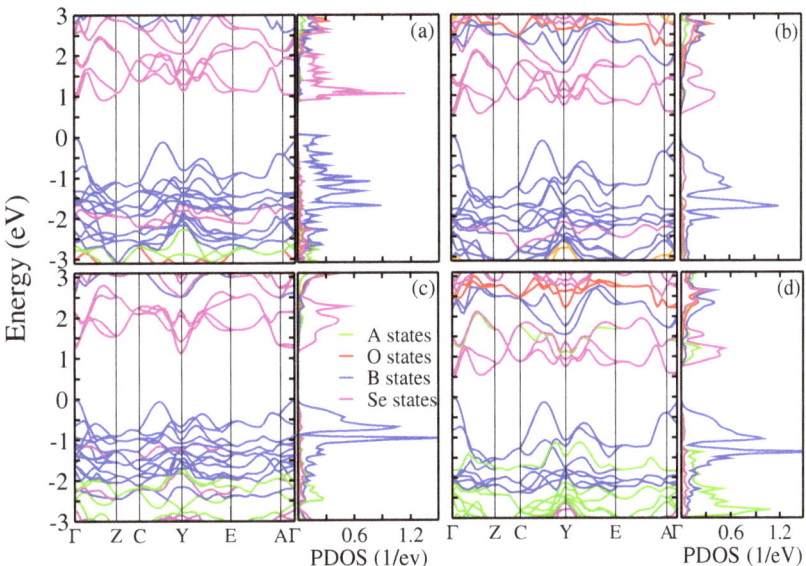

Figure 2. (color online) Calculated band structures and partial density of states for (**a**) $Sr_2O_2Bi_2Se_3$, (**b**) $Sr_2O_2Sb_2Se_3$, (**c**) $Ba_2O_2Bi_2Se_3$, and (**d**) $Ba_2O_2Sb_2Se_3$. A refers to Sr or Ba and B replaces Bi or Sb.

3.4. Optical Properties

Optical response functions of the materials are described by the complex dielectric function ($\varepsilon(\omega) = \varepsilon_1(\omega) + i\varepsilon_2(\omega)$). It gives the optical response of the medium at all photon energies $E = \hbar\omega$ and it is closely related to the electronic structure of the material. The imaginary part of the dielectric function $\varepsilon_2(\omega)$ is calculated using the matrix elements of occupied and unoccupied wave functions; it is given as follows [28]:

$$\varepsilon_2(\omega) = \frac{Ve^2}{2\pi m^2 \hbar^2 \omega^2} \int d^3k \sum_{n,n'} |<kn|p|kn'>|^2 f(kn) \times (1 - f(kn'))\delta(E_{kn} - E_{kn'} - \hbar\omega),$$

where n and n' are the initial and final states, respectively, p is the momentum operator $(\hbar/i)\partial/\partial x$, $|kn>$ is a crystal wave function, $f(kn)$ is the Fermi function for the nth state, and $\hbar\omega$ is the energy of the incident photon. The real part $\varepsilon_1(\omega)$ of the dielectric function can be obtained from the imaginary part $\varepsilon_2(\omega)$ through the Kramers–Kroning relations [29,30],

$$\varepsilon_1(\omega) = 1 + \frac{2}{\pi} P \int_0^\infty \frac{\omega' \varepsilon_2(\omega')}{\omega'^2 - \omega^2} d\omega'.$$

P implies the principal value of the integral.

The optical reflectivity spectra $R(\omega)$, the refractive index $n(\omega)$, and the absorption coefficient $\alpha(\omega)$ are derived from the dielectric function as follows [30–32]:

$$R(\omega) = \left| \frac{\sqrt{\varepsilon(\omega)} - 1}{\sqrt{\varepsilon(\omega)} + 1} \right|^2,$$

$$n(\omega) = \left[\frac{\varepsilon_1(\omega) + (\varepsilon_1^2(\omega) + \varepsilon_2^2(\omega))^{1/2}}{2} \right]^{1/2}$$

and

$$\alpha(\omega) = \left[2\omega^2 \left(\sqrt{\varepsilon_1^2(\omega) + \varepsilon_2^2(\omega)} - \varepsilon_2(\omega) \right) \right]^{1/2}$$

We note that this part of the Brillouin zone integration was performed using the tetrahedron method with denser k-points in the irreducible part of the Brillouin zone without broadening. Moreover, all the calculations are along parallel and perpendicular directions to the directions of propagation.

Since the dielectric function and the absorption coefficient play crucial roles in the characterization and optical applications of materials, we discuss the optical properties of the $A_2O_2B_2Se_3$ (A = Sr, Ba; B = Bi, Sb) in this section.

The calculated imaginary part shows three different components of dielectrics, which predict the anisotropic nature of the materials. In Figure 3a, we present only imaginary part of the dielectric function for E∥a. The peaks were caused by transitions from the upper valence band to the lower conduction band. These correspond to transitions from Se-$4p$ valence states to Bi-$6p$/Sb-$5p$ conduction band. It is noted that the ε_2 of all studied materials have similar shapes but different amplitudes because of the similar band structures. Moreover, the ε_2 of $Sr_2O_2Bi_2Se_3$ and $Ba_2O_2Bi_2Se_3$ were much higher than that of the $Sr_2O_2Sb_2Se_3$ and $Ba_2O_2Sb_2Se_3$, showing remarkably enhanced absorption of the photons.

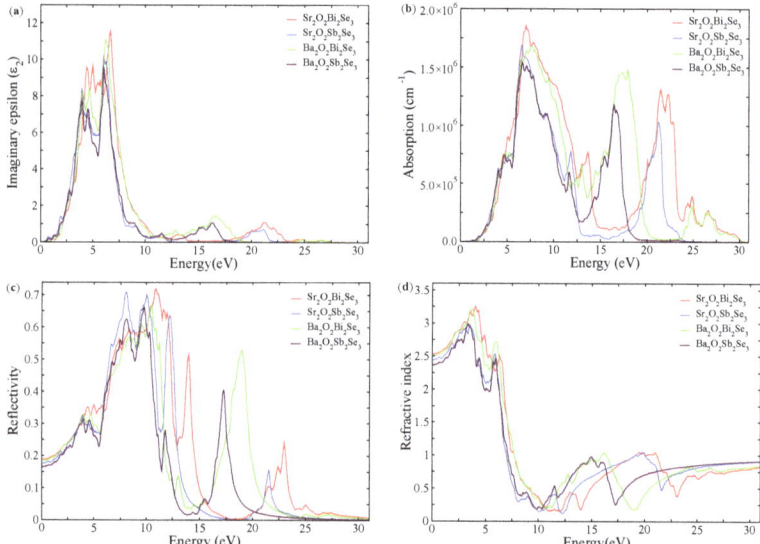

Figure 3. Optical spectrum of $A_2O_2B_2Se_3$ (A = Sr, Ba; B = Bi, Sb) compounds for E∥a. (**a**) Imaginary part of the dielectric function, (**b**) absorption coefficient, (**c**) reflection spectra, and (**d**) refractive index.

The absorption coefficient defines the region where a material absorbs energy. The energy dependence of the absorption spectrum of the present compounds for E∥a is given in Figure 3b. The absorption edge is away from 0 eV, which corresponds to the energy gaps. The absorption coefficient exhibits two prominent peaks indicating that they could absorb visible light. The first peaks of $Sr_2O_2Bi_2Se_3$ and $Ba_2O_2Bi_2Se_3$ are present in the same region at around 3 eV and extend to 12 eV, the second peaks are located at 16 eV and 24 eV for $Ba_2O_2Bi_2Se_3$ and $Sr_2O_2Bi_2Se_3$, respectively. For $Sr_2O_2Sb_2Se_3$ and $Ba_2O_2Sb_2Se_3$, the same trend is found but in a different region of adsorption. Notably, they exhibited two prominent peaks.

The reflection spectra of the studied compounds are presented in Figure 3c. We can see that the spectra are mainly in the areas between 5 eV and 25 eV, after that, the reflectivity falls sharply to low values (high transparency) for higher energy ranges. The peaks of the reflectivity correspond to the dielectric peaks, which is the macroscopic expression of the inter-band transition behavior. Several obvious peaks are identified, which correspond to the transition from the valence bands to the conduction bands located at 7, 10, 11, 14, 15, 18, 21, and 24 eV.

The calculated refractive index of the studied materials is shown in Figure 3d. The refractive index spectra of all studied materials show similar features, first reaching a maximum value of around 3.3 at around 4 eV, falling at intermediate energies, and then the curve vanishes at higher energies. Indeed, beyond certain energy, the considered material absorbs high-energy photons and cannot behave as transparent material. The extracted static refractive indices are 2.51, 2.48, 2.50, and 2.47 for $Sr_2O_2Bi_2Se_3$, $Ba_2O_2Bi_2Se_3$, $Sr_2O_2Sb_2Se_3$, and $Ba_2O_2Sb_2Se_3$, respectively.

3.5. Thermodynamic Properties

Thermodynamic properties of materials, such as specific heat, melting temperature, and thermal conductivity [33] are most suitably described in terms of the Debye temperature (θ_D). It is a fundamental and very important parameter that helps to obtain the thermodynamic properties and stability of crystals and, thus, design and develop new materials. At a low temperature, the average sound velocities (v_m) and the Debye temperature (θ) can be calculated from elastic constants taking into account the fact that the vibrational excitations arise solely from acoustic vibrations. We calculated the θ_D from the average sound velocity through the following equation v_m [34]:

$$\theta_D = \frac{h}{k_B}\left[\frac{3n}{4\pi V_a}\right]^{1/3},$$

where h is Planck's constant, k_B is Boltzmann's constant, V_a is the atomic volume, n is the number of atoms per formula unit, and v_m can be obtained from [34]:

$$v_m = \left[\frac{1}{3}\left(\frac{2}{v_t^3} + \frac{1}{v_l^3}\right)\right]^{-1/3},$$

where v_t is transverse velocity, v_l is longitudinal velocity. v_t and v_l are calculated from Navier's equation [35]:

$$v_t = \left[\frac{4G+3B}{3\rho}\right]^{1/2} \text{ and } v_l = \left[\frac{G}{\rho}\right]^{1/2}.$$

The calculated results of Debye temperatures, transverse, longitudinal, and average sound velocities are given in Table 4. $Sr_2O_2Sb_2Se_3$ has a higher Debye temperature; thus, it has a greater micro-hardness. According to References [36–38], a larger θ suggests a higher normal vibration, which is associated with better thermal conductivity. Meanwhile, the Debye temperature can characterize the strength of the covalent bond for the solid. Unfortunately, to the best of our knowledge, there are no other theoretical or measured data available in the literature to compare with our results. Nevertheless, we hope that our results can support further experiments.

Table 4. Calculated Debye temperature θ_D (K), average elastic wave velocities v_m (ms^{-1}), longitudinal v_l (ms^{-1}), and transverse v_t (ms^{-1}) for $A_2O_2B_2Se_3$ (A = Sr, Ba, and B = Bi, Sb) compounds.

	θ_D	v_l	v_t	v_m
$Sr_2O_2Bi_2Se_3$	260.55	4000.36	2219.89	2472.64
$Sr_2O_2Sb_2Se_3$	293.61	4463.99	2482.98	2765.16
$Ba_2O_2Bi_2Se_3$	250.38	3918.86	2189.84	2437.81
$Ba_2O_2Sb_2Se_3$	255.84	3833.41	2245.95	2490.15

4. Conclusions

In summary, first-principles calculations were performed on the $A_2O_2B_2Se_3$ (A = Sr, Ba; B = Bi, Sb) compounds, including structural parameters, band structure, the density of state, elastic constants, and optical parameters. Our study shows that all studied materials are mechanically stable. By analyzing the band structures and the DOS, we show that the $Sr_2O_2Bi_2Se_3$, $Sr_2O_2Sb_2Se_3$, and $Ba_2O_2Sb_2Se_3$ compounds have direct band gaps while $Ba_2O_2Bi_2Se_3$ possesses an indirect band gap. In addition, this paper also shows that the $Sr_2O_2Bi_2Se_3$, $Sr_2O_2Sb_2Se_3$, and $Ba_2O_2Bi_2Se_3$ are ductile while $Ba_2O_2Sb_2Se_3$ is brittle. Moreover, the calculated refractive indices n(0) for all compounds are in the range of 2.47–2.51, making them good candidates for use as waveguides. Finally, this new family offers a unique chance to research the impact of dimensionality on superconductivity due to the structural and electronic similarities between them and the $LnOBiX_2$ (X = S, Se, and Ln = La, Nd, Ce, Pr, Yb) superconductors.

Author Contributions: Conceptualization: A.M., M.D., M.H.D. and B.B.; data curation: A.M., M.D. and M.H.D.; formal analysis: M.D., M.H.D. and B.B.; Investigation: A.M., M.D. and M.H.D.; resources: M.D.; validation: A.M. and M.D.; writing—original draft: A.M., M.D. and M.H.D. All authors have read and agreed to the published version of the manuscript.

Funding: Qassim University, represented by the Deanship of Scientific Research, on the material support for this research under the number (10244-cos-2020-1-3-I).

Data Availability Statement: Data can be requested from the authors.

Acknowledgments: The authors gratefully acknowledge Qassim University, represented by the Deanship of Scientific Research, on the material support for this research under the number (10244-cos-2020-1-3-I) during the academic year 1442AH/2020AD.

Conflicts of Interest: The authors declare no conflict of interest.

References

1. Chen, H.; McClain, R.; Shen, J.; He, J.; Malliakas, C.D.; Spanopoulos, I.; Zhang, C.; Zhao, C.; Wang, Y.; Li, Q.; et al. Christopher Wolverton, and Mercouri G. Kanatzidis. *Inorg. Chem.* **2022**, *61*, 8240.
2. Njema, H.; Debbichi, M.; Boughzala, K.; Said, M.; Bouzouita, K. Structural, electronic and thermodynamic properties of britholites $Ca_{10-x}La_x (PO4)_{6-x}(SiO4)_xF_2$ ($0 \leq x \geq 6$): Experiment and theory *Mater. Res. Bull.* **2014**, *51*, 216. [CrossRef]
3. Kamihara, Y.; Watanabe, T.; Hirano, M.; Hosono, H. Iron-Based Layered Superconductor $La[O_{1-x}F_x]FeAs$ (x = 0.05–0.12) with T_c = 26 K. *J. Am. Chem. Soc.* **2008**, *130*, 3296–3297. [CrossRef] [PubMed]
4. Zagorac, D.; Doll, K.; Zagorac, J.; Jordanov, D.; Matović, B. Barium sulfide under pressure: discovery of metastable polymorphs and investigation of electronic properties on ab initio level. *Inorg. Chem.* **2017**, *56*, 10644–10654. [CrossRef]
5. Wang, R.; Zhao, Y.; Zhang, X.; Huang, F. Structural dimension modulation in a new oxysulfide system of $Ae_2Sb_2O_2S_3$ (Ae = Ca and Ba). *Inorg. Chem. Front.*, **2022**, *9*, 3552–3558. [CrossRef]
6. Zhu, W.J.; Hor, P.H.; Jacobson, A.J.; Crisci, G.; Albright, T.A.; Wang, S.-H.; Vogt, T. $A_2Cu_2CoO_2S_2$ (A= Sr, Ba), a novel example of a square-planar CoO_2 layer. *J. Am. Chem. Soc.* **1997**, *119*, 12398–12399. [CrossRef]
7. Zhu, W.J.; Hor, P.H. Unusual layered transition-metal oxysulfides: $Sr_2Cu_2MO_2S_2$ (M = Mn, Zn). *J. Solid State Chem.* **1997**, *130*, 319–321. [CrossRef]
8. Zhao, J.; Islam, S.M.; Kontsevoi, O.Y.; Tan, G.; Stoumpos, C.C.; Chen, H.; Li, R.K.; Kanatzidis, M.G. The Two-Dimensional $A_xCd_xBi_{4-x}Q_6$ (A = K, Rb, Cs; Q = S, Se): Direct Bandgap Semiconductors and Ion-Exchange Materials. *J. Am. Chem. Soc.* **2017**, *139*, 6978–6987. [CrossRef]
9. Pacquette, A.L.; Hagiwara, H.; Ishihara, T.; Gewirth, A.A. Fabrication of an oxysulfide of bismuth Bi_2O_2S and its photocatalytic activity in a Bi_2O_2S/In_2O_3 composite. *J. Photochem. Photobiol. A* **2014**, *277*, 27–36. [CrossRef]
10. Ruleova, P.; Drasar, C.; Lostak, P.; Li, C.-P.; Ballikaya, S.; Uher, C. Thermoelectric properties of Bi_2O_2Se. *Mater. Chem. Phys.* **2010**, *119*, 299–302. [CrossRef]
11. Yazici, D.; Huang, K.; White, B.D.; Chang, A.H.; Friedman, A.J.; Maple, M.B. Superconductivity of F-substituted Ln $OBiS_2$ (Ln = La, Ce, Pr, Nd, Yb) compounds. *Philos. Mag.* **2013**, *93*, 673–680. [CrossRef]
12. Panella, J.R.; Chamorro, J.; McQueen, T.M. Synthesis and Structure of Three New Oxychalcogenides: $A_2O_2Bi_2Se_3$ (A = Sr, Ba) and $Sr_2O_2Sb_2Se_3$. *Chem. Mater.* **2016**, *28*, 890–895. [CrossRef]
13. Giannozzi, P.; Baroni, S.; Bonini, N.; Calandra, M.; Car, R.; Cavazzoni, C.; Ceresoli, D.; Chiarotti, G.L.; Cococcioni, M.; Dabo, I.; et al. QUANTUM ESPRESSO: A modular and open-source software project for quantum simulations of materials. *J. Phys. Condens. Mat.* **2009**, *21*, 395502. [CrossRef]

14. Perdew, J.P.; Burke, K.; Ernzerhof, M. Generalized gradient approximation made simple. *Phys. Rev. Lett.* **1996**, *77*, 3865. [CrossRef]
15. Wisesa, P.; McGill, K.A.; Mueller, T. Efficient generation of generalized Monkhorst-Pack grids through the use of informatics. *Phys. Rev. B* **2016**, *93*, 155109. [CrossRef]
16. Yates, J.R.; Pickard, C.J.; Mauri, F. Calculation of NMR chemical shifts for extended systems using ultrasoft pseudopotentials. *Phys. Rev. B* **2007**, *76*, 024401. [CrossRef]
17. Methfessel, M.; Paxton, A.T. High-precision sampling for Brillouin-zone integration in metals. *Phys. Rev. B* **1989**, *40*, 3616. [CrossRef]
18. Baerends, E.J. Perspective on "Self-consistent equations including exchange and correlation effects". *Theor. Chem. Acc.* **2000**, *103*, 265–269. [CrossRef]
19. Mokhtari, A.; Ribeiro, A. Global Convergence of Online Limited Memory BFGS. *J. Mach. Learn. Res.* **2015**, *16*, 3151–3181.
20. Iuga, M.; Neumann, G.S.; Meinhardt, J. Ab-initio simulation of elastic constants for some ceramic materials. *Eur. Phys. J. B* **2007**, *58*, 127–133. [CrossRef]
21. Golesorkhtabar, R.; Pavone, P.; Spitaler, J.; Puschnig, P.; Draxl, C. ElaStic: A tool for calculating second-order elastic constants from first principles. *Comp. Phys. Commun.* **2013**, *184*, 1861–1873. [CrossRef]
22. Nye, J.F. *Physical Properties of Crystals*; Oxford University Press: New York, NY, USA, 1985.
23. Debbichi, M.; Alresheedi, F. First-principles calculations of mechanical, electronic and optical properties of a new imidooxonitridophosphate. *Chem. Phys.* **2020**, *538*, 110917. [CrossRef]
24. Voigt, W. *Lehrbuch der Kristallphys*; Teubner Press: Leipzig, Germany, 1928.
25. Reuss, A. Berechnung der Fließgrenze von Mischkristallen auf Grund der Plastizitätsbedigung für Einkristalle. *Z. angew. Math. Mech.* **1929**, *9*, 49–58. [CrossRef]
26. Hill, R. The elastic behaviour of a crystalline aggregate. *Proc. Phys. Soc. Lond.* **1952**, *65*, 349.
27. Pugh, S.F. XCII. Relations between the elastic moduli and the plastic properties of polycrystalline pure metals. *Lond. Edinb. Dublin Philosoph. Mag. J. Sci.* **1954**, *45*, 823–843. [CrossRef]
28. Saha, S.; Sinha, T.P.; Mookerjee, A. Electronic structure, chemical bonding, and optical properties of paraelectric $BaTiO_3$. *Phys. Rev. B* **2000**, *62*, 8828. [CrossRef]
29. Wooten, F. *Optical Properties of Solids*; Academic: New York, NY, USA, 1972.
30. Yu, Y.P.; Cardona, M. *Fundamentals of Semiconductors: Physics and Materials Properties*, 2nd ed.; Springer: Berlin/Heidelberg, Germany, 1999.
31. Ravindran, P.; Delin, A.; James, P.; Johansson, B.; Wills, J.M.; Ahuja, R.; Eriksson, O. Magnetic, optical, and magneto-optical properties of MnX (X = As, Sb, or Bi) from full-potential calculations. *Phys. Rev. B* **1999**, *59*, 15680. [CrossRef]
32. Fox, M. *Optical Properties of Solids*; Oxford University Press: New York, NY, USA, 2001.
33. Fan, Q.Y.; Wei, Q.; Yan, H.Y.; Zhang, M.G.; Zhang, Z.X.; Zhang, J.Q. Elastic and electronic properties of Pbca-BN: First-principles calculations. *Comput. Mater. Sci.* **2014**, *85*, 80–87. [CrossRef]
34. Anderson, O.L. A simplified method for calculating the Debye temperature from elastic constants. *J. Phys. Chem. Solids* **1963**, *24*, 909–917. [CrossRef]
35. Screiber, E.; Anderson, O.L.; Soga, N. *Elastic Constants and Their Measurement*; McGraw-Hill: New York, NY, USA, 1973.
36. Duan, Y.H.; Huang, B.; Sun, Y.; Peng, M.J.; Zhou, S.G. Stability, elastic properties and electronic structures of the stable Zr–Al intermetallic compounds: A first-principles investigation. *J. Alloys Compd.* **2014**, *590*, 50–60. [CrossRef]
37. Hu, W.C.; Liu, Y.; Li, D.J.; Zeng, X.Q.; Xu, C.S. First-principles study of structural and electronic properties of C14-type Laves phase Al_2Zr and Al_2Hf. *Comput. Mater. Sci.* **2014**, *83*, 27–34. [CrossRef]
38. Li, J.; Zhang, M.; Luo, X. Theoretical investigations on phase stability, elastic constants and electronic structures of $D0_{22-}$ and $L1_2$-Al_3Ti under high pressure. *J. Alloys Compd.* **2013**, *556*, 214–220.

Disclaimer/Publisher's Note: The statements, opinions and data contained in all publications are solely those of the individual author(s) and contributor(s) and not of MDPI and/or the editor(s). MDPI and/or the editor(s) disclaim responsibility for any injury to people or property resulting from any ideas, methods, instructions or products referred to in the content. [CrossRef]

Article

Structural, Mechanical, and Piezoelectric Properties of Janus Bidimensional Monolayers

Abdulrahman Mallah [1], Mourad Debbichi [2], Mohamed Houcine Dhaou [3,*] and Bilel Bellakhdhar [4]

[1] Department of Chemistry, College of Science, Qassim University, Buraydah 51452, Almolaydah, Saudi Arabia
[2] Laboratoire de la Matière Condensée et Nanosciences, Département de Physique, Faculté des Sciences de Monastir, Monastir 5019, Tunisia
[3] Department of Physics, College of Science, Qassim University, Buraydah 51452, Almolaydah, Saudi Arabia
[4] Jeddah College of Technology, Jeddah 17608, Saudi Arabia
* Correspondence: m.dhaou@qu.edu.sa

Abstract: In the present work, the noncentrosymmetric 2D ternary Janus monolayers $Al_2XX'(X/X' = S,$ Se, Te and O), $Si_2XX'(X/X' = P, As, Sb$ and Bi), and $A_2PAs(A = Ge, Sn$ and Pb) have been studied based on first-principles calculations. We find that all the monolayers exhibit in-plane d_{12}, and out-of-plane d_{13} piezoelectric coefficients due to the lack of reflection symmetry with respect to the central A atoms. Moreover, our calculations show that $Al_2OX(T = S, Se, Te)$ chalcogenide monolayers have higher absolute in-plane piezoelectric coefficients. However, the highest out-of-plane values are achieved in the Si_2PBi monolayer, larger than those of some advanced piezoelectric materials, making them very promising transducer materials for lightweight and high-performance piezoelectric nanodevices.

Keywords: first-principles; elastic; piezoelectric; 2D material, Janus monolayer

Citation: Mallah, A.; Debbichi, M.; Dhaou, M.H.; Bellakhdhar, B. Structural, Mechanical, and Piezoelectric Properties of Janus Bidimensional Monolayers. *Crystals* 2023, 13, 126. https://doi.org/10.3390/cryst13010126

Academic Editors: Alexei A. Bokov, Khitouni Mohamed and Joan-Josep Suñol

Received: 25 December 2022
Revised: 3 January 2023
Accepted: 6 January 2023
Published: 10 January 2023

Copyright: © 2023 by the authors. Licensee MDPI, Basel, Switzerland. This article is an open access article distributed under the terms and conditions of the Creative Commons Attribution (CC BY) license (https://creativecommons.org/licenses/by/4.0/).

1. Introduction

During the last 10 years, the study of two-dimensional (2D) materials has received a lot of attention as a result of the successful exfoliation of a graphene monolayer and the revealing of its special properties [1–3]. This class of materials can have significantly different, and sometimes unexpected properties compared to their bulk counterparts [4,5].

Among these, the Janus materials, which are characterized by two faces with two different local environment, have received rapidly increasing attention in recent years [6–9]. This new type of 2D material is successfully predicted by using first-principles calculations and is exfoliated mechanically from its bulk. The Janus-type two-dimensional (2D) monolayers have been studied extensively both experimentally and theoretically. They have many new physical properties that are not present in bulk structures or other conventional 2D materials. They also possess many exceptional physical properties, making them good candidates for many fields like electronics, optoelectronics, and catalysis. For instance, the potential of these stable 2D In_2X_2X' (X and X' = S, Se, and Te) for photocatalytic and piezoelectric applications have been predicted by first-principles calculations [10]. Tuan et al. have predicted by first-principles calculations a novel stable Janus group III chalcogenide monolayers Al_2XY_2 (X/Y = S, Se, Te) suitable for applications in high-performance electronic nanodevices [8]. It has also been demonstrated by using first-principles calculations that the Janus Si_2XY (X,Y = P, As, Sb, Bi) monolayers have the potential for applications in spintronic devices [11]. Very recently, Yungang et al. [9] demonstrated that Nb_3SBr_7 and Ta_3SBr_7 bilayers are promising photocatalysts for water splitting due to their experimental feasibility and their distinct characteristic such as robust coexistence of intrinsic charge separations, ultrahigh solar-to-hydrogen (STH) efficiencies, and strong absorptions.

Piezoelectricity is a particularly interesting and useful property that has attracted tremendous interest because it allows for energy conversion between electrical and mechanical energy or vice versa. The growing demands for nanoscale and diverse functional piezoelectric devices have oriented researchers to explore low-dimensional piezoelectric materials. However, the piezoelectric effect is an electromechanical interaction between stresses and strains, and polarizations and electric fields in noncentrosymmetric semiconductors and insulators [12–14]. It has been revealed that the lack of mirror symmetry in Janus structures has resulted in many new physical effects that are not present in symmetric structures such as the piezoelectric effect. This class of piezoelectric materials has numerous promising applications in sensors, transducers, actuators, active flexible electronics, and energy conversion devices [15,16]. As shown by Yonghu et al. [17] the coupling of topology and piezoelectricity in Janus MTeS (M = Ga and In) monolayers may offer a new platform for novel spintronic and piezotronic device applications. Additionally, the monolayer Fe_2IX (X = Cl and Br) becomes a viable platform for multifunctional spintronic applications with a large gap and high Curie temperature, due to the combination of piezoelectricity, topology, and the ferromagnetic ordering [18]. Furthermore, the coexistence of piezoelectricity and magnetism and their interaction in 2D materials can be utilized for making piezoelectric-based multifunctional nanodevices [19,20]. Despite the existence of many works about piezoelectricity properties based on binary 2D materials, few works are down for 2D ternary. However, the search for new materials with large piezoelectric coefficients remains a challenge for nanogenerators, ultrasensitive mechanical detectors, and consumer touch-sensor applications. Here, by using the density-functional perturbation theory, we have predicted the piezoelectric coefficients of some stable Janus monolayers, including $Al_2XX'(X/X'=S, Se, Te$ and $O)$, $Si_2XX'(X/X'=P, As, Sb$ and $Bi)$ and $A_2PAs(A=Ge, Sn$ and $Pb)$. These Janus monolayers are distinguished by the lack of mirror symmetry. We find that all the monolayers exhibit an in-plane d_{12} and out-of plane d_{13} piezoelectric coefficients. Our first-principles calculations show $Al_2OX(T = S, Se, Te)$ chalcogenide monolayers have larger absolute in-plane piezoelectric coefficients. However, the highest out-of-plane value is achieved in the Si_2PBi monolayer, higher than those of some advanced piezoelectric materials.

2. Computational Details and Methods

Our DFT calculations are performed by using the Vienna ab initio simulation package (VASP) [21] and the projector-augmented wave method (PAW) with a cutoff energy of 600 eV. For the exchange-correlation potential, the generalized gradient approximation (GGA) within the Perdew–Burke–Ernzerhof (PBE) formalism is employed [22]. The Brillouin zone integration is sampled by using a Γ-centered $16 \times 16 \times 1$ k-point grid. For the all Janus A_2XX' monolayers, a vacuum spacing higher than 20 Å along the the direction perpendicular to the plane is included to avoid interactions between two neighboring images.

Elastic stiffness was calculated, including ionic relaxations by using the finite differences method [23]. However, the piezoelectric stress coefficients were calculated by employing the density functional perturbation theory (DFPT) method [23]. For more accurate results, a dense k-point mesh $25 \times 25 \times 1$ is used. All the structures are fully relaxed by using 10^{-6} eV and 10^{-3} eV/Å as convergence criteria for total energy and Hellmann–Feynman force, respectively. The localization of electrons in one unit cell with one monolayer is estimated by using the electron localization function (ELF) analysis, which was introduced in quantum chemistry to identify regions of space that can be associated with electron pairs [24].

3. Results and Discussion

3.1. Crystal Structures and Symmetry

In this paper, 15 possible models of A_2XX' monolayer, including $Al_2XX'(X/X' = S, Se, Te$ and $O)$, $Si_2XX'(X/X' = P, As, Sb$ and $Bi)$ and $A_2PAs(A = Ge, Sn$ and $Pb)$ are modeled. Figure 1 shows the top and side views of the optimized lattice structure of the Janus A_2XX' monolayers. The unit cell has a hexagonal symmetry with a C_{3v} space group, and are made up of one X atom, one X' atom, and two A atoms layers sandwiched between X and X' atomic layers in the sequence X-A-A-X'. The absence of inversion symmetry distinguishes these monolayers from their parent binary structure.

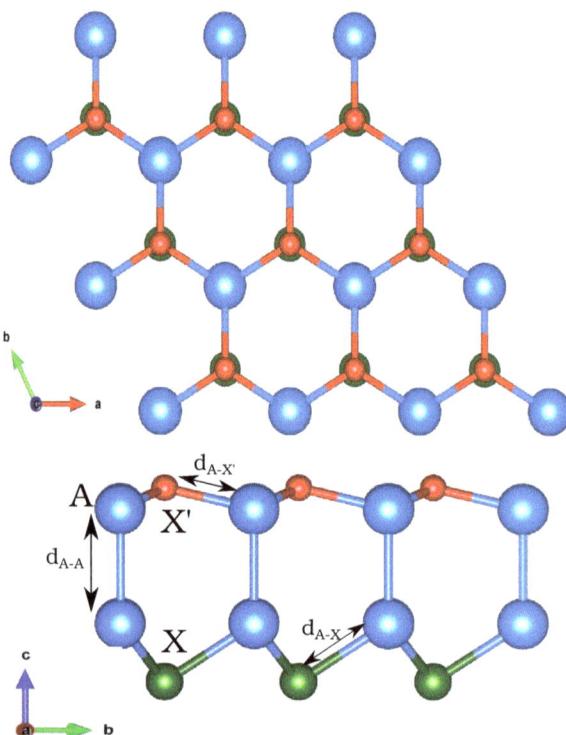

Figure 1. Top and side views of the prototype structure of the A_2XX' monolayers.

It was proven by previous theoretical works [11,25–28] based on static energy, phonon spectrum, and ab initio molecular dynamics simulations that all the Al_2XX' monolayers have thermal and kinetic stability. Meanwhile, DFT calculations that use different exchange-correlation functionals, such as PBE and HSE06, have been used to report the band structures of these monolayers. The calculated structural parameters of Al_2XX' monolayers are listed in Table 1. The obtained results are in good agreement with the available data [11,25–27].

Table 1. The lattice constant (a), A-A (d_{A-A}), A-X (d_{A-X}), A-X'$d_{A-X'}$ bond lengths and the work function difference $\Delta\Phi$ of 15 different structures of A_2XX' monolayers.

Monolayer	a (Å)	d_{A-A} (Å)	d_{A-X} (Å)	$d_{A-X'}$ (Å)	$\Delta\Phi$ (eV)
Al_2SO	3.36	2.60	2.26	2.00	0.30
Al_2SeO	3.42	2.59	2.38	2.03	0.42
Al_2TeO	3.49	2.63	2.61	2.06	0.24
Al_2TeS	3.85	2.58	2.41	2.63	0.16
Al_2TeSe	3.96	2.57	2.53	2.65	0.59
Al_2SeS	3.67	2.60	2.44	2.35	0.06
Si_2PAs	3.62	2.37	2.37	2.31	0.47
Si_2PSb	3.76	2.36	2.54	2.36	0.57
Si_2PBi	3.81	2.34	2.63	2.38	0.39
Si_2AsSb	3.84	2.34	2.57	2.45	0.26
Si_2AsBi	3.92	2.34	2.65	2.49	0.79
Si_2SbBi	4.08	2.35	2.69	2.64	0.26
Ge_2PAs	3.73	2.50	2.45	2.39	0.29
Sn_2PAs	4.03	2.90	2.59	2.65	0.087
Pb_2PAs	4.18	3.04	2.68	2.74	0.270

3.2. Elastic Theory and Properties

According to the symmetry group of our 2D compounds, only four elastic constants are nonzero, C_{11}, C_{22}, C_{12}, and C_{66}. Due to the symmetry of structures, we have $C_{11} = C_{22}$ et $C_{66} = \frac{1}{2}(C_{11}-C_{12})$. The calculated elastic constants C_{ij} are listed in Table 2. By checking the Born–Huang stability criteria [25]: $C_{11} > C_{12}$, $C_{22} > 0$, $C_{66} > 0$, and $C_{11}^2 - C_{21}^2 > 0$, we show that all the monolayers satisfy the stability condition.

Table 2. The 2D elastic constants C_{ij}(N/m), Young modulus Y^{2D} (N/m), Poisson ratio v^{2D}, Shear modulus G^{2D} (N/m), and layer modulus γ^{2D} (N/m) of the Janus monolayers A_2XX'.

Monolayer	C_{11}	C_{12}	C_{66}	Y^{2D}	v^{2D}	G^{2D}	γ^{2D}
Al_2SO	96.54	24.68	35.92	90.23	0.25	35.92	60.61
Al_2SeO	84.45	26.80	28.82	75.94	0.31	28.82	55.62
Al_2TeO	47.24	15.85	15.69	41.92	0.33	15.69	31.55
Al_2TeS	65.77	13.01	26.37	63.19	0.19	26.37	39.39
Al_2TeSe	62.20	13.23	24.48	59.38	0.21	24.48	37.72
Al_2SeS	75.50	15.58	29.96	72.29	0.20	29.96	45.54
Si_2PAs	112.95	16.96	47.99	110.40	0.15	47.99	64.95
Si_2PSb	94.02	13.96	40.03	91.95	0.14	40.03	53.99
Si_2PBi	52.36	24.72	13.82	40.69	0.47	13.82	38.54
Si_2AsSb	88.94	16.01	36.46	86.05	0.18	36.46	52.48
Si_2AsBi	80.65	18.74	30.95	76.30	0.23	30.95	49.70
Si_2SbBi	70.54	15.95	27.29	66.93	0.22	27.29	43.24
Ge_2PAs	98.80	20.54	39.12	94.52	0.20	39.12	59.67
Sn_2PAs	73.34	17.80	27.77	69.02	0.24	27.77	45.57
Pb_2PAs	44.19	6.57	18.80	43.21	0.14	18.80	25.38

The elastic properties of the A_2XX' monolayers are examined in terms of the in-plane Young modulus and the Poisson ratios. Due to hexagonal symmetry, A_2XX' monolayers are mechanically isotropic.

In terms of these elastic constants, the layer modulus is

$$\gamma^{2D} = \frac{1}{2}(C_{11} + C_{12}). \tag{1}$$

The angular dependence of the in-plane Poisson's ratio ($\nu^{2D}(\theta)$) and Young's modulus ($Y^{2D}(\theta)$) are obtained from the following formulas [1]:

$$\nu^{2D}(\theta) = \frac{C_{12}S^4 - BS^2C^2 + C_{12}C^4}{C_{11}S^4 + AS^2C^2 + C_{22}C^4} \quad (2)$$

$$Y^{2D}(\theta) = \frac{C_{11}C_{22} - C_{12}^2}{C_{11}S^4 + AS^2C^2 + C_{22}C^4} \quad (3)$$

where $S = sin(\theta)$, $C = cos(\theta)$, $A = \frac{C_{11}C_{22} - C_{12}^2}{C_{66}} - 2C_{12}$ and $B = C_{11} + C_{12} - \frac{C_{11}C_{22} - C_{12}^2}{C_{66}}$.

The orientation-dependent values for the all monolayers reveal strong isotropy of Young's modulus as well as Poisson's ratio. As an example, we show in Figure 2 the angular dependence of the Poisson's ratio (a) and Young's modulus of Al$_2$SSe monolayer. We find that ν^{2D} and Y^{2D} plots are perfect circulars, implying that these monolayers have highly isotropic elasticity due to their 2D isotropic atomic structures. The corresponding computed values are listed in Table 2. For the all monolayers, our calculated values of the elastic stiffness are in concordance with the available data [11,25,26]. The Young's moduli values are obviously smaller than those of other well-known 2D materials, such as graphene, hexagonal boron nitrite layer and MoS$_2$ [29–31], demonstrating their mechanical flexibility and can resist significantly to the mechanical strain. The calculated ν^{2D} values of all the monolayers except Si$_2$PBi are less than 0.33, which implies that these monolayers are brittle based on the Frantsevich rule [32,33].

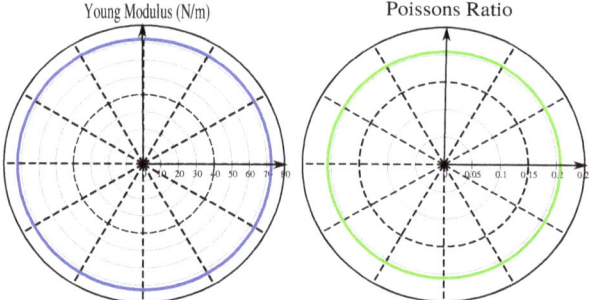

Figure 2. Orientation dependence on the in-plane Poisson's ratio (ν^{2D}) and Young's modulus (Y^{2D}) of Al$_2$SSe monolayer.

3.3. Piezoelectric Properties

In A$_2$XX', the difference in atom size and electronegativity, as well as the different bond types between Al-X (d$_{A-X}$) and Al-X'(d$_{A-X'}$) all contribute to unequal charge distributions in the systems as shown in the inset of the Figure 3, resulting in noncentrosymmetric materials. As an example, the planar average of the electrostatic potential energy of Al$_2$SSe is shown in Figure 3. As can clearly be seen, a significant potential difference between the two sides of the monolayer, reflecting the formation of an internal electric field and surface work function difference ($\Delta\Phi$). For the other compounds, the planar average of the electrostatic potential energy is calculated and the extracted work function difference is regrouped in Table 1, which is proportional to the magnitude of the dipole moment according to the Helmholtz equation [34].

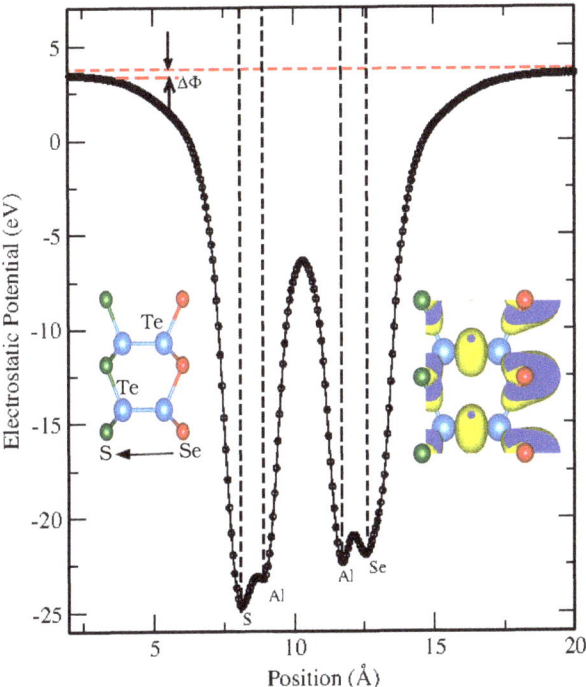

Figure 3. Planar average of the electrostatic potential energy for the Al$_2$SSe. In the inset, the electron localization function (ELF) of the Al$_2$SSe. The arrow indicates the direction of the local electric field.

The aforementioned properties, such as the lack of inversion symmetry and the intrinsic polar electric field, are two possible causes for the emergence of piezoelectricity in the materials. The relaxed-ion third-rank piezoelectric tensors c_{ijk} and d_{ijk}, which are the sum of ionic and electronic contributions, can be evaluated by

$$e_{ijk} = e_{ijk}^{ion} + e_{ijk}^{elc} = \frac{\partial P_i}{\partial \varepsilon_{ij}} \tag{4}$$

and

$$d_{ijk} = d_{ijk}^{ion} + d_{ijk}^{elc} = \frac{\partial P_i}{\partial \sigma_{ij}}, \tag{5}$$

where ε_{ij}, σ_{ij} and P_i represent the strain, stress, and polarization tensors, respectively. For 2D materials, $\varepsilon_{ij} = \sigma_{ij} = 0$ for i = 3 [16].

By using the Voigt notation, (1 = xx, 2 = yy, 3 = zz, 4 = yz, 5 = zx, and 6 = xy) [35]. The second-rank piezoelectric tensors e_{ij} and d_{jk} are related via the elastic stiffness tensor by

$$e_{ij} = d_{ik} C_{jk}. \tag{6}$$

For our Janus monolayers the point-group symmetry belongs to 3m, and the nonzero piezoelectric stress tensors, e_{ij} are given as

$$e_{ij} = \begin{pmatrix} \cdot & \cdot & \cdot & \cdot & e_{15} & e_{12} \\ e_{12} & -e_{12} & \cdot & e_{15} & \cdot & \cdot \\ e_{31} & e_{31} & e_{31} & \cdot & \cdot & \cdot \end{pmatrix}.$$

For 2D materials $e_{ij}^{2D} = d_{ik}^{2D} C_{jk}$, where $M_{ij}^{2D} = M/l_z$, (M = e or d), and l_z is the length of the unit cell along the z direction.

Based on Equation (6), the unique in-plane and out-of-plane piezoelectric coefficients e_{12}^{2D}, d_{12}^{2D} and e_{31}^{2D}, d_{31}^{2D}, respectively, are nonzero. The corresponding piezoelectric tensors matrix can be written as

$$\begin{pmatrix} 0 & 0 & e_{12}^{2D} \\ e_{12}^{2D} & -e_{12}^{2D} & 0 \\ e_{31}^{2D} & e_{31}^{2D} & 0 \end{pmatrix} = \begin{pmatrix} 0 & 0 & 2d_{12}^{2D} \\ d_{12}^{2D} & -d_{12}^{2D} & 0 \\ d_{31}^{2D} & d_{31}^{2D} & 0 \end{pmatrix} \times \begin{pmatrix} C_{11} & C_{12} & 0 \\ C_{12} & C_{11} & 0 \\ 0 & 0 & C_{66} \end{pmatrix}.$$

The d_{22} and d_{31} can be calculated by

$$d_{12}^{2D} = \frac{e_{12}^{2D}}{C_{11} - C_{22}} \tag{7}$$

and

$$d_{31}^{2D} = \frac{e_{31}^{2D}}{C_{11} + C_{22}}. \tag{8}$$

In the following, we use e_{ij} and d_{ij} instead of the e_{ij}^{2D} and d_{ij}^{2D} symbol, respectively. In this work, we adapted the relaxed-ion method, which is the sum of electronic and ionic parts to calculate the piezoelectric coefficients, which is the more reliable method compared to that of the clamped-ion one [36]. To verify the reliability of the applied method, we have first computed piezoelectric stress coefficient e_{11} for 1H-MoS$_2$ monolayer and found a predicted value as higher as $\sim 2.27 \times 10^{-10}$ C/m, in excellent agreement with the experimental value and the reported theoretical studies [12,37]. By using the above procedures, we derive the piezoelectric coefficients e_{ij} of the all monolayers. The results are shown in Figure 4. More significantly, these Janus monolayers with broken mirror symmetry possess, in addition to in-plane e_{12}/d_{12} nonzero out-of-plane piezoelectric coefficients e_{13}/d_{13}. The minus sign in calculated values indicates the direction of polarization.

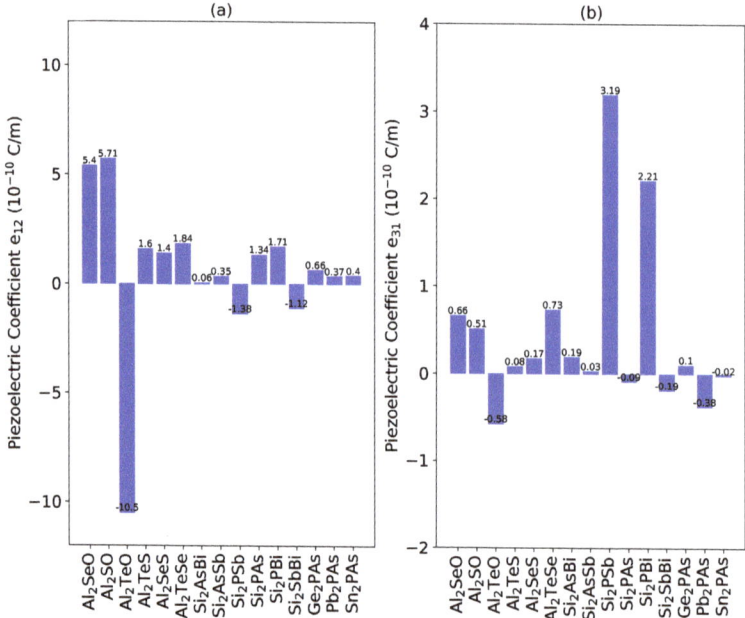

Figure 4. Piezoelectric coefficients (**a**) e_{12} and (**b**) e_{13} of 15 different structures of A$_2$XX' monolayers.

As shown in Figures 4a and 5a for a given metal element A, the monolayers containing heavier chalcogenide atoms (Te, Se and S) have the higher in-plane piezoelectric coefficient $|e_{12}|/|d_{12}|$ values. Compared with other 2D piezoelectric materials such as MoX_2 (X = S, Se, Te) and MoTO (T = S, Se, Te) with a value of 3.64–5.43 pm/V [30,38], the Janus Al_2OX(T = S, Se, Te) chalcogenide monolayers have larger absolute in-plane piezoelectric coefficients by several folds. More noticeably, $|d_{12}|$ attains 18.20 and 17.42 pm/V for Al_2TeO and Al_2SeO monolayers, respectively, in same order of magnitude as the Janus M_2SeX (M = Ge, Sn; X = S, Te) monolayers [13], which makes them appropriate for 2D piezoelectric sensors and nanogenerators. Some materials, such as the monolayers Al_2TeSe and Si_2PBi, have large d_{12} values but small e_{12} values because their Young's moduli are small, limiting the amount of force applied in electric field-induced deformations.

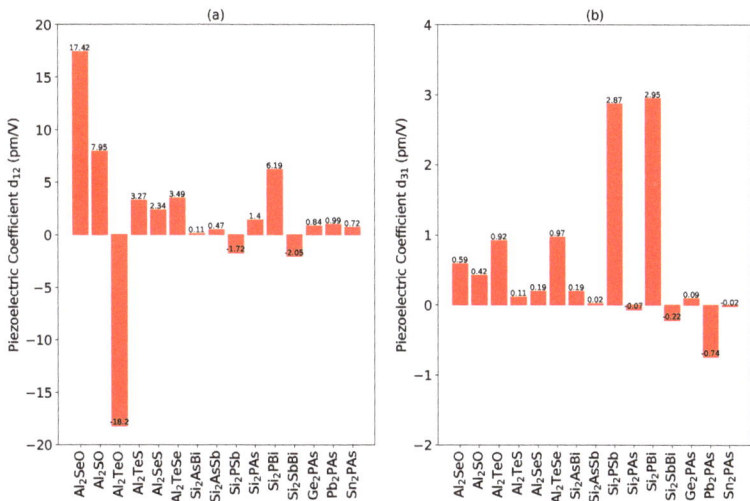

Figure 5. Piezoelectric coefficients (**a**) d_{12} and (**b**) d_{13} of 15 different structures of A_2XX' monolayers.

The noncentrosymmetric crystal structure of the A_2XX' monolayers in the out-of-plane direction gives rise to the finite out-of-plane piezoelectric constant. For all the monolayers presented in this work, the value of the out-of-plane strain piezoelectric coefficient d_{31} (Figure 5b) is about two orders of magnitude lower than the in-plane coefficient d_{12}. This means that vertical piezoelectric polarization due to vertical strain is much stronger than that of the in-plane strain. But these values are still comparable to other 2D materials such as Te_2Se, MoSTe, In_2SSe, $TiNX_{0.5}Y_{0.5}$(X, Y = Cl, Br, F) and $1T-MX_2$ (M = Zr and Hf; X = S, Se, and Te) [16,39,40]. The highest d_{31} value (2.95 pm/V) is achieved in the Si_2PBi and Si_2PSb monolayers, higher to those of some advanced piezoelectric materials such as MoSTe, GaN wirtzite, M_2XX'(M = Ga, In; X, X' = S, Se, Te and X≠X') and $MM'X_2$ (M, M' = Ga, In and M'≠M; X = S, Se, Te) [16,36]. This value is also comparable to that of the Janus Te_2Se multilayers with antiparallel orientations, $CrF_{1.5}I_{1.5}$, Sb_2Te_2Se, Sb_2Se_2Te, TePtS and TePtSe [14,16,41,42]. This significant out-of-plane piezoelectric effect would give these Janus monolayers a variety of functions in piezoelectric applications.

By comparing the d_{31} of the six 2D Janus Si_2XX' monolayers, we conclude that the absolute value of the out-of-plane piezoelectric coefficient d_{31} increases with the electronegativity difference between the atoms on both sides of the monolayer, X and X'. This finding is also valid for the Al_2XX' monolayers. This is understandable given that the difference in atomic sizes of X and X' breaks the reflection symmetry along the vertical direction, resulting in vertical piezoelectric polarization, which becomes stronger when the electronegativity difference between atoms increases.

4. Conclusions

On the basis of first-principles calculations, we have systematically studied the piezoelectric properties of some Janus 2D monolayers, Al_2XX' (X/X' = S, Se, Te and O), Si_2XX' (X/X' = P, As, Sb and Bi) and A_2PAs (A = Ge, Sn and Pb). The absence of inversion symmetry in these Janus structures gives rise to in-plane and out-of-plane piezoelectric coefficients. Our calculations, by using the DFPT method, reveal that the monolayers containing heavier chalcogenide atoms (Te, Se and S) have a higher in-plane piezoelectric coefficient larger than those of the widely studied 1H-MoX_2 (X = S, Se, Te) monolayers. Moreover, all the monolayers are characterized by out-of-plane piezoelectric coefficients e_{31}/d_{31}, due to the lack of reflection symmetry with respect to the central A atoms. The highest values of d_{31} (~2.9 pm/V) are achieved for the Si_2PBi and S_2PSb monolayers due to their mechanical flexibility.

Author Contributions: Conceptualization: A.M., M.D., M.H.D. and B.B.; Data curation: A.M., M.D. and M.H.D.; Formal analysis: M.D., M.H.D. and B.B.; Funding acquisition: A.M. and M.H.D.; Investigation: A.M. and M.D.; Resources: M.D.; Validation: A.M. and M.D.; Writing-original draft: A.M., M.D. and M.H.D. All authors have read and agreed to the published version of the manuscript.

Funding: Qassim University, represented by the Deanship of Scientific Research, on the material support for this research under the number (10244-cos-2020-1-3-I).

Data Availability Statement: Data can be requested from the authors.

Acknowledgments: The authors gratefully acknowledge Qassim University, represented by the Deanship of Scientific Research, on the material support for this research under the number (10244-cos-2020-1-3-I) during the academic year 1442AH/2020AD.

Conflicts of Interest: On behalf of all authors, the corresponding author states that there is no conflict of interest.

References

1. Debbichi, M.; Alhodaib, A. Stability, electronic and magnetic properties of the penta-CoAsSe monolayer: A first-principles and Monte Carlo study. *Phys. Chem. Chem. Phys.* **2022**, *24*, 5680. [CrossRef] [PubMed]
2. Debbichi, M.; Mallah, A.; Dhaou, M.H.; Lebègue, S. First-Principles Study of Monolayer penta-CoS_2 as a Promising Anode Material for Li/Na-ion Batteries. *Phys. Rev. Appl.* **2021**, *16*, 024016. [CrossRef]
3. Debbichi, M.; Said, H.; Garbouj, H.; El Hog, S.; An Dinh, V. A new ternary Pentagonal Monolayer based on Bi with large intrinsic Dzyaloshinskii-Moriya interaction. *J. Phys. D Appl. Phys.* **2021**, *55*, 015002. [CrossRef]
4. Debbichi, M.; Debbichi, L.; Lebègue, S. Tuning the magnetic and electronic properties of monolayer chromium tritelluride through strain engineering. *Phys. Lett. A* **2020**, *384*, 126684. [CrossRef]
5. Debbichi, M.; Debbichi, L.; Lebègue, S. Controlling the stability and the electronic structure of transition metal dichalcogenide single layer under chemical doping. *Phys. Lett. A* **2019**, *383*, 2922–2927. [CrossRef]
6. Sa, B.; Hu, R.; Zheng, Z.; Xiong, R.; Zhang, Y.; Wen, C.; Zhou, J.; Sun, Z. High-Throughput Computational Screening and Machine Learning Modeling of Janus 2D III–VI van der Waals Heterostructures for Solar Energy Applications. *Chem. Mater.* **2022**, *34*, 6687–6701. [CrossRef]
7. Zhang, L.; Gu, Y.; Du, A. Two-Dimensional Janus Antimony Selenium Telluride with Large Rashba Spin Splitting and High Electron Mobility. *ACS Omega* **2021**, *6*, 31919–31925. [CrossRef]
8. Vu, T.V.; Hieu, N.N. Novel Janus group III chalcogenide monolayers Al_2XY_2 (X/Y = S, Se, Te): first-principles insight onto the structural, electronic, and transport properties. *J. Phys. Condens. Matter* **2021**, *34*, 115601. [CrossRef]
9. Zhou, Y.; Zhou, L.; He, J. 2D Nb_3SBr_7 and Ta_3SBr_7: Experimentally Achievable Janus Photocatalysts with Robust Coexistence of Strong Optical Absorption, Intrinsic Charge Separation, and Ultrahigh Solar-to-Hydrogen Efficiency. *ACS Appl. Mater. Interfaces* **2022**, *14*, 1643–1651. [CrossRef]
10. Wang, P.; Liu, H.; Zong, Y.; Wen, H.; Xia, J.B.; Wu, H.B. Two-Dimensional In_2X_2X' (X and X' = S, Se, and Te) Monolayers with an Intrinsic Electric Field for High-Performance Photocatalytic and Piezoelectric Applications. *ACS Appl. Mater. Interfaces* **2021**, *13*, 34178–34187. [CrossRef]
11. Babaee Touski, S.; Ghobadi, N. Structural, electrical, and Rashba properties of monolayer Janus Si_2XY (X,Y =P, As, Sb, and Bi). *Phys. Rev. B* **2021**, *103*, 165404. [CrossRef]
12. Blonsky, M.N.; Zhuang, H.L.; Singh, A.K.; Hennig, R.G. Ab Initio Prediction of Piezoelectricity in Two-Dimensional Materials. *ACS Nano* **2015**, *9*, 9885–9891. [CrossRef] [PubMed]

13. Qiu, J.; Zhang, F.; Li, H.; Chen, X.; Zhu, B.; Guo, H.; Ding, Z.; Bao, J.; Yu, J. Giant Piezoelectricity of Janus M_2SeX (M = Ge, Sn; X = S, Te) Monolayers. *IEEE Electron Device Lett.* **2021**, *42*, 561–564. [CrossRef]
14. Guo, S.D.; Guo, X.S.; Cai, X.X.; Mu, W.Q.; Ren, W.C. Intrinsic piezoelectric ferromagnetism with large out-of-plane piezoelectric response in Janus monolayer $CrBr_{1.5}I_{0.5}$. *J. Appl. Phys.* **2021**, *129*, 214301. [CrossRef]
15. Wu, W.; Wang, Z.L. Piezotronics and piezo-phototronics for adaptive electronics and optoelectronics). *Nat. Rev. Mater.* **2016**, *1*, 16031. [CrossRef]
16. Chen, Y.; Liu, J.; Yu, J.; Guo, Y.; Sun, Q. Symmetry-breaking induced large piezoelectricity in Janus tellurene materials. *Phys. Chem. Chem. Phys.* **2019**, *21*, 1207–1216. [CrossRef] [PubMed]
17. Wang, Y.; Lei, S.; Huang, Q.; Wan, N.; Xu, F.; Yu, H.; Li, C.; Chen, J. Coexistence of the Piezoelectricity and Intrinsic Quantum-Spin Hall Effect in GaTeS and InTeS Monolayers: Implications for Spintronic Devices. *ACS Appl. Nano Mater.* **2022**, *5*, 11037–11044. [CrossRef]
18. Guo, S.D.; Mu, W.Q.; Xiao, X.B.; Liu, B.G. Intrinsic room-temperature piezoelectric quantum anomalous hall insulator in Janus monolayer Fe_2IX (X = Cl and Br). *Nanoscale* **2021**, *13*, 12956–12965. [CrossRef] [PubMed]
19. Guo, S.D.; Zhu, Y.T.; Qin, K.; Ang, Y.S. Large out-of-plane piezoelectric response in ferromagnetic monolayer NiClI. *Appl. Phys. Lett.* **2022**, *120*, 232403. [CrossRef]
20. Noor-A-Alam, M.; Nolan, M. Large piezoelectric response in ferroelectric/multiferroelectric metal oxyhalide MOX_2 (M = Ti, V and X = F, Cl and Br) monolayers. *Nanoscale* **2022**, *14*, 11676–11683. [CrossRef]
21. Kresse, G.; Hafner, J. Ab initio molecular dynamics for liquid metals. *Phys. Rev. B* **1993**, *47*, 558–561. [CrossRef] [PubMed]
22. Perdew, J.P.; Burke, K.; Ernzerhof, M. Generalized Gradient Approximation Made Simple. *Phys. Rev. Lett.* **1996**, *77*, 3865. [CrossRef] [PubMed]
23. Wu, X.; Vanderbilt, D.; Hamann, D.R. Systematic treatment of displacements, strains, and electric fields in density-functional perturbation theory. *Phys. Rev. B* **2005**, *72*, 035105. [CrossRef]
24. Becke, A.D.; Edgecombe, K.E. A simple measure of electron localization in atomic and molecular systems. *J. Chem. Phys.* **1990**, *92*, 5397–5403. [CrossRef]
25. Liu, M.Y.; Gong, L.; He, Y.; Cao, C. Intraband Lifshitz transition and Stoner ferromagnetism in Janus PA_2As (A=Si,Ge,Sn, and Pb) monolayers. *Phys. Rev. B* **2021**, *104*, 035409. [CrossRef]
26. Demirtas, M.; Varjovi, M.J.; Cicek, M.M.; Durgun, E. Tuning structural and electronic properties of two-dimensional aluminum monochalcogenides: Prediction of Janus Al_2XX'(X/X':O,S,Se,Te) monolayers. *Phys. Rev. Mater.* **2020**, *4*, 114003.
27. Kumar, V.; Jung, J. Two-dimensional Janus group-III ternary chalcogenide monolayer compounds B_2XY, Al_2XY, and $BAlX_2$ (X, Y = S, Se, Te) with high carrier mobilities. *Bull. Korean Chem. Soc.* **2022**, *43*, 138–146. [CrossRef]
28. Huang, A.; Shi, W.; Wang, Z. Optical Properties and Photocatalytic Applications of Two-Dimensional Janus Group-III Monochalcogenides. *J. Phys. Chem. C* **2019**, *123*, 11388–11396. [CrossRef]
29. Lee, C.; Wei, X.; Kysar, J.W.; Hone, J. Measurement of the Elastic Properties and Intrinsic Strength of Monolayer Graphene. *Science* **2008**, *321*, 385–388. [CrossRef]
30. Duerloo, K.A.N.; Ong, M.T.; Reed, E.J. Intrinsic Piezoelectricity in Two-Dimensional Materials. *J. Phys. Chem. Lett.* **2012**, *3*, 2871–2876.
31. Song, L.; Ci, L.; Lu, H.; Sorokin, P.B.; Jin, C.; Ni, J.; Kvashnin, A.G.; Kvashnin, D.G.; Lou, J.; Yakobson, B.I.; et al. Large Scale Growth and Characterization of Atomic Hexagonal Boron Nitride Layers. *Nano Lett.* **2010**, *10*, 3209–3215. [CrossRef] [PubMed]
32. Frantsevich, I.; Voronov, F.; Bokuta, S. *Elastic Constants and Elastic Moduli of Metals and Insulators Handbook*; Naukova Dumka: Kiev, Ukraine, 1983; pp. 60–180.
33. Debbichi, M.; Alresheedi, F. First-principles calculations of mechanical, electronic and optical properties of a new imidooxonitridophosphate. *Chem. Phys.* **2020**, *538*, 110917. [CrossRef]
34. Xiao, W.Z.; Xu, L.; Xiao, G.; Wang, L.L.; Dai, X.Y. Two-dimensional hexagonal chromium chalco-halides with large vertical piezoelectricity, high-temperature ferromagnetism, and high magnetic anisotropy. *Phys. Chem. Chem. Phys.* **2020**, *22*, 14503–14513. [CrossRef]
35. Yin, H.; Gao, J.; Zheng, G.P.; Wang, Y.; Ma, Y. Giant Piezoelectric Effects in Monolayer Group-V Binary Compounds with Honeycomb Phases: A First-Principles Prediction. *J. Phys. Chem. C* **2017**, *121*, 25576–25584. [CrossRef]
36. Guo, Y.; Zhou, S.; Bai, Y.; Zhao, J. Enhanced piezoelectric effect in Janus group-III chalcogenide monolayers. *Appl. Phys. Lett.* **2017**, *110*, 163102. [CrossRef]
37. Zhu, H.; Wang, Y.; Xiao, J.; Liu, M.; Xiong, S.; Wong, Z.J.; Ye, Z.; Yin, X.; Zhang, X. Observation of Piezoelectricity in Free-standing Monolayer Molybdenum Disulfide. *Nat. Nanotech.* **2015**, *10*, 151–155. [CrossRef]
38. Li, Y.Q.; Wang, X.Y.; Zhu, S.Y.; Tang, D.S.; He, Q.W.; Wang, X.C. Active Asymmetric Electron-Transfer Effect on the Enhanced Piezoelectricity in MoTO (T = S, Se, or Te) Monolayers and Bilayers. *J. Phys. Chem. Lett.* **2022**, *13*, 9654–9663. [CrossRef]
39. Shi, X.; Yin, H.; Jiang, S.; Chen, W.; Zheng, G.P.; Ren, F.; Wang, B.; Zhao, G.; Liu, B. Janus 2D titanium nitride halide $TiNX_{0.5}Y_{0.5}$ (X, Y = F, Cl, Br and X≠Y) monolayers with giant out-of-plane piezoelectricity and high carrier mobility. *Phys. Chem. Chem. Phys.* **2021**, *23*, 3637. [CrossRef]
40. Jena, N.; Rawat, A.; Ahammed, R.; Mohanta, M.K.; De Sarkar, A. Emergence of high piezoelectricity along with robust electron mobility in Janus structures in semiconducting Group IVB dichalcogenide monolayers. *J. Mater. Chem. A* **2018**, *6*, 24885.

41. Qiu, J.; Li, H.; Chen, X.; Zhu, B.; Guo, H.; Zhang, F.; Ding, Z.; Lang, L.; Yu, J.; Bao, J. Piezoelectricity of Janus Sb_2Se_2Te monolayers: A first-principles study. *J. Appl. Phys.* **2021**, *129*, 125109. [CrossRef]
42. Kahraman, Z.; Kandemir, A.; Yagmurcukardes, M.; Sahin, H. Single-Layer Janus-Type Platinum Dichalcogenides and Their Heterostructures. *J. Phys. Chem. C* **2019**, *123*, 4549–4557. [CrossRef]

Disclaimer/Publisher's Note: The statements, opinions and data contained in all publications are solely those of the individual author(s) and contributor(s) and not of MDPI and/or the editor(s). MDPI and/or the editor(s) disclaim responsibility for any injury to people or property resulting from any ideas, methods, instructions or products referred to in the content.

Article

Effect of Thermal Exposure on Mechanical Properties of Al-Si-Cu-Ni-Mg Aluminum Alloy

Fanming Chen, Chengwen Liu *, Lijie Zuo *, Zhiyuan Wu, Yiqiang He, Kai Dong, Guoqing Li and Weiye He

School of Mechanical Engineering, Jiangsu Ocean University, Lianyungang 222005, China
* Correspondence: chwl2014@163.com (C.L.); zuoliji@126.com (L.Z.)

Abstract: The microstructure morphology and evolution of mechanical properties are investigated in this study. The results show that the phases displayed no clear change after thermal exposure at 250 °C for 200 h. The tensile strength of the as-cast alloy showed a downward trend in different degrees with the increase in the tensile temperature, while the influence of elongation was opposite to the tensile strength. In addition, the tensile strength tended to be stable after thermal exposure at 250 °C for 100 h. The main creep mechanism of the as-cast alloy at a low temperature and low stress (T ≤ 250 °C; σ ≤ 40 MPa) is grain-boundary creep. The Monkman–Grant empirical formula was used to fit the relationship between the creep life and the minimum creep rate, and the fitting results are: $t_r \cdot \dot{\varepsilon}_{min}^{0.95} = 0.207$.

Keywords: thermal exposure; casting alloys; mechanical properties; creep mechanism

1. Introduction

Because of its good specific strength and excellent mechanical properties, heat-resistant aluminum alloys have been widely used in aerospace engineering, vehicle engineering, and marine engineering. Al-Si-Cu-Ni-Mg alloy is a typical cast heat-resistant aluminum alloy, which is used to produce engine pistons. Therefore, the alloy usually works at a high temperature and alternating load. The service life of the piston is determined by the high-temperature endurance performance of the alloy [1–5].

With the increase in engine power, new challenges are presented to the high-temperature stability and endurance strength of piston materials. The development of a heat-resistant aluminum alloy with better high-temperature performance is imminent [6,7]. Thermal exposure experiments have been widely used by domestic and foreign scholars to verify the high-temperature stability of heat-resistant aluminum alloys. For instance, the eutectic silicon and Al_3Ni phases of Al_3Zr/Al-8Si-2Ni composites were spheroidized after being exposed at 450 °C for 150 h. The elongation of the Al_3Zr/Al-8Si-2Ni composites was positively correlated with the heat exposure time, and the tensile strength tended to stabilize after decreasing [8]. A study reported that the average number of fatigue cycles before failure of the alloy was 2.24×10^7 after thermal exposure at 425 °C/100 h, which was 3 times higher than the 8.21×10^6 fatigue life of T6 heat treatment. The δ-Al_3CuNi phase and γ-Al_7Cu_4Ni were thermally stable intermetallic compounds, which significantly improved the high-temperature fatigue life of Al-Si piston alloys [9]. Zhang et al. [10] observed that the steady-state creep rate of the studied alloy decreased threefold when the loading changed from 70 MPa to 90 MPa with a constant temperature of 90 °C. Under a constant stress of 90 MPa, the steady creep rate of the studied alloy increased from 1.94×10^{-7} to 1.48×10^{-6} with the temperature rising from 90 °C to 150 °C. The grain size was basically unchanged under different compressive creep stresses. Zhao et al. [11] reported that the increase in YS was attributed to the precipitation of β'', Q', and θ' nanophases in an α-Al matrix after aging at 175 °C for 4 h. After thermal exposure at 350 °C for 100 h, the nanophases were mainly dissolved in the α-Al matrix, and the ultra-fine eutectic silicon

network was fractured and coarsened, resulting in the hardness of the alloy decreasing from 110 HV to 61 HV and the YS dropping to 117 MPa. The current literature reports that the high-temperature mechanical properties after thermal exposure are positively correlated with Ti and Cu content, which is attributed to the increase in the number of precipitated particles. The high strength value of the alloy after thermal exposure at 350 °C for 0.5 h is attributed to the number of nano sizes. The high heat resistance of T($Al_{20}Cu_2Mn_3$) and Fe($Al_{15}(FeMn)_3Si_2$) makes the strength stable after 100 h thermal exposure [12].

At present, the research on the high-temperature endurance strength of heat-resistant aluminum alloys mainly focuses on fatigue failure, and high-temperature creep performance is another key factor in the high-temperature service life of alloys [13–17]. Therefore, the aim of this study is to find out the thermal stability of the high-temperature endurance strength of Al-Si-Cu-Ni-Mg piston alloys, including the change law of the creep properties of the alloys when exposed to high temperature, and then guide subsequent study of heat-resistant aluminum alloy.

In this work, the microstructures of the cast alloy before and after thermal exposure were characterized using optical microscopy and scanning electron microscopy. The different temperature tensile properties of the cast alloy were investigated. The evolution law of the tensile strength during thermal exposure, as well as the creep mechanism of the alloy before and after thermal exposure, were investigated in this study. This study provides a basis for subsequent study on the influence of the high-temperature properties of Al-Si-Cu-Ni-Mg alloy.

2. Experimental Procedures

The chemical composition of the designed alloys is shown in Table 1. The alloys were prepared by melting pure Mg, Al, and Al-20Si (wt. %), Al-50Cu (wt. %), and Al-10Ni (wt. %) master alloys in an electric smelting furnace. Before melting the alloy, the crucible and mold were kept at 200 °C for 2 h. The Al-50Cu and Al-10Ni were kept at 150 °C for 2 h to remove moisture in the raw materials. First, pure Al and master alloys were added to the crucible and melted at 740 °C. Then pure Mg wrapped in aluminum foil was added to the melt at approximately 720 °C. The melt was refined using the refining agent C_2Cl_6 at 740 °C. During the process of refining, the melt was stirred to ensure the refining effect and homogeneous chemical composition.

Table 1. Composition of cast Al-Si-Cu-Ni-Mg alloy.

Elements	Si	Cu	Ni	Mg	Mn	Zn	Ti	Fe	Ce	Al
wt. %	12.34	3.81	1.97	0.79	0.53	0.26	0.23	0.18	0.23	Bal

The metallographic samples were ground with different levels of sandpapers, polished with a PG-2B metallographic polishing machine for final polishing, and then etched with Keller's reagent (a mixed solution of 2.5%HNO_3, 1.5%HCl, 1%HF, and 95% H_2O) for 10 s. The fracture surface of the tensile specimen was observed with NOVA NanoSEM 230 scanning electron microscopy (Hong Kong, China). The tensile test was carried out on a WDW-10S (Jinan, China) universal tensile testing machine with a tensile rate of 0.4 mm/min at room temperature or 250 °C. The thermal exposure experiment was carried out in a tubular furnace with a temperature error of ± 2 °C. The thermal exposure temperature was set at 250 °C, and the time was 0~200 h. Each sample was measured after holding for 15 min at the set temperature, and the results are the average value of three parallel samples. The creep test was carried out on a CSS-3902 creep testing machine (Nanjing, China). The test temperature was 200–300 °C, and the stress was 20 MPa and 28 MPa. The sample was kept at the preset temperature for 1 h before loading. Two creep samples were tested for each condition, and no repeatability test was carried out if there was no abnormality. The geometric sizes of the tensile creep sample and the creep testing device are shown in Figures 1 and 2, respectively.

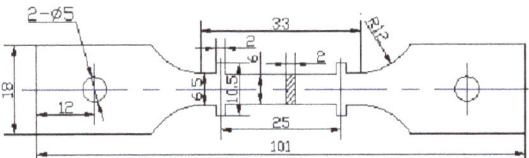

Figure 1. The geometric dimensions of tensile creep samples [18].

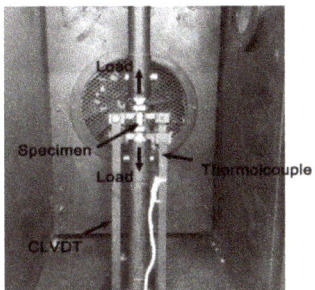

Figure 2. The tension creep testing setup.

3. Results and Discussion

3.1. Microstructure and Morphology of Alloys in Different States

Figure 3 depicts the microstructure morphology of Al-Si-Cu-Ni-Mg alloy under different characterization methods. Figure 3a shows that the α-Al was in the form of coarse dendrites, and most of the eutectic silicon was in the shape of short rods. Uniformly distributed black needle-like strengthening phases and some fishbone-like AlSiFeNiCu phases were detected. The characteristics of the fishbone-like phases were consistent with those reported in the related literature [19]. Figure 3b presents that primary silicon and eutectic silicon were randomly distributed on the Al matrix. The fishbone-like phase is observable in Figure 3a, and the network-like precipitation phase can be found in Figure 3b.

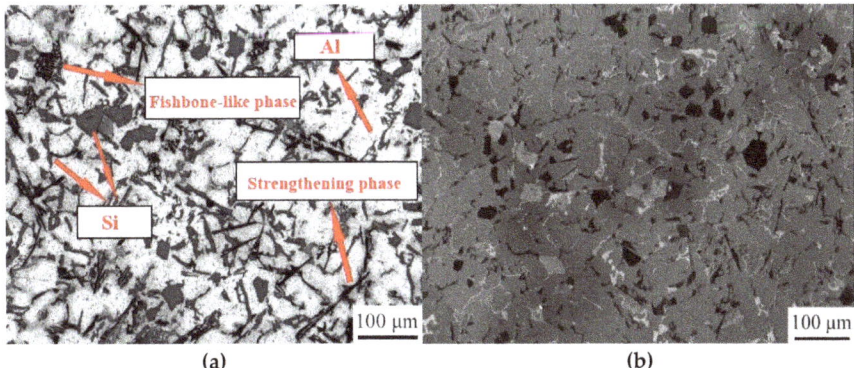

Figure 3. Microstructure of Al-Si-Cu-Ni-Mg alloy under (**a**) metallography and (**b**) scanning electron microscopy.

Compared with the as-cast alloy in Figure 4, the elements in the thermal exposed alloy occurred with different degrees of segregation and aggregation as shown in Figure 5. With the temperature increase, the solute atoms of the precipitation phase diffused, which promoted the nucleation and growth of Q-$Al_5Cu_2Mg_8Si_6$, displayed with the Mg element obviously aggregated. The distribution of Cu and Ni elements was more uniform, whereas there was an increase in the distribution of slightly roughened rod-shaped eutectic Si. In conclusion, after 200 h thermal exposure at 250 °C, the microstructures at the micro scale demonstrated no significant changes.

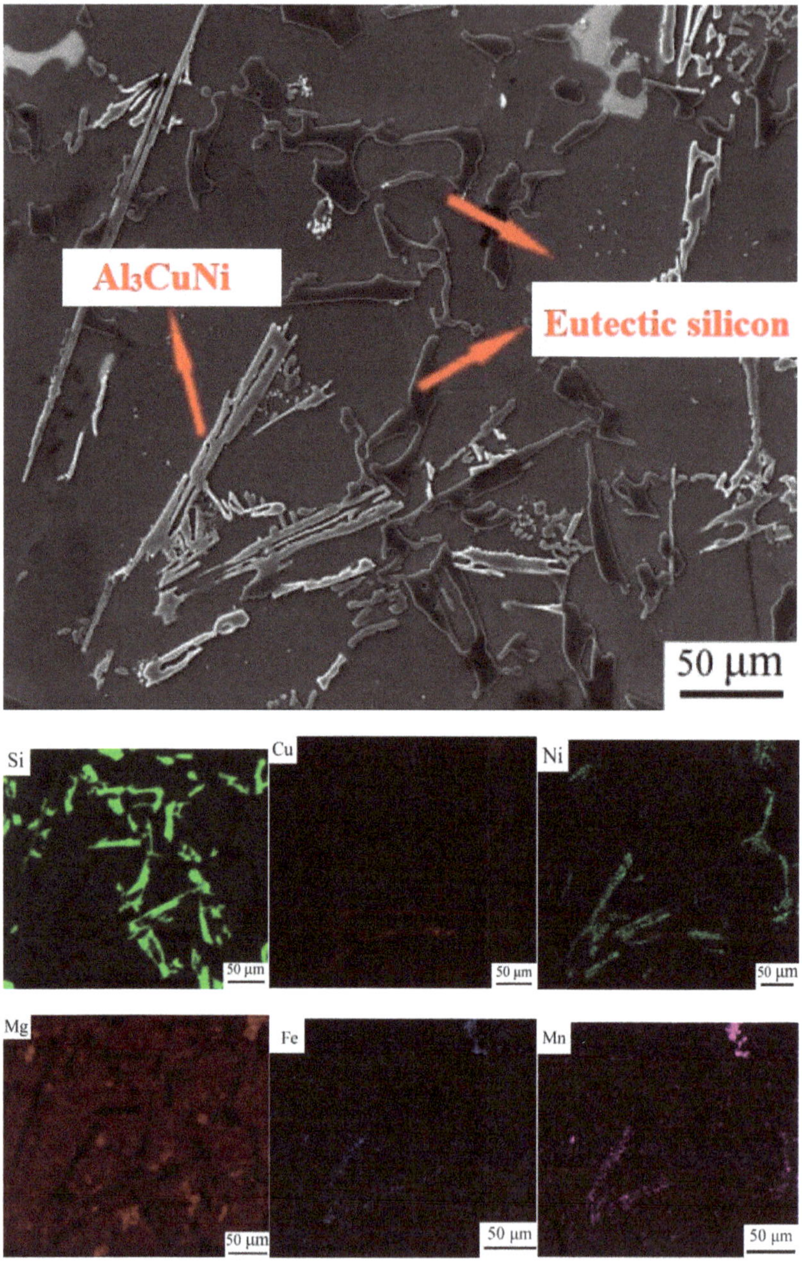

Figure 4. SEM morphology of as-cast Al-Si-Cu-Ni-Mg alloy and distribution of alloy constituent elements.

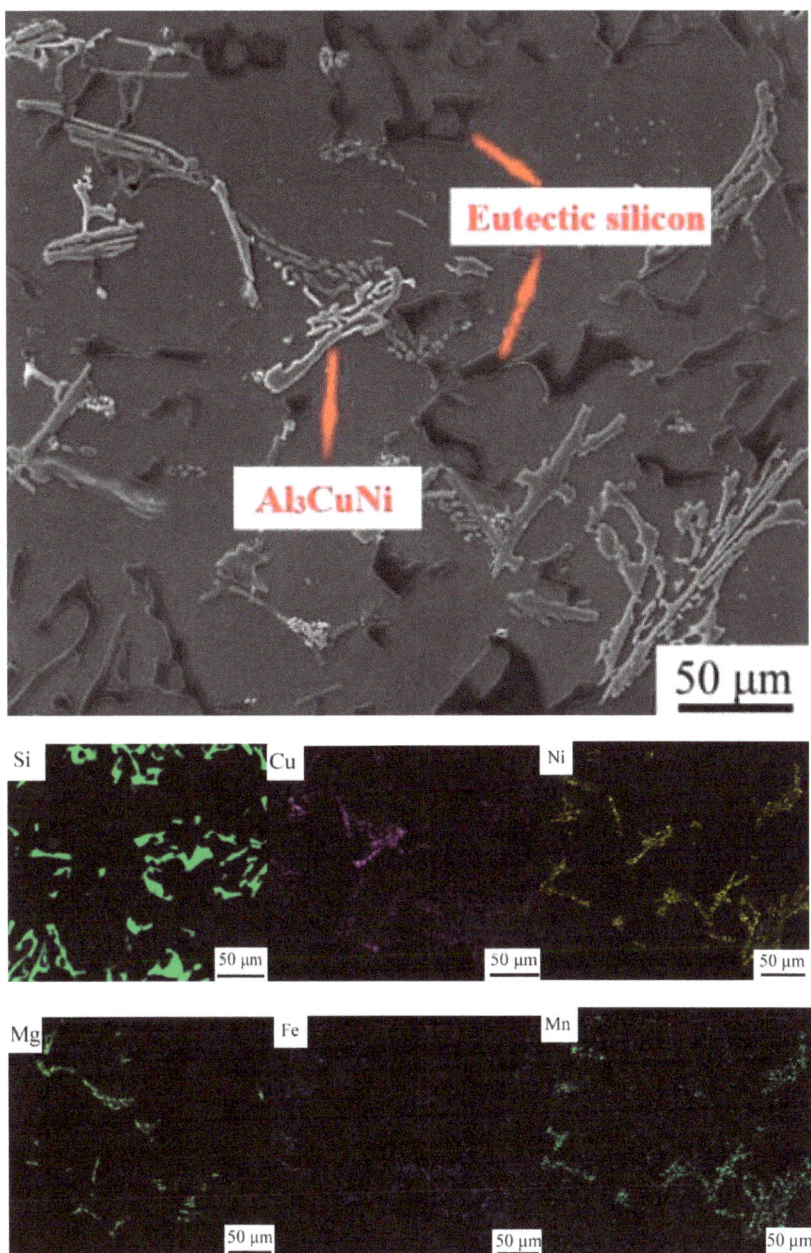

Figure 5. SEM morphology of thermally exposed Al-Si-Cu-Ni-Mg alloy and the distribution of alloy constituent element.

3.2. Evolution of Tensile Properties during Thermal Exposure

Figure 6 shows the typical stress–strain curves of the Al-Si-Cu-Ni-Mg alloy under different conditions tested at 250 °C. The ultimate tensile strength (UTS) of the alloy measured at 250 °C decreased from 202.8 MPa to 120.2 MPa after thermal exposure at 250 °C for 200 h, which is a 40.7% decrease. The elongation of the alloy measured at 250 °C increased from 0.33% to 0.95% after thermal exposure.

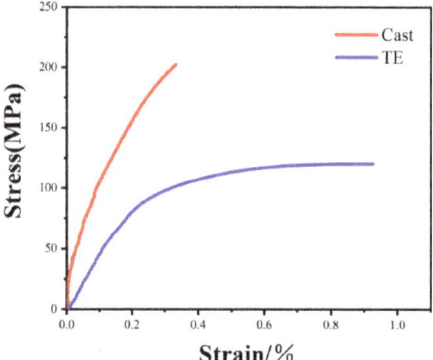

Figure 6. The typical stress–strain curves of Al-Si-Cu-Ni-Mg alloy under different conditions measured at 250 °C (TE represents thermal exposure at 250 °C for 200 h).

The tensile properties of the alloy with different states measured at different temperatures are displayed in Figure 7. The UTS of the as-cast alloy decreased from 266.8 MPa to 202.8 MPa when the temperature was increased from 25 °C to 250 °C, which is a drop of about 24%. The UTS of the TE alloy decreased from 266.8 MPa to 170.2 MPa at room temperature, with a decrease of about 36.2%. The elongation of the as-cast alloy increased slightly when the temperature was increased from 25 °C to 250 °C.

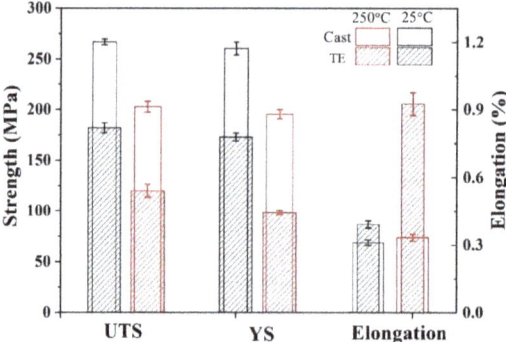

Figure 7. Tensile properties of the Al-Si-Cu-Ni-Mg alloy measured at different states (TE represents thermal exposure at 250 °C for 200 h).

The UTS of the TE alloy dropped from 170.2 MPa to 120.2 MPa when the temperature was increased from 25 °C to 250 °C, which is a significant drop of 29.4%. The elongation of the TE alloy increased greatly from 0.39% to 0.92% when the temperature was increased from 25 °C to 250 °C. With a test temperature of 250 °C, dislocations can climb to a certain extent under the action of stress, which provides more possibilities for alloy cross-slip and reduces the resistance of the cross-slip process. The climbing of a dislocation leads to a decrease in mechanical properties, as can be seen in the tensile strength decreasing accordingly.

As illustrated in Figure 8, the UTS of the alloy was tested at room temperature after thermal exposure at 250 °C for different times. The UTS of the studied alloys decreased greatly in the initial stage and then basically tended to be stable when the exposure temperature reached 32 h [12]. The UTS of the alloy was 173 MPa after thermal exposure at 250 °C for 200 h. Compared with the as-cast alloy, the UTS decreased by about 35%. The relation between the UTS (σ) at room temperature and the thermal exposure time (t) at 250 °C fitted using ExpAssoc is shown as follows:

$$\sigma_t^{250} = \sigma_0^{250} - 38(1 - Exp\left(-\frac{t}{0.84}\right)) - 56.8(1 - Exp\left(-\frac{t}{45.2}\right)), \; R^2 = 0.98 \quad (1)$$

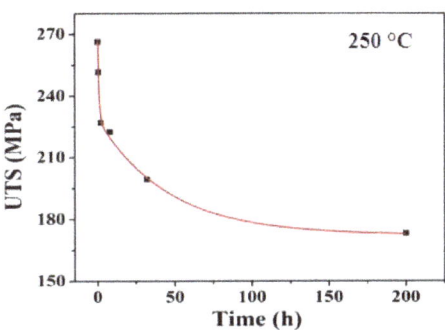

Figure 8. The ultimate tensile strength of the Al-Si-Cu-Ni-Mg alloy at room temperature fitted with exposure time at 250 °C for different times.

3.3. Creep Behavior before and after Thermal Exposure

The typical creep curve and creep-rate curve of the alloy are displayed in Figure 9. The creep rate decreased sharply in the first stage as a result of the work-hardening phenomenon. The creep stage where work hardening and dynamic recovery were in equilibrium was called the second stage of creep. The main feature of this stage was that the creep rate was constant and minimum. The creep rate of the material increased rapidly until fracturing, called the accelerated creep stage.

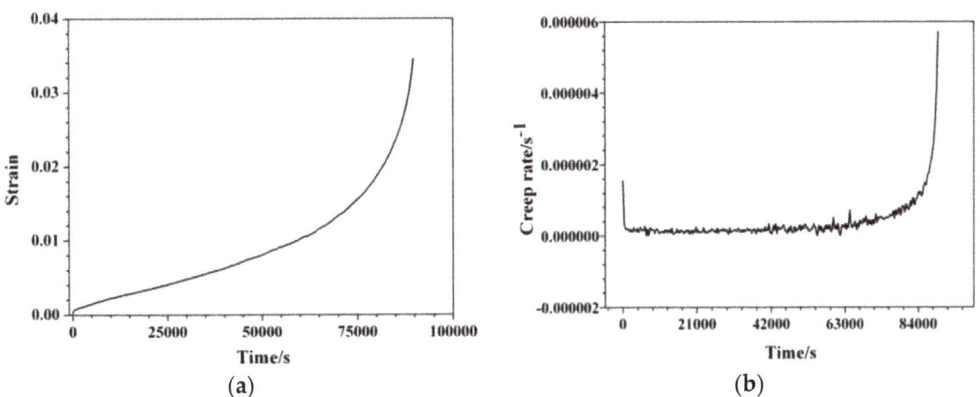

Figure 9. Typical creep curve (**a**) and creep-rate curve (**b**) of Al-Si-Cu-Ni-Mg alloy.

The minimum creep rates of the alloys at different states are presented in Table 2. As shown in Table 2, the minimum creep rates of the as-cast and thermally exposed alloys were similar at 250 °C. However, the minimum creep rate increased by almost an order of magnitude when the creep stress increased from 20 MPa to 28 MPa at the same temperature.

Table 2. Part of the minimum creep rate of Al-Si-Cu-Ni-Mg alloy before and after heat exposure under different creep conditions.

Alloy State	Temperature	Creep Rate (s^{-1})	
		20 MPa	28 MPa
As-cast alloy	250 °C	6.69×10^{-9}	1.02×10^{-8}
Thermal exposed alloy	250 °C	3.78×10^{-9}	1.10×10^{-8}

Table 2 also displays the minimum creep rate of the alloy after thermal exposure at 250 °C for 200 h. At 20 MPa/250 °C, the minimum creep rate of the alloy was around $10^{-9} \cdot s^{-1}$. However, the minimum creep rate of the alloy under the same conditions was lower than the minimum creep rate of the as-cast state, indicating that the creep resistance of the alloy at a low temperature was slightly improved after the thermal exposure. Compared with the minimum creep rate of the as-cast alloy, the minimum creep rate of the thermal exposed alloy was increased by an order of magnitude, in which the creep stress increased from 20 MPa to 28 MPa. Therefore, the creep resistance of the alloy significantly deteriorated.

The creep fracture morphology of the alloy at 20 MPa and 28 MPa is displayed in Figure 10. In contrast with Figure 10a, the quantity of dimples in Figure 10b significantly decreases. Apparently, Figure 10b has almost no dimples and a large area of cleavage planes.

Owing to the interactions of dislocation creating long-range stress fields in the alloy, the dislocations relied on shear stress to overcome these stress fields and move. Hence, the alloy with the higher applied stress load had a faster creep rate and worse creep resistance. The applied load condition of 28 MPa offered a larger shear stress, which provided more possibilities for the dislocation of the alloy. The larger applied load made the high-strength, hard and brittle phase more prone to brittle fracture, and the α-Al was less stressed. The large-scale precipitates had already initiated cracks and led to cracking when plastic deformation occurred, so a mass of dimples formed after plastic deformation, which is observable in Figure 10a but not in Figure 10b. Under the interaction of the stress field and high temperature, the strengthening phases shown in Figure 10b became globular more rapidly than those shown in Figure 10a, and the ability to deform was also decreased, which is explained by the phenomenon of a large-area cleavage surface occurring and no dimples forming after fracturing.

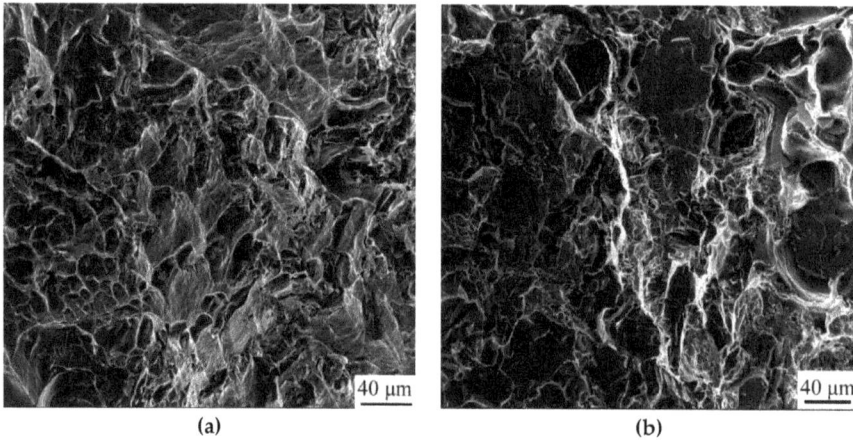

Figure 10. The creep fracture morphology of Al-Si-Cu-Ni-Mg alloy at (**a**) 20 MPa/250 °C and (**b**) 28 MPa/250 °C.

With 20 MPa of applied stress loaded, dislocation was hindered as a result of less shear stress distribution, and the tendency of stress concentration was less. The forces distributed in the α-Al and strengthening phase were relatively balanced. Compared with 28 MPa, the coarsening rate of the strengthening phase was lower with the combined action of the small stress field and temperature. The deformation performance declined slowly, and the strengthening phase reached the stress limit only after the plastic deformation of the α-Al produced cracks, manifesting as plenty of dimples and the cleavage surface being small. This could explain why the creep resistance deteriorated signally with the creep stress increasing from 20 MPa to 28 MPa at the same creep temperature.

The minimum creep rate and creep stress of materials are closely related to the creep temperature, and their corresponding relationship is expressed by the classical power law Equation (2) [16]:

$$\dot{\varepsilon} = A \cdot \sigma^n \cdot \exp\left(-\frac{Q}{RT}\right) \quad (2)$$

where $\dot{\varepsilon}$ is the minimum creep rate, A is the materials constant, σ is the creep stress, R is the gas constant (R = 8.31 J/mol), T is the absolute temperature, n is the creep stress exponent, and Q is the creep activation energy. Stress exponent n and activation energy Q are often used to infer the creep mechanism of alloys. The creep mechanism corresponding to stress exponent n and activation energy Q are summarized simply as follows: When the grain-boundary creep $n \leq 2$, which is divided into grain-boundary diffusion and grain-boundary slip, the creep activation energy Q corresponding to this creep mechanism is 82 kJ/mol. When N ≈ 3, this indicates the dislocation slip mechanism. When n is 4–6, this indicates a dislocation climbing mechanism, and the minimum creep rate is related to the diffusion of vacancy.

Using Equation (2), the creep stress exponent of the alloy at different temperatures was obtained. Figure 11 displays that the creep stress exponent of the as-cast alloy was about 1–2, and it is speculated that the creep mechanism of the material under this creep condition was grain-boundary creep. After thermal exposure at 250 °C for 200 h, the creep stress exponent of the alloy at 250 °C was 1.54, so the creep mechanism was grain-boundary creep.

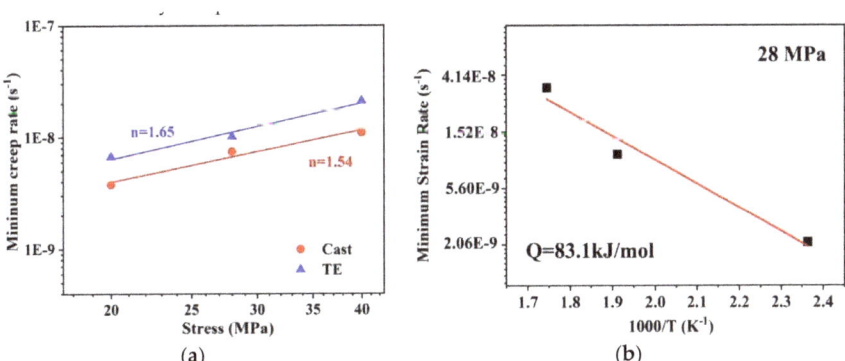

Figure 11. Al-Si-Cu-Ni-Mg alloy (**a**) creep stress exponent and (**b**) creep activation energy (TE represents thermal exposure at 250 °C for 200 h).

With the condition of invariable stress, the creep activation energy of the material at a temperature of 250 °C was obtained by changing the creep temperature; the result is shown in Figure 11b. Figure 11b displays that the creep activation energy of the alloy at 250 °C is 83.1 kJ/mol, which is very close to the theoretical grain-boundary diffusion activation energy of 82 kJ/mol. It also illustrates that the creep mechanism of the Al-Si-Cu-Ni-Mg alloy at 250 °C is grain-boundary creep.

After the steady-state creep, the accelerated creep stage occurred until the final failure of the material. The Monkman–Grant model is the most widely used among many material creep failure prediction models, which are summarized as the following Equation (3):

$$t_r \cdot \dot{\varepsilon}_{min}^m = C \quad (3)$$

where t_r is the creep life, $\dot{\varepsilon}_{min}$ is the minimum creep rate, and m and C are both material constants. In order to further understand the influence of material constants m and C on the creep life of the material, let the minimum creep rate = 1.10×10^{-8} s^{-1}, and the results of the influence of the material constants m and C on the creep life are presented in Figure 12.

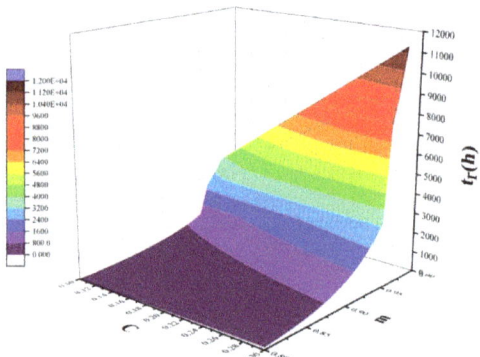

Figure 12. The influence of C and m constants in the Monkman–Grant model on the fracture life under the condition of constant minimum creep rate.

As shown in Figure 12, under the condition of constant m, creep life and material constant C increase linearly, and the increase is inconspicuous, while under the condition of constant C, creep life and material constant m increase exponentially, and the closer the m value is to 1, the more obvious its effect on creep life is.

The fitting results are displayed in Figure 13, showing that the data fitting degree is as high as 99.12%, and the material constant m is also close to 1, indicating the good creep resistance of the material to a certain extent. The Monkman–Grant Equation (4) is written as follows:

$$t_r \cdot \dot{\varepsilon}_{min}^{0.95} = 0.207 \quad (4)$$

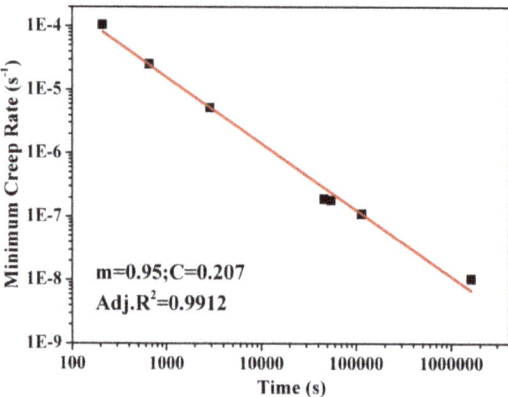

Figure 13. Monkman–Grant fitting of creep life of Al-Si-Cu-Ni-Mg alloy.

From Equation (4), it was calculated that for the creep life to reach 1000 h, the minimum creep rate should be lower than 2.4×10^{-8} s^{-1}. The relationship between the specific creep life and the corresponding minimum creep rate limit value are shown in Table 3.

Table 3. Correspondence table of creep life and minimum creep rate.

t_r (h)	10	100	1000
$\dot{\varepsilon}$ (s^{-1})	3.0×10^{-6}	2.7×10^{-7}	2.4×10^{-8}

4. Conclusions

The effect of thermal exposure on the mechanical properties of Al-Si-Cu-Ni-Mg aluminum alloy has been investigated in the present work. The main conclusions of this work are summarized as follows:

1. The main phases at the micro scale of the Al-Si-Cu-Ni-Mg alloy were stable during thermal exposure at 250 °C.
2. The UTS of the studied alloys decreased greatly in the initial stage and then basically tended to be stable after thermal exposure at 250 °C for about 100 h.
3. The creep resistance deteriorated signally when the creep stress was increased from 20 MPa to 28 MPa at the same creep temperature.
4. The main creep mechanism of the as-cast Al-Si-Cu-Ni-Mg alloy at a low temperature and low stress (T ≤ 250 °C; σ ≤ 40 MPa) was grain-boundary creep. The Monkman–Grant empirical formula was used to fit the relationship between the creep life and the minimum creep rate of the cast Al-Si-Cu-Ni-Mg alloy, and the fitting results were: $t_r \cdot \dot{\varepsilon}_{min}^{0.95} = 0.207$.

Author Contributions: Conceptualization, F.C.; methodology, C.L.; software, L.Z.; validation, Z.W.; formal analysis, Y.H.; investigation, K.D.; resources, Z.W.; data curation, L.Z.; writing—original draft preparation, G.L.; writing—review and editing, F.C.; visualization, W.H.; supervision, L.Z.; project administration, L.Z.; funding acquisition, Y.H. All authors have read and agreed to the published version of the manuscript.

Funding: This research was funded by the Natural Science Foundation of the Jiangsu Higher Education Institutions of China (Grant No. 20KJB430015) and the Natural Science Foundation of Jiangsu Province (Grant No. BK20201467).

Data Availability Statement: Not applicable.

Acknowledgments: This work was supported by the Natural Science Foundation of the Jiangsu Higher Education Institutions of China (Grant No. 20KJB430015). This work was supported by the Natural Science Foundation of Jiangsu Province (Grant No. BK20201467).

Conflicts of Interest: The authors declare no conflict of interest.

References

1. Mikhaylovskaya, A.V.; Kishchik, A.A.; Tabachkova, N.Y.; Kotov, A.D.; Cheverikin, V.V.; Bazlov, A.I. Microstructural Characterization and Tensile Properties of Al-Mg-Fe-Ce Alloy at Room and Elevated Temperatures. *JOM* **2020**, *72*, 1619–1626. [CrossRef]
2. Hu, K.; Xu, Q.; Ma, X.; Sun, Q.; Gao, T.; Liu, X. A novel heat-resistant Al-Si-Cu-Ni-Mg base material synergistically strengthened by Ni-rich intermetallics and nano-AlNp microskele-tons. *Mater. Sci. Technol.* **2019**, *35*, 306–312. [CrossRef]
3. Mahmoud, E.R.; Shaharoun, A.; Gepreel, M.A.; Ebied, S. Microstructure and Mechanical Properties of Fe-Mn-Ni-Cr-Al-Si High Entropy Alloys. *Metals* **2022**, *12*, 1164. [CrossRef]
4. Silva, C.; Barros, A.; Vida, T.; Garcia, A.; Cheung, N.; Reis, D.A.; Brito, C. Assessing Microstructure Tensile Properties Relationships in Al-7Si-Mg Alloys via Multiple Regression. *Metals* **2022**, *12*, 1040. [CrossRef]
5. Zhang, J.; Feng, J.; Zuo, L.; Ye, B.; Kong, X.Y.; Jiang, H.; Ding, W. Effect of Sc microalloying addition on microstructure and mechanical properties of as-cast Al-12Si alloy. *Mater. Sci. Eng.* **2019**, *766*, 138343.1–138343.4. [CrossRef]
6. Liu, H.Q.; Pang, J.C.; Wang, M.; Li, S.X.; Zhang, Z.F. Effect of temperature on the mechanical properties of Al-Si-Cu-Mg-Ni-Ce alloy. *Mater. Sci. Eng. A* **2021**, *3*, 141762. [CrossRef]
7. Shlyaptseva, A.D.; Petrov, I.A.; Ryakhovsky, A.P.; Medvedeva, E.V.; Tcherdyntsev, V.V. Complex Structure Modification and Improvement of Properties of Aluminium Casting Alloys with Various Silicon Content. *Metals* **2021**, *11*, 1946. [CrossRef]

8. Sui, Y.; Wang, Q.; Liu, T.; Ye, B.; Jiang, H.; Ding, W. Influence of Gd content on microstructure and mechanical properties of cast Al-12Si-4Cu-2Ni-0.8Mg alloys. *J. Alloy. Compd.* **2015**, *622*, 572–579. [CrossRef]
9. Herve, E.; Dendievel, R.; Bonnet, G. Steady-state power-law creep in "inclusion matrix" composite materials. *Acta Metall.* **1995**, *43*, 4027–4034. [CrossRef]
10. Zhang, J.; Fan, J.; Chen, L.; Li, Y. Compressive creep aging behavior and microstructure evolution in extruded Al-Mg-Si alloy under different temperature and stress levels. *Mater. Today Commun.* **2022**, *33*, 104722. [CrossRef]
11. Zhao, B.; Ye, B.; Wang, L.; Bai, Y.; Yu, X.; Wang, Q.; Yang, W. Effect of ageing and thermal exposure on microstructure and mechanical properties of a HPDC Al–Si–Cu–Mg alloy. *Mater. Sci. Eng. A* **2022**, *849*, 143463. [CrossRef]
12. Lin, B.; Li, H.; Xu, R.; Zhao, Y.; Xiao, H.; Tang, Z.; Li, S. Thermal exposure of Al-Si-Cu-Mn-Fe alloys and its contribution to high temperature mechanical properties. *J. Mater. Res. Technol.* **2020**, *9*, 1856–1865. [CrossRef]
13. Yuan, J.; Wang, Q.; Yin, D.; Wang, H.; Chen, C.; Ye, B. Creep behavior of Mg-9Gd-1Y-0.5Zr (wt.%) alloy piston by squeeze casting. *Mater. Charact.* **2013**, *78*, 37–46. [CrossRef]
14. Dong, Z.-Q.; Wang, J.-G.; Guan, Z.-P.; Ma, P.-K.; Zhao, P.; Li, Z.-J.; Lu, T.-S.; Yan, R.-F. Effect of Short T6 Heat Treatment on the Thermal Conductivity and Mechanical Properties of Different Casting Processes Al-Si-Mg-Cu Alloys. *Metals* **2021**, *11*, 1450. [CrossRef]
15. Ning, Z.L.; Yi, J.Y.; Qian, M.; Sun, H.C.; Cao, F.Y.; Liu, H.H.; Sun, J.F. Microstructure and elevated temperature mechanical and creep properties of Mg-4Y-3Nd-0.5Zr alloy in the product form of a large structural casting. *Mater. Des.* **2014**, *60*, 218–225. [CrossRef]
16. Zafar, H.; Khushaim, M.; Ravaux, F.; Anjum, D.H. Scale-Dependent Structure-Property Correlations of Precipitation-Hardened Aluminum Alloys: A Review. *JOM* **2022**, *74*, 361–380. [CrossRef]
17. Kaja, S.S.T.; Gangadasari, P.R.; Ayyagari, K.P.R. Extending the Tolerance of Iron in Cast Al-Si Alloy. *JOM* **2021**, *73*, 2652–2657. [CrossRef]
18. *ISO 204:2009*; Metallic Materials—Uninterrupted Materials Uniaxial Creep Testingin Tension—Method of Test. International Organization for Standardization: Geneva, Switzerland, 2009.
19. Zuo, L.; Ye, B.; Feng, J.; Kong, X.; Jiang, H.; Ding, W. Effect of Q-$Al_5Cu_2Mg_8Si_6$ phase on mechanical properties of Al-Si-Cu-Mg alloy at elevated temperature. *Mater. Sci. Eng. A* **2017**, *693*, 26–32. [CrossRef]

Disclaimer/Publisher's Note: The statements, opinions and data contained in all publications are solely those of the individual author(s) and contributor(s) and not of MDPI and/or the editor(s). MDPI and/or the editor(s) disclaim responsibility for any injury to people or property resulting from any ideas, methods, instructions or products referred to in the content.

Article

Synthesis of a Novel Zinc(II) Porphyrin Complex, Halide Ion Reception, Catalytic Degradation of Dyes, and Optoelectronic Application

Soumaya Nasri [1,2,*], Mouhieddinne Guergueb [2], Jihed Brahmi [2], Youssef O. Al-Ghamdi [1], Frédérique Loiseau [3] and Habib Nasri [2]

[1] Department of Chemistry, College of Science Al-Zulfi, Majmaah University, Majmaah 11952, Saudi Arabia
[2] Laboratory of Physical Chemistry of Materials, Faculty of Sciences of Monastir, University of Monastir, Avenue de l'environnement, Monastir 5019, Tunisia
[3] Département de Chimie Moléculaire, 301 rue de la Chimie, Université Grenoble Alpes, CS 40700, CEDEX 9, 38058 Grenoble, France
* Correspondence: soumaya.n@mu.edu.sa

Abstract: This work describes the synthesis of a novel zinc(II) porphyrin complex, namely [*Meso*-4α-tetra-(1,2,3-triazolyl)phenylporphyrinato]zinc(II) symbolized by 4α-[Zn(TAzPP)] (**4**), using the click chemistry approach in the presence of copper iodide. All of the synthetic porphyrin species reported herein were fully characterized by elemental analysis, infrared spectroscopy, proton nuclear magnetic resonance, UV-visible spectroscopy, and fluorescence. To synthesize the 4α-[Zn(TAzPP)] complex (**4**), we produced 4α-*Meso*-tetra-o-nitrophenylporphyrin ($H_2T_{NO_2}PP$) and 4α-*meso*-tetra-o-aminophenylporphyrin (4α-H_2TNH_2PP) (**1**) using known classic literature methods. This 4α atropisomer was converted to 4α-*meso*-tetra-o-azidophenylporphyrin (4α-H_2TN_3PP) (**3**) by reaction with sodium nitrite and sodium azide, and then it was metalated by Zn(II), leading to [4α-*meso*-tetra(2-azidophenyl)porphyrinate]zinc(II) (4α-[Zn(TN_3PP)]) (**3**). The click chemistry synthetic method was finally used to prepare 4α-[Zn(TAzPP)] (**4**). This new tetracoordinated zinc(II) porphyrin complex was prepared and characterized in order to: (i) produce a receptor for anion recognition and sensing application for Cl$^-$ and Br$^-$; (ii) study the catalytic decomposition of rhodamine B (RhB) and methyl orange (MO) dyes; and (iii) determine the electronic characteristics as a photovoltaic device. Complex (**4**) formed 1:1 complex stoichiometric species with chloride and bromide halides and the average association constants of the 1.1 addicts were ~ 10^3. The photodecomposition of RhB and MO dyes in the presence of complex (**4**) as a catalyst and molecular oxygen showed that complex (**4**) presented a photodegradation yield of approximately 70% and could be reused for five successive cycles without any obvious change in its catalytic activity. The current-voltage characteristics and impedance spectroscopy measurements of complex (**4**) confirmed that our zinc(II) metalloporphyrin could be used as a photovoltaic device.

Keywords: zinc(II) porphyrins; click chemistry; optical anion sensing; UV-visible titration; photoelectronic degradation; photovoltaic devices

1. Introduction

Porphyrins are aromatic tetrapyrrolic macrocycles that are widely represented in living systems. They participate, in a metalated form, in many biological processes. This is the case of hemoglobin and myoglobin, which are built on the basis of the iron protoporphyrin IX complex (heme) and ensure the transport and storage of molecular oxygen [1]. Such natural macromolecules are also involved in the oxidation of substrates by cytochromes (especially cytochromes P450) [2] or in photosynthesis in plants and photosynthetic bacteria.

Unlike iron, cobalt, magnesium, and nickel metals present in natural metalloporphyrins, zinc(II) is not present in biological systems. Nevertheless, synthetic zinc(II) porphyrin complexes are actually widely used in a large number of fields, e.g., the manufacture of liquid crystals used in display devices for watches, computer screens, etc. [3,4]. These porphyrinic derivatives are also used in the design of biosensors [5] as well as in photoluminescent [6] and optoelectronic systems [7].

In is worth noting that anion sensing is a rapidly expanding area of research in supramolecular chemistry [8–14]. This stems from many fundamental roles that the anion plays in nature, with biological, chemical, biomedical, and environmental applications.

During the last two decades, many investigations have been devoted to the preparation, characterization, and study of new compounds to be used in the detection of ionic species. Studying the recognition and sensing of such ionic inorganic species is important for several reasons: (i) cationic and anionic inorganic compounds, such as cations of heavy metals (e.g., Cd^{2+}, Hg^{2+}, Pb^{2+}, Sb^{5+}) and many anions (e.g., CN^-, $Cr_2O_7^{2-}$, AsO_4^{3-}), are very toxic and must be removed from the environment; (ii) anions such as NO_3^- and PO_4^{2-} are present in agricultural fertilizers; and (iii) many ions, such as K^+, Na^+, and F^-, have very important roles in the functioning of biological systems [15,16]. Among compounds used for the detection and sensing of inorganic ions, metal-organic frameworks (MOFs) should be mentioned in the first place [17,18]. The other important species used as receptors and sensors of anionic and cationic inorganic compounds are the calixarenes, especially the calix [6], homooxacalix [3], and homoazacalix [3] arenes [19,20].

On the other hand, porphyrins and metalloporphyrins are very attractive hosts to use for anion recognition studies, as they are spectrophoto-electroactive, which enables the complexation of anions via several physical methods. It has been shown that the well-known *meso*-tetraphenylporphyrin (H_2TPP) does not have anion binding power alone [21,22]. This is due to the small size of the cavity of this porphyrin, which does not complex anions via hydrogen bonding interactions between the ion anion and the porphyrin N ··· H bonds.

In addition, the rigidity of the porphyrin backbone and the cavity also weaken the formation of anionic bonds. This gave rise to the expansion of the porphyrinic cavity. This is the case for urea porphyrins, also known as "picket fence porphyrins" [23], and metalloporphyrin-cage systems [24].

The sensing of anions by hosts that are zinc(II) porphyrin complexes can be monitored by UV-visible spectral titration studies, e.g., the detection of Cl^- and Br^- ions by the zinc(II) porphyrin complex [24]. On the other hand, recent developments with porphyrin-based solar cells exhibit a promising advance because they use low production cost materials, are easy to synthesize, have low toxicity, rigid geometry, and efficient electron transfer, etc. [25–33]. Moreover, porphyrin-based solar cells possess high molar absorption coefficients and exceptional light harvesting properties, which make them excellent sensitizers for dye-sensitized solar cells (DSSCs) [34–42].

In this work, a new *meso*-porphyrin, namely [4α-*meso*-tetra-(1,2,3-triazolyl)-phenylporphyrinato]zinc(II) symbolized by 4α-[Zn(TAzPP)] (**4**), was synthesized using the click chemistry method [43–45], and the ability of this new zinc(II) porphyrin complex to capture Cl^- and Br^- ions was studied. UV-visible, fluorescence, IR, and ^1H NMR spectroscopic characterization of (**4**) is described. The bonding of Cl^- and Br^- ions by complex (**4**), investigated by UV-visible titration, is also reported. Furthermore, the efficiency of the catalytic oxidative degradation and photocatalysis of rhodamine B (RhB) and methyl orange (MO) dyes using the triazole *meso*-arylporphyrin zinc complex were also investigated. Additionally, the current-voltage characteristics and impedance spectroscopy measurements of 4α-[Zn(TAzPP)] (**4**) were studied to determine their electronic properties.

2. Method and Materials

All commercially available reagents were used without further purification. All anions that were used for selectivity testing were in the form of tetrabutylammonium salt.

UV-visible absorption spectra and titration were recorded on a WinASPECT PLUS (SPECORD PLUS version 4.2 validation) scanning spectrophotometer. ^1H NMR spectroscopy was performed on a Bruker DPX 400 spectrometer and chemical shifts are reported in ppm below the internal tetramethylsilane (TMS) field. IR spectra with Fourier transformation were obtained using a PerkinElmer Spectrum Two FT-IR spectrometer. Emission spectra were recorded in dichloromethane at room temperature on a Horiba Scientific FluoroMax-4 spectrofluorometer. Samples were placed in 1 cm path length quartz cuvettes. Luminescence lifetime measurements were performed after irradiation at = 430 nm obtained by the second harmonic of a titanium: sapphire laser (Tsunami Spectra Physics 3950-M1BB picosecond laser + 39868-03 pulse doubler) at a repetition rate of 800 kHz. The luminescence decays were studied with FLUOFIT software (Picoquant). The emission quantum yields were calculated at room temperature in dichloromethane solutions using the optical dilution method. [Zn(TPP)] in air-equilibrated dichloromethane solution was chosen as the quantum yield standard (ϕ_f = 0.031) [46].

The oxidative degradation and photodegradation of MO and RhB dye experiments were performed at room temperature using 10 mg of the catalyst compound and 10 mL of an aqueous solution of the MO and RhB dyes (at pH = 6). Stirring was kept at 250 rpm. The resulting mixture was filtered, and the concentration was then recorded by measuring the absorption at 555 and 418 nm for MO and RhB dyes, respectively. The decolorization yields (R%) are given by the following relationship (Equation 1):

$$R\% = (A_o - A_t)/A_o.100 \qquad (1)$$

where A_o and A_t are the absorption at t = 0 and at the t instant, respectively.

3. Results and Discussion

3.1. Synthesis

4α-*meso*-tetra-*o*-nitrophenylporphyrin($H_2T_{NO2}PP$) was synthesized using the method described in the literature [47]. 4α-*meso*-tetra-*o*-aminophenylporphyrin (**1**) (4α-$H_2T_{NH2}PP$) was then prepared by the reduction of the nitro group of 4α-*meso*-tetra-*o*-nitrophenylporphyrin to the amine group, following the literature method [47] (Scheme 1). Separation was carried out using a one-column procedure that enriched the desired cis isomer (designated by α atropisomer), as described in the literature [48], leading to 4α-*meso*-tetra-*o*-aminophenylporphyrin (4α-$H_2T_{NH2}PP$) (**1**). 4α-*meso*-tetra-*o*-azidophenylporphyrin (**2**) (4α-$H_2T_{N3}PP$) was produced using sodium nitrite (NaNO$_2$) and sodium azide (NaN$_3$). Compound (**2**) was then metalated using Zn(OAc)$_2$·2H$_2$O, leading to [4α-meso-tetra(2-azidophenyl)porphyrinato]zinc(II) (**3**) (4α-[Zn(TN$_3$PP)]) Finally, using the click chemistry reaction [49], [4α-*meso*-tetra-(1,2,3-triazolyl)phenylporphyrinato]zinc(II) (**4**) (4α-[Zn(TAzPP)]) was synthesized.

3.2. Spectroscopic ^1H NMR and IR Data

For the 4α-$H_2T_{NH2}PP$ (**1**) and 4α-$H_2T_{N3}PP$ (**2**) free-base porphyrins, the characteristic types of protons were observed. Thus, the NH-pyrrolic protons, which are exchangeable and strongly shielded, appeared between −2.5 and −2.7 ppm. The eight β-pyrrolic protons of the porphyrin macrocycle resonated around 8.8 ppm. The phenyl protons of these two *meso*-porphyrins resonated in the range of 8.88 to 7.49 ppm. For the 4α-$H_2T_{NH2}PP$ porphyrin, a singlet was shown around 3.56 ppm, which corresponded to the amine protons (Figures S5 and S6).

The disappearance of the signal at −2.68 ppm, corresponding to NH-pyrrolic protons of compound (**2**), was an indication of the insertion of the Zn(II) ion into the porphyrin ring (Figure S6). The positions of the peaks of the Hβ-pyrrolic protons, as well as those of the phenyl protons of the 4α-[Zn(TN$_3$PP)] and 4α-[Zn(TAzPP)] complexes (**3**)–(**4**), underwent a slight shift compared to those of the 4α-$H_2T_{NH2}PP$ and 4α-$H_2T_{N3}PP$ free-base porphyrins (Figures S5–S8) [49].

The azide stretching vibration ν(N$_3$) was easily identified from the IR spectra of compounds (**2**) and (**3**), which appeared in the 2130–2068 and 2133–2098 cm^{-1} domains, respectively. The IR spectrum of 4α-[Zn(TAzPP)] (**4**) confirmed the formation of the triazole *meso*-arylporphyrin which showed a strong absorption band at 1731 cm^{-1} attributed to the ν(N==N) and ν(C==N) stretching vibrations of the triazole group (Figures S1–S4) [50].

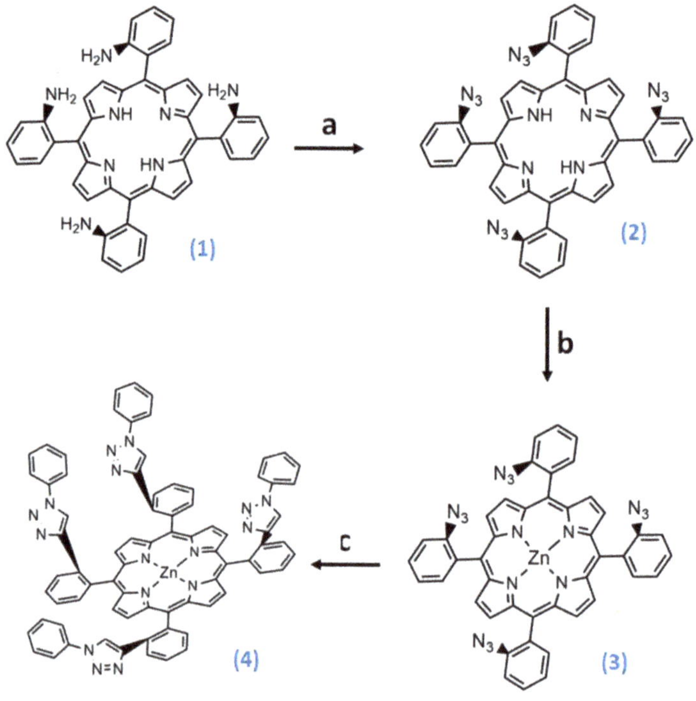

Scheme 1. General scheme for the synthesis of compounds (**1**)–(**4**): (**a**): (1) HCl, H$_2$O; (2) NaNO$_2$, HCl, H$_2$O; (3) NaN$_3$, HCl, H$_2$O, (**b**): Zn(OAc)·2H$_2$O, CHCl$_3$/C$_2$H$_5$O, (**c**): CuI, Et$_3$N, Phenylacetylene, in THF/Acetonitrile.

3.3. Optical Absorption

Figure 1 depicts the electronic absorption spectra of compounds (**1**)–(**4**), while the UV-visible data of these porphyrinic species are given in Table 1. 4α-H$_2$T$_{NH2}$PP and 4α-H$_2$T$_{N3}$PP free-base *meso*-porphyrins (**1**)–(**2**) presented similar UV-visible spectra in solution, with λ$_{max}$ values of the Soret band at ca. 424 nm and four Q bands at ca. 515, 550, 590, and 660 nm. The UV-visible spectra of the 4α-[Zn(T$_{N3}$PP)] (**3**) and 4α-[Zn(TAzPP)] (**4**) Zn(II) porphyrin complexes were slightly shifted compared to those of the corresponding free-base porphyrins, and the number of Q bands was reduced from four to two, which was indicative of the metalation of a porphyrin [51].

The optical gap (E$_g$-op) values of compounds (**1**)–(**4**) were 1.83, 1.92, 2.02, and 2.03 eV, respectively. In particular, the E$_g$-op values of Zn(II)-metalloporphyrins were close to 2.00 eV. It is worth mentioning that the optical gap values of the two zinc(II) metalloporphyrins indicated that these complexes could be used for the development of new optoelectronic organic semiconductor materials [52].

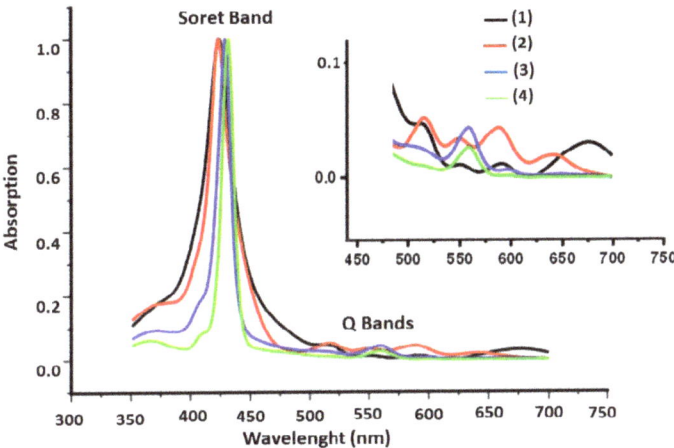

Figure 1. UV–visible spectra of compounds (**1**)–(**4**) recorded in dichloromethane at concentrations ~10^{-6} M. The inset shows the enlarged view of the Q bands region.

Table 1. UV-visible data of the free-base porphyrins and the *meso*-arylporphyrin zinc(II) tetracoordinated complexes. The spectra were recorded in dichloromethane.

Compound	λ_{max} (nm) ($\varepsilon \times 10^{-3} M^{-1}.cm^{-1}$)					$E_{gap-opt}$ (eV)	Ref
	Soret band	Q bands					
Free-base meso-arylporphyrins							
H$_2$(TPP) [a]	416(419)	513(20)	550(20)	590(6)	646(6)	1.89	[53]
H$_2$(TEBOP) [b]	422(295)	517(9)	554(8)	593(5)	651(7)	1.85	[54]
H$_2$(T$_{AzP}$-IVP) [c]	424(576)	520(46)	555(29)	595(24)	652(18)	1.86	[55]
H$_2$T$_{NH_2}$PP	424(545)	514(39)	552(41)	592(40)	677(35)	1.83	this work
H$_2$T$_{N_3}$PP	424(519)	516(39)	550(37)	594(38)	642(36)	1.92	this work
Zinc(II) meso-arylporphyrin complexes							
[Zn(TPP)]	421(524)	550(21)	591(25)			1.91	[43]
[Zn(T$_{AzP}$-IVP)]	424(530)	551(26)	592(10)			2.04	[50]
4α-[Zn(T$_{N_3}$PP)]	430(535)	560(410)	598(361)			2.02	this work
4α-[ZnTAzPP]	430(544)	561(394)	601(321)			2.03	this work

[a]: TTP = *meso*-tetratolylporphyrinato, [b]: TEBOP = *meso*-tetrakis(ethyl-4(4- butyryl)oxyphenyl)porphyrinato, [c]: TAzP-IVP = 4-((1-(4-iodinephenyl)-1H-1,2,3-triazol-4-yl)methoxy)-3-methoxyphenyl.

3.4. Photoluminescence Studies

Porphyrins and metalloporphyrins are known to exhibit two types of emissions. The first emission type, which is between the second excited state S$_2$ and the ground state S$_o$ (S$_2$→S$_o$), corresponds to the Q bands [Q (0,0) and Q(0,1)]. The second emission type is between the first exited state S$_1$ and the ground state S$_o$ (S$_1$→S$_o$), corresponding to the Soret band. The S$_2$→S$_o$ emission is very weak and negligible; only the S$_1$→S$_o$ emission is considered for porphyrins and metalloporphyrins.

As shown in Figure 2, we noticed a major hypochromic shift of approximately 50 nm of the Q(0,0) and Q(0.1) bands between free-base porphyrins (**1**) and (**2**) and their corresponding zinc porphyrins. The Q(0,0) and Q(0,1) emission bands of compounds (**3**) and (**4**) had wavelengths of about 600 and 665 nm, respectively. The quantum yield values of compounds (**1**)–(**4**) were 0.085, 0.078, 0.054, and 0.033, respectively. The decrease in the fluorescence quantum yield values (Φ_f) was due to the insertion of zinc(II) on the free-base porphyrins. The fluorescent lifetime values of compounds (**1**)–(**4**) were 8.61, 8.78, 3.1, and 1.91, respectively. The photophysical property values of the synthesized compounds showed that they could be used for various optoelectronic applications, a priori DSSC systems.

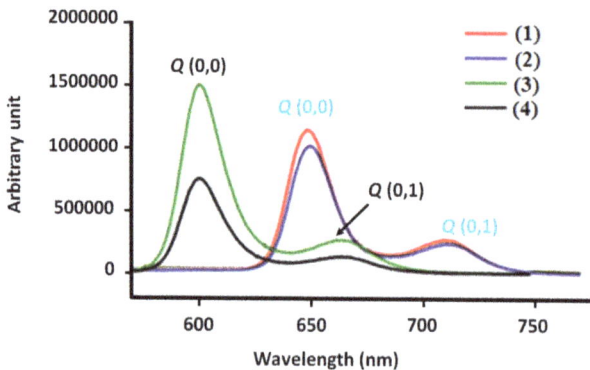

Figure 2. Emission spectra of compounds (**1**)–(**4**). The spectra were recorded in dichloromethane at concentrations ~10^{-6} M. The excitation wavelength value was 430 nm.

3.5. Anion Binding Studies

[4α-*meso*-tetra-(1,2,3-triazolyl)phenylporphyrinato]zinc(II) complex (**4**) (4α-[Zn(TAzPP)]) was tested as a detector of Cl⁻ and Br⁻ anions by UV-visible titration in dichloromethane. Anions were added as their salts of the non-complexing cation tetrabutylammonium (TBA).

The UV-visible titration spectra of complex (**4**) showed a clear change of the Soret and Q bands as the concentration of the Cl⁻ and Br- anions increased (Figure 3a,b). The titrations for Cl⁻ and Br⁻ ions on [Zn(TTP)] (TTP = *meso*-tolylporphyrin) used as a reference are shown in Figure 3c,d. Table 2 summarizes the values of the association constants K_{as} for [Zn(Porph)Cl] (Porph = TTP and TAzPP) and [Zn(Porph)Br] complexes. The K_{as} values obtained from the titration of the [Zn(PC)X] complex with the cage porphyrin (PC = 4α-*meso*-(tetrakis(2-azidoacetamidophenyl)porphyrinate [23] with Cl⁻ and Br⁻ ions are also shown in Table 2.

Upon successive addition of Cl⁻ to complex (**4**), the UV-visible titration study showed a bathochromic shift of the Soret band from 430 to 439 nm ($\Delta\lambda_{max}$ = 9 nm), with one distinct isosbestic point at 434 nm, thus proving the formation of a 1:1 coordination complex type [Zn(Porph)(L)] (L = axial ligand). A red shift was also observed for the Q(0,0) and Q(0,1) bands. Similar changes were also noted upon Br⁻ addition to a solution of 4α-[Zn(TAzPP)] (**4**), using the same concentrations, showing a red shift of the Soret and Q bands. As the titration progressed, an isosbestic point was also observed at 436 nm for the Soret band.

A UV-visible titration with zinc(II)-*meso*-tetratolylphenylporphyrin ([Zn(TTP)]) was also performed to compare the Cl⁻ and Br⁻ detecting properties of zinc complex (**5**) with those of the [Zn(TTP)] complex.

The association constants for the 1:1 complex, calculated using the so-called "strong interactions" method [56] (see the supplementary information for details), are summarized in Table 2. From this table, it can be seen that in the case of Cl-, the average K_{as} value of 4α-[Zn(TAzPP)] (**4**) was 0.301×10^3, which was higher than that of [Zn(TTP)] porphyrin with a K_{as} value equal to 0.063×10^3. On the other hand, these two values were far lower than that obtained with the cage porphyrin PC [23], with a value is equal to 1.220×10^4. For the bromide ion, the K_{as} values were 0.441×10^3 for porphyrin derivative (**4**) and 0.168×10^3 for the [Zn(TTP)] complex, while the association constant K_{as} value for [Zn(PC)Br] was equal to 0.005 [23]. These results showed that our synthetic zinc(II) porphyrin 4α-[Zn(TAzPP)] (**4**) was selective for Br− over Cl⁻ anions and that complex (**4**) presented a better binding affinity for Br⁻ than the cage porphyrin (PC). This could be explained by the fact that the cage porphyrin has a cavity which is not large enough to accommodate the large size of the bromide ion.

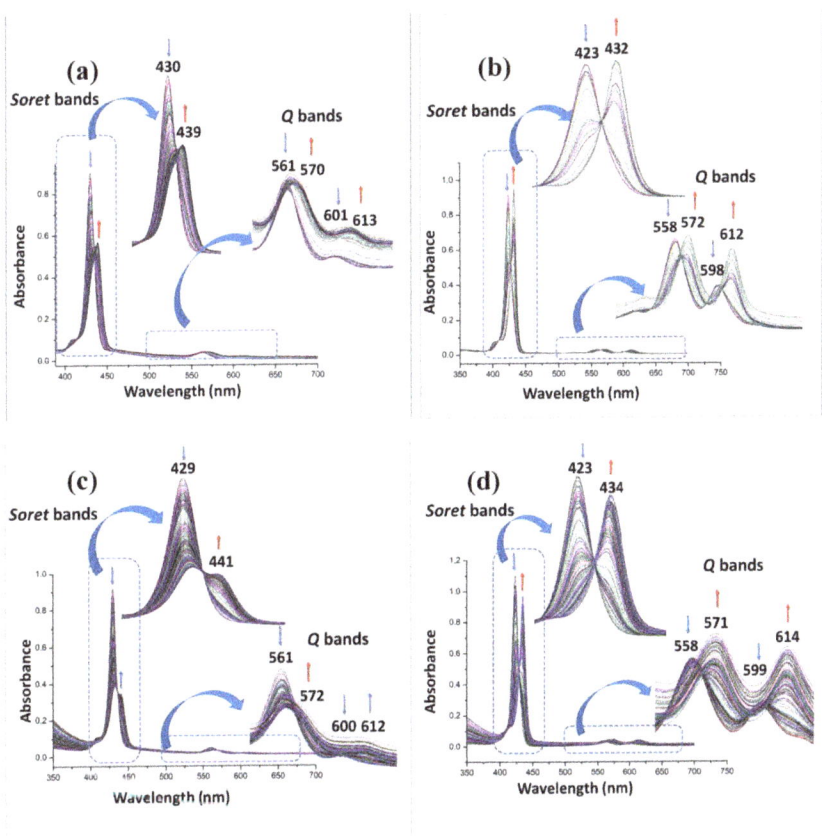

Figure 3. Evolution of the Q and Soret bands: (**a**) complex (**4**) as a function of the addition of the Cl⁻ anion, (**b**) [Zn(TTP)] as a function of the addition of the Cl⁻ anion, (**c**) complex (**4**) as a function of the addition of the Br⁻ anion, and (**d**) [Zn(TTP)] as a function of the addition of the Br⁻ anion.

Table 2. Values of association constants (K_{as}) and log(K_{as}) for our zinc(II) metalloporphyrins and other related complexes.

Complexes	log(K_{as})	(K_{as})	Ref.
[Zn(TAzPP)Cl]	2.407	0.301×10^3	this work
[Zn(TTP)Cl]	1.791	0.063×10^3	this work
[Zn(PC)Cl]	-	1.220×10^4	[23]
[Zn(TAzPP)Br]	1.299	0.441×10^3	this work
[Zn(TTP)Br]	1.789	0.168×10^3	this work
[Zn(PC)Br]	-	0.005	[23]

3.6. Degradation of Rhodamine B (RhB) and Methyl Orange (MO) Dyes

The ability of complex (**4**) to catalyze the degradation of RhB and MO dyes was tested using an aqueous hydrogen peroxide solution at room temperature. The optimal condition of this degradation was found to be as follows: mass of complex (**4**) was m = 10 mg, the H_2O_2 aqueous solution concentration was C_o = 20 mg.L^{-1}.

The oxidation of organic compounds by hydrogen peroxide catalyzed per metallic species is known to involve the radical •OH, leading to a formation of intermediate species. In our case, the disappearance rate of the RhB and MO dyes could be obtained through the following equation (Equation (2)):

$$\frac{dC}{dt} = -k.C.[OH^\cdot] \qquad (2)$$

where C is the concentration of the MO and RhB dyes at time t and k is defined as the second order rate constant of the MO and Rh B dyes reacting with •OH. The equation can be further simplified if one considers that the concentration of •OH is constant, assuming the steady state situation for the net formation rate of these intermediates. Thus, the degradation rate of the MO and RhB dyes due to the combination of hydrogen peroxide is finally given by Equation (3):

$$\frac{dC}{dt} = -k_o.C \qquad (3)$$

where k_o (in min^{-1}) is the pseudo-first order rate constant, and C_t and C_o are the concentrations at time t and the initial concentration, respectively. Figure 4 shows the curves C_t/C_o versus time. The degradation yield ($R\%$) is given by the following relation (Equation (4)):

$$R(\%) = \left(\frac{C_o - C_t}{C_o}\right).100 \qquad (4)$$

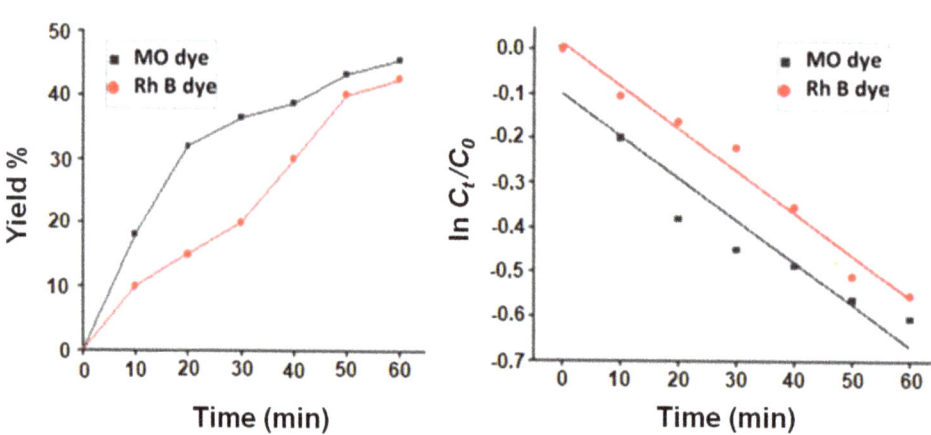

Figure 4. (Left) Changes in C_t/C_o versus time for the following conditions: H_2O_2 + MO + complex (**4**) and H_2O_2 + Rh B + complex (**4**). (Right) Kinetics of complex (**4**)-catalyzed degradation of MO and Rh B in aqueous solution.

As shown in Figure 4, when we used only MO and Rh B dyes with the H_2O_2 aqueous solution, there was no degradation of the organic dyes. The use of an aqueous solution of H_2O_2 (C_o = 10 mg.L^{-1}) led to degradation yields of 45.5% and 42.3% for the MO and RhB dyes, respectively, after 60 min of reaction. The k_o values of the pseudo-first order rate constant of the degradation concerning the MO or RhB dye-H_2O_2-complex (**4**) systems were 0.01×10^{-2} min^{-1} (R^2 = 0.9017) and 0.011×10^{-2} min^{-1} (R^2 = 0.9776), respectively.

3.7. Photodegradation of MO and RhB Dyes

First, complex (**4**) was utilized to degrade MO and RhB dyes under visible light illumination (λ > 400 nm) for the sake of exploring further photocatalytic transformations.

In a typical trial, an aqueous suspension (50 mL) containing MO or RhB dye (20 mg/L) and 10 mg of complex (**4**) was placed in the reactor under visible light irradiation. The suspension was stirred in the dark for 30 min before illumination to ensure the adsorption/desorption balance was established.

At defined time intervals, an appropriate amount of suspension was centrifuged and filtered through a filter membrane to remove solid particles and collect the filtrate for further analysis. The maximum absorption wavelength of the MO and RhB dyes (λ_{max}) were 418 and 555 nm, respectively.

As shown in Figure 5, complex (**4**) showed effective degradation of MO dye. More than 75% of the RhB dye was degraded after irradiation for 60 min, while the percentage of degradation of the MO dye was 63%.

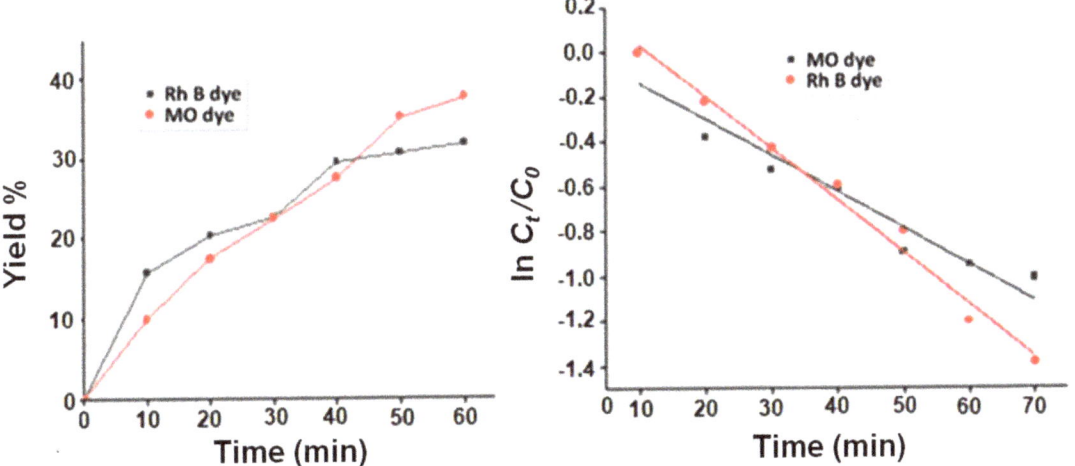

Figure 5. (**Left**) Changes in C_t/C_o versus time for the following conditions: MO + complex (**4**) and RhB + complex (**4**). (**Right**) Kinetics of complex (**4**)-catalyzed photo degradation of MO and Rh B dyes in aqueous solution.

The kinetics of the degradation reaction can be described using a first order model for low concentrations of the MO and RhB dye solutions. The pseudo first-order kinetics equation is expressed as follows (Equation (5)):

$$\ln(C_o/C_t) = k_o t \qquad (5)$$

where C_t is the MO or Rh B dye concentration in aqueous solution at time t (mg/L), C_o is the initial MO or RhB dye concentration (mg/L), and k_o is the apparent pseudo-first-order kinetic constant (min^{-1}). The plots $\ln(C_t/C_o)$ as a function of time are shown in Figure 5. The calculated values of k_o were 1.6×10^{-2} min^{-1} ($R^2 = 0.9344$) and 2.3×10^{-2} min^{-1} ($R^2 = 0.9856$) for the MO and RhB dyes, respectively. The excellent fitting indicated that the photoreaction followed first-order reaction kinetics.

The principle of heterogeneous photocatalysis (Figure 6) is based on the activation of complex (**4**) by a supply of light energy $h\nu \geq E_g$ (E_g = band gap energy). During this activation step, an electron (e−)/hole (h+) pair is created, which results from the passage of an electron from the valence band to the conduction band. The electron will react with the oxygen adsorbed on the surface of our porphyrinic compounds, while the hole h+, reacts with the surface of the OH ions to form highly oxidizing hydroxyl radicals (OH·), which is responsible for the degradation of pollutants.

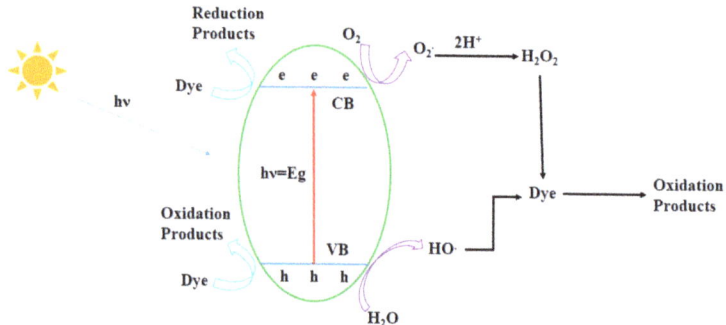

Figure 6. Pictorial representation of indirect dye degradation process.

Complex (**4**), repeatedly used, exhibited properties identical to those of the initial complex, with no obvious drop in photocatalytic efficacy even after five cycles, achieving photodegradation efficiency of 60% and 73% for the MO and RhB dyes, respectively (Figure 7).

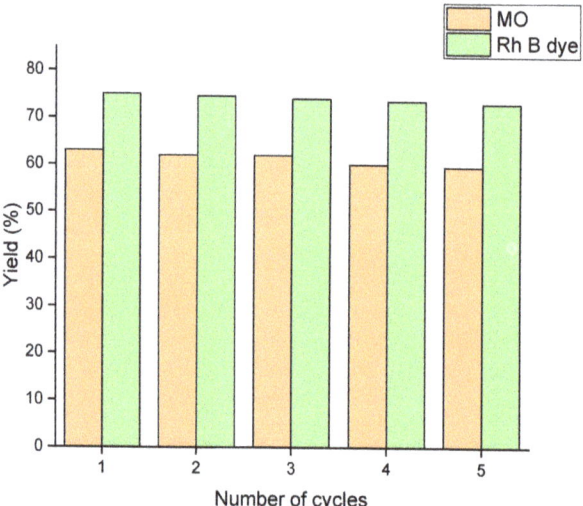

Figure 7. Redegradation efficiency of complex (**4**) for the MO and RhB dyes.

4. Electronic Study on Complex (4)

To prepare the thin film containing complex (**4**), ITO glass slides were washed in an ultrasonic bath containing acetone, then in an isopropyl alcohol bath. Subsequently, the clean substrates were dried with a nitrogen gas flow. A 15 mg sample of complex (**4**) was dissolved in 10 mL of dichloromethane. Afterwards, the solution containing the zin(II) coordination compound was deposited on an indium tin oxide (ITO) glass slides by spin coating at 2000 rpm for 25 s. The aluminum (Al) electrodes were deposited by thermal evaporation. To obtain the best quality images of the film surface, AFM (atomic force microscopy) was employed, which showed the homogeneity of the film with a coherent structure (1.86 nm).

Owing to the interesting value of the gap energy of complex (**4**), which was in the range of semiconductor materials, we carried out electrical and dielectric tests on this new zinc(II) porphyrin compound in order to study its electronic properties.

The I-V measurements were obtained using a Keithley 236 instrument and the spectroscopic impedance measurement was performed using an impedance analyzer (Solartron 1260). The electronic properties of the ITO/complex (**4**)/Al system can provide information about the transmission properties in organic materials. The current-voltage curve measured at room temperature of the ITO/Complex (**4**)/Al system is shown in Figure 8. The curve shown in this figure presents a similar behavior to that of electronic devices, such as diodes, indicating complex (**4**) could be used as a photosensitizer in DSSCs (dye-sensitized solar cells) [57–60]. The threshold voltage was approximately 0.64 V.

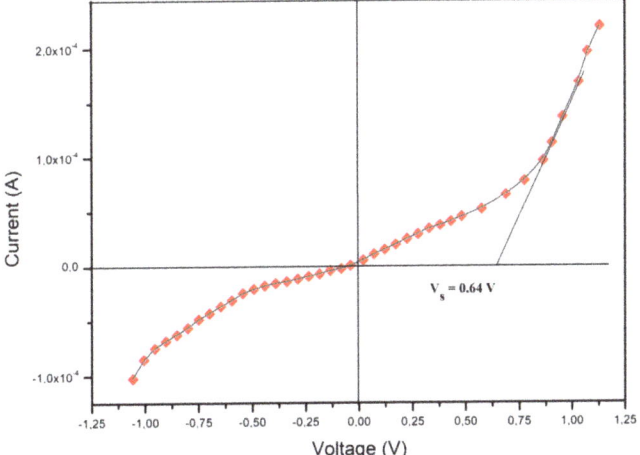

Figure 8. Current–voltage curves of ITO/complex (**4**).

Indeed, we observed an asymmetric curve for high voltage, which was related to the difference in the injection of electrons and holes of the anode (ITO) and cathode (Al). At low voltage, the I-V curve indicated symmetric behavior. This behavior is explained by the theory that the localized state with defects provides the localized gap states.

Notably, recent work [61] has shown a useful current hysteresis behavior for some porphyrin species. By introducing a triazole group, the hysteresis behavior was eliminated and consequently, we completely changed the electronic properties of our tested molecule.

The semi-logarithmic scale of the I-V curve of complex (**4**) indicated that the value of the barrier height of this species was approximately 1.3083 V and the saturation current value was 5.97×10^{-6} A.

In addition, we studied the mechanism of electrical conduction through the junction by presenting the I-V characteristics in double logarithmic plot (Figure 9). The I-V plot of the ITO/complex (**4**)/Al system showed the presence of different parts in which the current depended mainly on the applied voltage. At low voltage, the first part of the curve corresponded to a value of the slope in the order of 1.2, indicating prevention of the charge injection due to the presence of a small amount of interface barrier. For this part of the curve, which defines the ohmic region, the amount of the heat-activated charge carriers was too small and the trap levels were vacant. The current density equation is as follows (Equation (6)):

$$J_\Omega = q.p_0.\mu.\frac{V}{d} \qquad (6)$$

where μ represent the charge mobility, q defines the electronic charge, d refers to the film thickness, and p_0 is the free carrier density.

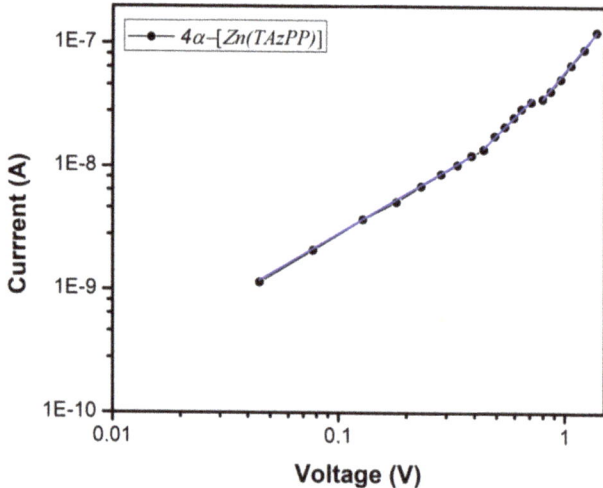

Figure 9. Log-log curve of complex (**4**) structure.

In the second part of the curve, where the voltage was moderate, the value of the slope was close to 2.2. This can be explained by the dependence of the voltage according to the power law (I-V), which is associated with the space charge limited current mechanism (SCLC) [62]. In addition, the applied voltage increased and passed through the transition voltage, which reflected the increase in the density of charges injected by the electrodes. The charge density injected will govern the transport ability of the layer of complex (**4**). The current density varies following equation (Equation (7)):

$$J_{SCLC} = \frac{9}{8} \varepsilon \cdot \mu_{eff} \cdot \frac{V^2}{d^3} \tag{7}$$

where d is the film thickness, V is the applied voltage, ε is the material permittivity, and μ_{eff} is the effective carrier mobility.

Based on the SCLC model (Equation (7)), the μ_{eff} in the film of complex (**4**) had a value of 0.45 (10^{-5} cm^2/Vs). In the third part of the curve, where the voltage is high, the value of the slope was approximately 3.4. This represented the trapped charge limit current (TCLC) area where the distribution of traps changed exponentially. However, the transition between SCLC and TCLC mechanisms is affected by the trapping levels. This transition occurs when the quantity of injected carriers surpasses the density of free carriers [63].

4.1. Impedance Spectroscopy

To investigate the dielectric characteristics of complex (**4**) and determine the participation associated with the volume and interface, we carried out an impedance spectroscopy study [64–66]. Equation (8) describes the impedance Z(ω) of complex (**4**) as a function of frequency:

$$Z(\omega) = Re(Z) + jIm(Z) = Z'(\omega) + jZ''(\omega) \tag{8}$$

This equation shows that the complex impedance Z(ω) is composed of two parts: the first part is the real part (Re (Z) = Z') and the second part is the imaginary part (Im (Z) = Z''). The semicircular spectrum present in the impedance spectrum (Nyquist plot) of the complex (**4**) structure suggested the homogeneity of the electrode-organic interface (Figure 10).

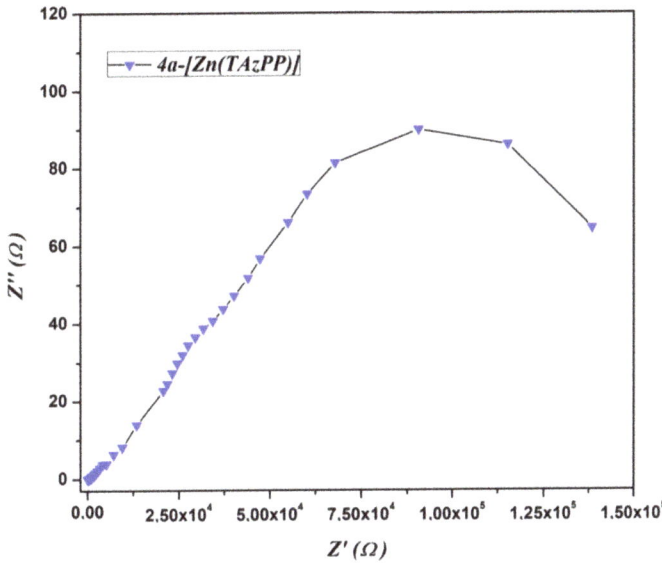

Figure 10. Impedance plot spectrum of complex (**4**).

4.2. Conductance

Figure 11 shows two regimes of conductance of complex (**4**), which depended essentially on the frequency applied. The first regime was observed at low frequency, where the conductance increased with increasing frequency until reaching a maximum at a frequency of approximately 1.4 Hz, which indicated a disordered system. However, the second regime observed at high frequency indicated that the conductance tended toward zero, where the dipoles neglected the frequency. This phenomenon was associated with the jump transport mechanism, where the dipoles will be guided by the applied field, which will lead to an increase in the charge hopping process (Figure 11) [67].

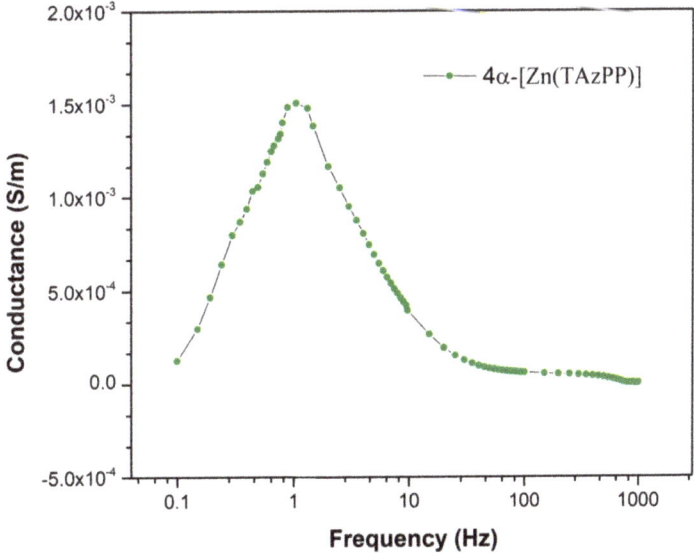

Figure 11. Conductance characteristics of complex (**4**).

5. Conclusions

We successfully synthesized a new zinc(II) *meso*-arylporphyrin coordination compound: [*meso*-4α-tetra-(1,2,3-triazolyl)phenylporphyrinato]zinc(II) (4) with the formula 4α-[Zn(TAzPP)]. This new Zn(II) metalloporphyrin was characterized by ^{1}H NMR and infrared, UV-visible, and fluorescence spectroscopies. This coordination compound was able to make 1:1 stoichiometric complexes with Cl$^-$ and Br$^-$ ions, with average association constant values of 0.30×10^3 and 0.44×10^3, respectively, which were higher than those of the related [Zn(TPP)] (TPP = *meso*-tetraphenylporphyrinate) complex. In addition, complex (4) was tested as a catalyst in the degradation reaction of rhodamine B (RhB) and methyl orange (MO) dyes, using both photodegradation and degradation by aqueous hydrogen peroxide solution. The photodegradation yield values of the MO and RhB dyes using complex (4) were close to 63% and 75%, respectively, while the degradation yield values using aqueous dye solutions, H_2O_2, and complex (4) were 45.5% and 42.3% for MO and RhB, respectively. Notably, the use of this complex several times without variation in the degradation yield of the MO and RhB dyes indicated that complex (4) was a good catalyst for such reactions. Furthermore, our new Zn(II)-porphyrin species was used in the ITO/complex (4)/Al system for current-voltage and impedance spectroscopy measurements. The I-V curve of this system exhibited a similar behavior to that of diodes, with a threshold voltage of approximately 0.64 V. The impedance spectrum (Nyquist plot) of complex (4) presented a semicircular spectrum that suggested the homogeneity of the electrode-organic interface. Finally, the conductance properties of complex (4) were investigated, indicating the presence of two regimes of conductance depending essentially on the frequency applied.

Supplementary Materials: The following supporting information can be downloaded at: https://www.mdpi.com/article/10.3390/cryst13020238/s1, Figure S1: IR spectrum of the free base porphyrin (4α-H$_2$TNH$_2$PP) (1); Figure S2: IR spectra of the free base porphyrin 4α-H$_2$TN$_3$PP (2); Figure S3: IR spectra of the complex 4α-[Zn(TN$_3$PP)]) (3); Figure S4: IR spectra of complex 4α-[Zn(TAzPP)] (4); Figure S5: ^{1}H NMR spectrum of the free base porphyrin 4α-H2TNH2PP (1) (400 MHz, CDCl$_3$); Figure S6: ^{1}H NMR spectrum of free base porphyrin 4α-H2TN3PP (2) (400 MHz, CDCl$_3$); Figure S7: ^{1}H NMR spectrum of 4α-[Zn(TN$_3$PP)] (3) (400 MHz, CDCl$_3$); Figure S8: ^{1}H NMR spectrum of α4-[Zn(TAZPP)] compound (4) (400 MHz, CDCl$_3$).

Author Contributions: Conceptualization, S.N., M.G., J.B., Y.O.A.-G. and H.N.; Methodology, J.B. and F.L.; Software, Y.O.A.-G. and F.L.; Validation, J.B. and H.N.; Formal analysis, F.L.; Investigation, M.G. and F.L.; Writing–original draft, S.N., M.G. and J.B.; Writing–review & editing, H.N.; Visualization, H.N.; Project administration, S.N., J.B and Y.O.A.-G. All authors have read and agreed to the published version of the manuscript.

Funding: The authors extend their appreciation to the deputyship for Research & Innovation, Ministry of Education in Saudi Arabia for funding this research work through the project number (IFP-2020-05).

Data Availability Statement: Not applicable.

Conflicts of Interest: The authors declare no conflict of interest.

References

1. Perutz, M. Regulation of Oxygen Affinity of Hemoglobin: Influence of Structure of the Globin on the Heme Iron. *Ann. Rev. Biochem.* **1979**, *48*, 327. [CrossRef] [PubMed]
2. Milgrom, L.R. *The Colors of Life: An Introduction to the Chemistry of Porphyrins and Related Compounds*; Oxford University Press: New York, NY, USA, 1997; p. 249.
3. Campbell, W.M.; Jolley, K.W.; Wagner, P.; Wagner, K.; Walsh, P.J.; Gordon, K.C.; Schmidt-Mende, L.; Nazeeruddin, M.K.; Wang, Q.; Grätzel, M.; et al. Highly Efficient Porphyrin Sensitizers for Dye-Sensitized Solar Cells. *J. Phys. Chem. C. Lett.* **2007**, *111*, 11760. [CrossRef]
4. Gregg, B.A.; Fox, M.A.; Bard, A.J. Functionalized Porphyrin Discotic Liquid Crystals. *J. Am. Chem. Soc.* **1989**, *111*, 3024. [CrossRef]
5. Garg, K.; Singh, A.; Majumder, C.; Nayak, S.K.; Aswal, D.K.; Gupta, S.K.; Chattopadhyay, S. Room temperature ammonia sensor based on jaw like bis-porphyrin molecules. *Org. Electron.* **2013**, *14*, 14189. [CrossRef]

6. Kadish, K.M.; Smith, K.M.; Guilard, R.; Harvey, P.D. *The Porphyrin Handbook*; Academic Press: San Diego, CA, USA, 2003; Volume 18, p. 63.
7. Lammi, R.K.; Ambroise, A.; Balasubramanian, T.; Wagner, R.W.; Bocian, D.F.; Holten, D.; Lindsey, J.S. One-step synthesis and characterization of difunctionalized N-confused tetraphenylporphyrins. *J. Am. Chem. Soc.* **2000**, *122*, 7579. [CrossRef]
8. Izatt, R.M.; Pawlak, K.; Bradshaw, J.S. Thermodynamic and kinetic data for macrocycle interaction with cations, anions, and neutral molecules. *Chem. Rev.* **1995**, *95*, 2529. [CrossRef]
9. Dietrich, B. Design of anion receptors: Applications. *Pure Appl. Chem.* **1993**, *65*, 1457. [CrossRef]
10. Kavallieratos, K.; de Gala, S.R.; Austin, D.J.; Crabtree, R. A readily available non-preorganized neutral acyclic halide receptor with an unusual nonplanar binding conformation. *J. Am. Chem. Soc.* **1997**, *119*, 2325. [CrossRef]
11. Davis, A.P.; Perry, J.J.; Williams, R.P. Anion recognition by tripodal receptors derived from cholic acid. *J. Am. Chem. Soc.* **1997**, *119*, 1793. [CrossRef]
12. Berger, M.; Schmidtchen, F. Electroneutral artificial hosts for oxoanions active in strong donor solvents. *J. Am. Chem. Soc.* **1996**, *118*, 8947. [CrossRef]
13. Kral, V.; Furuta, H.; Shreder, K.; Lynch, V.; Sessler, J.L. Protonated sapphyrins. Highly effective phosphate receptors. *J. Am. Chem. Soc.* **1996**, *118*, 1595. [CrossRef]
14. Gale, P.A.; Sessler, J.L.; Kral, V.; Lynch, V. Calix [4] pyrroles: Old yet new anion-binding agents. *J. Am. Chem. Soc.* **1996**, *118*, 5140. [CrossRef]
15. Andrews, P.A.; Mann, S.C.; Huynh, H.H.; Albright, K.D. Role of the Na+,K+-Adenosine Triphosphatase in the Accumulation of cis-Diamminedichloroplatinum(II) in Human Ovarian Carcinoma Cells. *Cancer Res.* **1991**, *51*, 3677.
16. Zhou, Y.; Dong, X.; Zhang, Y.; Tong, P.; Qu, J. Highly selective fluorescence sensors for the fluoride anion based on carboxylate-bridged diiron complexes. *Dalton Trans.* **2016**, *45*, 6839. [CrossRef] [PubMed]
17. Mandal, T.N.; Karmakar, A.; Sharma, P.; Ghosh, S.K. Metal-Organic Frameworks (MOFs) as Functional Supramolecular Architectures for Anion Recognition and Sensing. *Chem. Rec.* **2018**, *18*, 154. [CrossRef] [PubMed]
18. Rozveh, Z.S.; Kazemi, S.; Karimi, M.; Ali, G.A.; Safarifard, V. Photocatalytic aerobic oxidative functionalization (PAOF) reaction of benzyl alcohols by GO-MIL-100(Fe) composite in glycerol/K2CO3 deep eutectic solvent. *Polyhedron* **2020**, *183*, 113514.
19. Dieleman, C.B.; Matt, D.; Neda, I.; Schmutzler, R.; Harriman, A.; Yaftian, R. Hexahomotrioxacalix [3] arene: A scaffold for a C 3-symmetric phosphine ligand that traps a hydrido-rhodium fragment inside a molecular funnel. *Chem. Commun.* **1999**, 1911. [CrossRef]
20. Kumar, R.; Sharma, A.; Singh, H.; Suating, P.; Kim, H.S.; Sunwoo, K.; Shim, I.; Gibb, B.C.; Kim, J.S. Hexahomotrioxacalix [3] arene: A scaffold for a C 3-symmetric phosphine ligand that traps a hydrido-rhodium fragment inside a molecular funnel. *Chem. Rev.* **2019**, *119*, 9657. [CrossRef] [PubMed]
21. Sessler, J.L.; Cyr, M.; Furuta, H.; Kral, V.; Mody, T.; Morishima, T.; Shionoya, M.; Weghorn, S. Anion binding: A new direction in porphyrin-related research. *Pure. Appl. Chem.* **1993**, *65*, 393. [CrossRef]
22. Sessler, J.L.; Burrell, A.K. Sapphyrins and heterosapphyrins. *Top. Curr. Chem.* **1992**, *161*, 177. [CrossRef]
23. Shionoya, M.; Furuta, H.; Harriman, A.; Sessler, J.L. Diprotonated sapphyrin: A fluoride selective halide anion receptor. *J. Am. Chem. Soc.* **1992**, *114*, 5714. [CrossRef]
24. Gilday, L.C.; White, N.G.; Beer, D. Halogen-and hydrogen-bonding triazole-functionalised porphyrin-based receptors for anion recognition. *Dalton Trans.* **2013**, *42*, 15766. [CrossRef] [PubMed]
25. Imahori, H.; Umeyama, T.; Ito, S. Large π-aromatic molecules as potential sensitizers for highly efficient dye-sensitized solar cells. *Acc. Chem. Res.* **2009**, *42*, 1809. [CrossRef] [PubMed]
26. Radivojevic, I.; Varotto, A.; Farley, C.; Drain, C.M. Commercially viable porphyrinoid dyes for solar cells. *Energy. Environ. Sci.* **2010**, *3*, 1897. [CrossRef]
27. Martinez-Diaz, M.V.; Torre, G.; Torres, T. Lighting porphyrins and phthalocyanines for molecular photovoltaics. *Chem. Commun.* **2010**, *46*, 7090. [CrossRef]
28. Walter, M.G.; Rudine, A.B.; Wamser, C.C. Porphyrins and phthalocyanines in solar photovoltaic cells. *J. Porph. Phthalocyanines* **2010**, *14*, 759. [CrossRef]
29. Griffith, M.J.; Sunahara, K.; Wagner, P.; Wagner, K.; Wallace, G.G.; Officer, D.L.; Furube, A.; Katoh, R.; Mori, S.; Mozer, A.J. Porphyrins for dye-sensitised solar cells: New insights into efficiency-determining electron transfer steps. *Chem. Commun.* **2012**, *48*, 4145. [CrossRef] [PubMed]
30. Imahori, H.; Umeyama, T.; Kurotobi, K.; Takano, Y. Self-assembling porphyrins and phthalocyanines for photoinduced charge separation and charge transport. *Chem. Commun.* **2012**, *48*, 4032. [CrossRef]
31. MPanda, K.; Ladomenou, K.; Coutsolelos, A.G. Porphyrins in bio-inspired transformations: Light-harvesting to solar cell. *Coord. Chem. Rev.* **2012**, *256*, 2601.
32. Hasobe, T.; Imahori, H.; Kamat, P.V.; Ahn, T.K.; Kim, S.K.; Kim, D.; Fujimoto, A.; Hirakawa, T.; Fukuzumi, S. Photovoltaic cells using composite nanoclusters of porphyrins and fullerenes with gold nanoparticles. *J. Am. Chem. Soc.* **2005**, *127*, 1216. [CrossRef]
33. Yella, A.; Lee, H.W.; Tsao, H.N.; Yi, C.; Chandiran, A.K.; Nazeeruddin, M.K.; Diau, E.W.; Yeh, C.Y.; Zakeeruddin, S.M.; Grätzel, M. Porphyrin-sensitized solar cells with cobalt (II/III)–based redox electrolyte exceed 12 percent efficiency. *Science* **2011**, *334*, 629. [CrossRef] [PubMed]

34. Kay, A.; Grätzel, M. Artificial photosynthesis. 1. Photosensitization of titania solar cells with chlorophyll derivatives and related natural porphyrins. *J. Phys. Chem.* **1993**, *97*, 6272. [CrossRef]
35. Cherian, S.; Wamser, C.C. Adsorption and Photoactivity of Tetra(4-carboxyphenyl)porphyrin (TCPP) on Nanoparticulate TiO_2. *J. Phys. Chem. B.* **2000**, *104*, 3624. [CrossRef]
36. Nazeeruddin, M.K.; Humphry-Baker, R.; Officer, D.L.; Campbell, W.M.; Burrell, A.K.; Grätzel, M. Conformation and π-conjugation of olefin-bridged acceptor on the pyrrole β-carbon of nickel tetraphenylporphyrins: Implicit evidence from linear and nonlinear optical properties. *Langmuir* **2004**, *20*, 6514. [CrossRef]
37. Wang, Q.; Campbell, W.M.; Bonfantani, E.E.; Jolley, K.W.; Officer, D.L.; Walsh, P.J.; Gordon, K.; Humphry-Baker, R.; Nazeeruddin, M.K.; Grätzel, M. Efficient Light Harvesting by Using Green Zn-Porphyrin-Sensitized Nanocrystalline TiO2 Films. *J. Phys. Chem. B.* **2005**, *109*, 15397. [CrossRef]
38. Park, J.K.; Lee, H.R.; Chen, J.; Shinokubo, H.; Osuka, A.; Kim, D. Photoelectrochemical Properties of Doubly β-Functionalized Porphyrin Sensitizers for Dye-Sensitized Nanocrystalline-TiO2 Solar Cells. *J. Phys. Chem. C* **2008**, *112*, 16691. [CrossRef]
39. Bessho, T.; Zakeeruddin, S.M.; Yeh, C.Y.; Diau, E.W.G.; Grätzel, M. Highly efficient mesoscopic dye-sensitized solar cells based on donor–acceptor-substituted porphyrins. *Angew.Chem. Int. Ed.* **2010**, *49*, 6646. [CrossRef]
40. Campbell, W.M.; Burrell, A.K.; Officer, D.L.; Jolley, K.W. Efficient Light Harvesting by Using Green Zn-Porphyrin-Sensitized Nanocrystalline TiO2 Films. *Coord. Chem. Rev.* **2004**, *248*, 1363. [CrossRef]
41. He, H.; Gurung, A.; Si, L. 8-Hydroxyquinoline as a strong alternative anchoring group for porphyrin-sensitized solar cells. *Chem. Commun.* **2012**, *48*, 5910. [CrossRef]
42. Kolb, H.C.; Finn, M.G.; Sharpless, K.B. Click chemistry connections for functional discovery. *Angew. Chem. Int. Ed.* **2001**, *40*, 2004. [CrossRef]
43. Rostovtsev, V.V.; Green, L.G.; Fokin, V.V.; Sharpless, K.B. A Stepwise Huisgen Cycloaddition Process: Copper(I)-Catalyzed Regioselective "Ligation" of Azides and Terminal Alkynes. *Angew. Chem. Int. Ed.* **2002**, *41*, 2596. [CrossRef]
44. Shetti, V.S.; Ravikanth, M. Synthesis and studies of Thiacorroles. *Eur. J. Org. Chem.* **2010**, *75*, 4172–4182. [CrossRef] [PubMed]
45. Chatterjee, S.; Sengupta, K.; Bhattacharyya, S.; Nandi, A.; Samanta, S.; Mittra, K.; Dey, A. Photophysical and ligand binding studies of metalloporphyrins bearing hydrophilic distal superstructure. *J. Porph. Phthalocyanines* **2013**, *17*, 210. [CrossRef]
46. Samanta, S.; Mittra, K.; Sengupta, K.; Chatterjee, S.; Dey, A. Second Sphere Control of Redox Catalysis: Selective Reduction of O_2 to O_2^- or H_2O by an Iron Porphyrin Catalyst. *Inorg. Chem.* **2013**, *52*, 1443. [CrossRef] [PubMed]
47. Mittra, K.; Chatterjee, S.; Samanta, S.; Sengupta, K.; Bhattacharjee, H.; Dey, A. A hydrogen bond scaffold supported synthetic heme Fe III–O 2− adduct. *Chem. Comm.* **2012**, *48*, 10535. [CrossRef] [PubMed]
48. Samanta, S.; Sengupta, K.; Mittra, K.; Bandyopadhyay, S.; Dey, A. Selective four electron reduction of O 2 by an iron porphyrin electrocatalyst under fast and slow electron fluxes. *Chem. Comm.* **2012**, *48*, 7631. [CrossRef]
49. Mandal, A.K.; Taniguchi, M.; Diers, J.R.; Niedzwiedzki, D.M.; Kirmaier, C.; Lindsey, J.S.; Bocian, D.F.; Holten, D. Photophysical Properties and Electronic Structure of Porphyrins Bearing Zero to Four meso-Phenyl Substituents: New Insights into Seemingly Well Understood Tetrapyrroles. *J. Phys. Chem. A* **2016**, *120*, 9719. [CrossRef]
50. Collman, J.P.; Gagne, R.R.; Halbert, T.R.; Marchon, J.C.; Reed, C.A. Reversible oxygen adduct formation in ferrous complexes derived from a picket fence porphyrin. Model for oxymyoglobin. *J. Am. Chem. Soc.* **1973**, *95*, 7868. [CrossRef]
51. Hartle, M.D.; Prell, J.S.; Plut, M.D. Spectroscopic investigations into the binding of hydrogen sulfide to synthetic picket-fence porphyrins. *Dalton Trans.* **2016**, *45*, 4843. [CrossRef]
52. Lindsey, J. Increased yield of a desired isomer by equilibriums displacement on binding to silica gel, applied to meso-tetrakis (o-aminophenyl) porphyrin. *J. Org. Chem.* **1980**, *45*, 5215. [CrossRef]
53. Gorlitzer, K.; Huth, S.; Jones, P.G. Color reaction of chlorhexidine and proguanil with hypobromite. *Pharmazie* **2005**, *60*, 269.
54. Guergueb, M.; Brahmi, J.; Nasri, S.; Loiseau, F.; Aouadi, K.; Guerineau, V.; Nasri, H. Zinc (II) triazole meso-arylsubstituted porphyrins for UV-visible chloride and bromide detection. Adsorption and catalytic degradation of malachite green dye. *RSC Adv.* **2020**, *10*, 22712. [CrossRef] [PubMed]
55. Chen, H.B.; Zeng, J.B.; Deng, X.; Chen, L.; Wang, Y.Z. Block phosphorus-containing poly (trimethylene terephthalate) copolyester via solid-state polymerization: Retarded crystallization and melting behaviour. *CrystEngComm.* **2013**, *15*, 2688–2698. [CrossRef]
56. Polster, J.; Lachmann, H. Spectrometric Titrations. Analysis of Chemical Equilibria. *Verlag Chemie* **1989**, *12*, 292.
57. Dumoulin, F.; Ahsen, V. Design and conception of photosensitisers. *J. Porph. Phthalocyanines* **2011**, *15*, 481. [CrossRef]
58. Brahmi, J.; Nasri, S.; Saidi, H.; Nasri, H.; Aouadi, K. Synthesis of new porphyrin complexes: Evaluations on optical, electrochemical, electronic properties and application as an optical sensor. *Chem. Select* **2019**, *14*, 1350. [CrossRef]
59. Ceyhan, T.; Altindal, A.; Erbil, M.; Bekaroglu, O. Synthesis, characterization, conduction and gas sensing properties of novel multinuclear metallo phthalocyanines (Zn, Co) with alkylthio substituents. *Polyhedron* **2006**, *25*, 7. [CrossRef]
60. Xue, X.; Tan, G. Effect of bivalent Co ion doping on electric properties of Bi0. 85Nd0. 15FeO3 thin film. *J. Alloys Compd.* **2013**, *575*, 90. [CrossRef]
61. Ghataka, S.; Ghosh, A. Observation of trap-assisted space charge limited conductivity in short channel MoS2 transistor. *App. Phys. Lett.* **2013**, *103*, 122103. [CrossRef]
62. Brahmi, J.; Nasri, S.; Saidi, H.; Aouadi, K.; Sanderson, M.R.; Winter, M.; Cruickshank, D.; Najmudin, S.; Nasri, H. Optical and photoelectronic properties of a new material: Optoelectronic application. *Comptes. Rendus. Chimie.* **2020**, *23*, 403. [CrossRef]

63. Al Mogren, M.M.; Ahmed, M.N.; Hasanein, A.A. Molecular modeling and photovoltaic applications of porphyrin-based dyes: A review. *J. Saudi. Chem. Soc.* **2020**, *24*, 303. [CrossRef]
64. Aloui, W.; Ltaief, A.; Bouazizi, A. Dielectrical properties of PET-MWCNT/P3HT: PC70BM/Al device: Impedance spectroscopy analysis. *Microelectron. Eng.* **2014**, *129*, 96–99. [CrossRef]
65. Mahmood, A.; Hu, J.Y.; Xiao, B.; Tang, A.; Wang, X.; Zhou, E. Recent progress in porphyrin-based materials for organic solar cells. *J. Mater. Chem. A.* **2018**, *6*, 16769. [CrossRef]
66. Opeyemi, O.; Louis, H.; Opara, C.; Funmilayo, O.; Magu, T. Porphyrin and Phthalocyanines-Based Solar Cells: Fundamental Mechanisms and Recent Advances. *Adv. J. Chem. Sect. A* **2019**, *2*, 21.
67. Fishchuk, I.I.; Kadashchuk, A.; Ullah, M.; Sitter, H.; Pivrikas, A.; Genoe, J.; Bassler, H. Electric field dependence of charge carrier hopping transport within the random energy landscape in an organic field effect transistor. *Phys. Rev. B.* **2012**, *86*, 045207. [CrossRef]

Disclaimer/Publisher's Note: The statements, opinions and data contained in all publications are solely those of the individual author(s) and contributor(s) and not of MDPI and/or the editor(s). MDPI and/or the editor(s) disclaim responsibility for any injury to people or property resulting from any ideas, methods, instructions or products referred to in the content.

Article

Room Temperature Ferromagnetic Properties of $Ga_{14}N_{16-n}Gd_2C_n$ Monolayers: A First Principle Study

Shijian Tian [1], Libo Zhang [2,*], Yuan Liang [1,*], Ruikuan Xie [3], Li Han [2], Shiqi Lan [1], Aijiang Lu [1], Yan Huang [4,*], Huaizhong Xing [1,*] and Xiaoshuang Chen [4]

[1] Department of Optoelectronic Science and Engineering, Donghua University, Shanghai 201620, China
[2] College of Physics and Optoelectronic Engineering, Hangzhou Institute for Advanced Study, University of Chinese Academy of Sciences, No. 1, Sub-Lane Xiangshan, Xihu District, Hangzhou 310024, China
[3] State Key Laboratory of Structural Chemistry, Fujian Institute of Research on the Structure of Matter, Chinese Academy of Sciences (CAS), Fuzhou 350002, China
[4] State Key Laboratory of Infrared Physics, Shanghai Institute of Technical Physics, Chinese Academy of Sciences, Shanghai 200083, China
* Correspondence: zhanglibo@ucas.ac.cn (L.Z.); yliang@dhu.edu.cn (Y.L.); yhuang@mail.sitp.ac.cn (Y.H.); xinghz@dhu.edu.cn (H.X.)

Abstract: Electronic and magnetic properties of $Ga_{14}N_{16-n}Gd_2C_n$ monolayers are investigated by means of the first principle calculation. The generalized gradient approximation (GGA) of the density functional theory with the on-site Coulomb energy U was considered (GGA + U). It is found that the total magnetic moment of a $Ga_{14}N_{16}Gd_2$ monolayer is 14 μ_B with an antiferromagnetic (AFM) phase. C atom substitutional impurity can effectively change the magnetic state of $Ga_{14}N_{16-n}Gd_2C_n$ monolayers to ferromagnetic phases (FM), and the magnetic moment increases by $1\mu_B/1C$. The stable FM phase is due to the p-d coupling orbitals between the C-2p and Gd-5d states. Moreover, Curie temperature (T_C) close to room temperature (T_R, 300 K) is observed in the $Ga_{14}N_{16}Gd_2C_2$ monolayer, and the highest value can reach 261.46 K. In addition, the strain effect has a significant positive effect on the T_C of the $Ga_{14}N_{16-n}Gd_2C_n$ monolayer, which is much higher than the T_R, and the highest value is 525.50 K. This provides an opportunity to further explore the application of two-dimensional magnetic materials in spintronic devices.

Keywords: density functional theory; GaN:Gd monolayer; ferromagnetic property; strain effect; p-d coupling

1. Introduction

Diluted magnetic semiconductors (DMSs), as an important part of spintronics, have attracted much attention in terms of harnessing the spin and charge of electrons [1–3]. Scientists modulate the ferromagnetism and T_C by controlling the doping atoms to investigate the potential applications of DMSs. It opens a new gateway for extending future classes of materials. A magnetic dopant was used to substitute cations in the host compound semiconductors and observed distinct properties [4–6]. The room-temperature FM (T_R-FM) phase has been observed in Cr, Mn, Fe, Co-doped TiO_2, ZnO and GaN [7–9].

The metal nitrides (MNs), including group IIIA nitrides and nitride MXene, exhibit unique electronic and magnetic characteristics [10–12]. In recent decades, scientists have paid much attention to the magnetic properties and T_C of GaN materials doped with transition metal (TM) [13–15], alkali metal, alkaline earth metal, etc. [16–18]. The low solubility of TM atoms in GaN materials restrains their potential application in T_R-FM spintronic devices. In addition, rare-earth elements such as Sm, Dy and Gd with large magnetic moments have attracted a lot of attention [19–21]. Nobuaki found that the T_C of GaN:Gd materials achieves a T_C of 400 K, far above T_R [22]. The T_R-FM coupling of the

GaN:Gd system was also detected by Asahi [23]. Dhar's group found a colossal magnetic moment and T_R-FM phase at a low concentration of Gd atoms [24,25]. It is well known that defects are one of the important reasons for FM and AFM coupling in GaN-based systems. The spintronic properties of GaN-based materials can be modulated not only by n-type (interstitial O, N and C dopants) but also by p-type defects (Ga vacancies and transition metal doping) [26–29]. Dalpian found that the FM phase in n-type GaN is mainly derived from s-f orbital hybridization [30]. It was found that the 4f orbitals of Gd are usually far from the Fermi energy [31]. Therefore, the s-f coupling effect near the Fermi energy level will be weak. Xie found T_R-FM in the GaN:Gd nanowires doped with C atoms, which are strongly influenced by hybridized p-d coupling [32]. Therefore, it is necessary to select a suitable material doped with GaN:Gd and to explore the origin of the ferromagnetic mechanism in-depth and definitively.

Until now, there has been a lot of research, but the research on GaN:Gd monolayers doped with C atoms is sparse. In the presented paper, the electronic and magnetic properties of Gd-pair-doped GaN ($Ga_{14}N_{16}Gd_2$) monolayers with and without C atoms are studied by employing the first principle calculation. This paper is organized as follows. In Section 2, we present the details of computational methods. In Section 3A, the geometric structure, band structure, partial density of states (PDOS), magnetic properties and T_C of $Ga_{14}N_{16}Gd_2$ monolayers doped with and without C atoms are determined. In Section 3 B, the biaxial strain effect in $Ga_{14}N_{16-n}Gd_2C_n$ monolayers are investigated. In Section 4, the results are briefly concluded.

2. Computational Method

All calculations are based on the density function theory (DFT) of the exchange-correlation potential. The computational work is conducted by using Vienna ab initio simulation packages (VASP) [33]. The Perdew–Burke–Ernzerhof (PBE) formalism of the generalized gradient approximation (GGA) is used to deal with electron exchange and correlation energies by using projection-enhanced waves (PAW) to understand the interactions between electrons and ions [34]. The cutoff energy of the plane wave basis set is kept at 500 eV [16]. When the structure is optimized, the atomic force and the energy convergence are kept at 0.01 eV/Å and 10^{-5} eV [13], respectively. A vacuum space above 12 Å is created to eliminate the effects of interactions between neighboring layers along the z-direction. The sample of k points in the Brillouin zone is set as $5 \times 5 \times 1$ [3]. The valence electron configurations of Ga, N, Gd and C atoms are described as: $3d^{10}4s^24p^1$, $2s^22p^3$, $4f^75d^16s^2$ and $2s^22p^2$, respectively. A self-consistent formulation of on-site Coulomb interaction for the Gd-4f orbitals is computed. The Coulomb repulsion energy U and the exchange parameter J are set to 6.7 eV and 0.7 eV, respectively [30,31].

The structural stability is studied by the binding energy (E_b), which is expressed as [35]:

$$E_b = \frac{E_{Total} - 14E_{Ga} - (16-n)E_N - 2E_{Gd} - nE_C}{32} \quad (1)$$

where E_{total} represents the total energy of $Ga_{14}N_{16-n}Gd_2C_n$, E_{Ga}, E_N, E_{Gd} and E_C corresponding to the energy of isolated Ga, N, Gd and C atoms, respectively.

We performed calculations of critical temperature to fully characterize the magnetic properties of $Ga_{14}N_{16-n}Gd_2C_n$ monolayers. The Heisenberg model based on the mean-field approximation theory is used to estimate the T_C^{MFA} [16]:

$$\frac{3}{2}K_B T_C^{MFA} = \frac{\Delta E_{AFM-FM}}{n} \quad (2)$$

where K_B and n are the Boltzmann constant and the number of Gd atoms, respectively. This temperature is often overestimated by the mean field approximation, thus an empirical relationship is used [36]:

$$\frac{T_C}{T_C^{MFA}} = 0.61 \quad (3)$$

3. Results and Discussion

The top and side views of atomic structures of free-standing GaN monolayers ($Ga_{16}N_{16}$) are relaxed to a flat honeycomb structure which is stripped from the (0001) plane of wurtzite GaN structure. The band structure and PDOS of the pure GaN monolayer in Figure S1a,e (shown in Supplementary Materials) indicate that the spin-up and spin-down channels are degenerated, indicating a non-magnetic semiconductor material. It possesses a wide band gap of 2.32 eV with a Ga-N bond length of about 1.84 Å, which agrees with previous studies [35,37,38]. Figure 1a,b shows the top and side views of optimized $Ga_{14}N_{16}Gd_2$ monolayers. The bulges in the doping position show that the Gd atom have tendency to break away from the monolayer. It is also shown in Figure S1 and Table S1 that different concentrations of Gd have a weak effect on the electronic structure and magnetic properties of the GaN:Gd monolayer. Thus, $Ga_{14}N_{16}Gd_2$ monolayers are used as the main research subject. The substitution sites have a great influence on the monolayer, so we explored the effects of different doping sites on the electronic properties, magnetic properties and T_C (the details are placed in Table S1, Figures S2 and S3). The topic of this article revolves around the optimal structures, with doping sites (M), (M, 8) and (M, 5, 7), respectively.

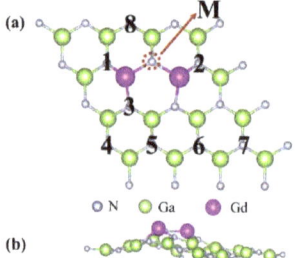

Figure 1. (**a**) Top and (**b**) side views of the optimized structures of $Ga_{14}N_{16}Gd_2$ monolayers.

3.1. The Structural, Magnetic Properties and T_C

The charge density differences of $Ga_{14}N_{16-n}Gd_2C_n$ monolayers are depicted in Figure 2, which is used to effectively investigate the accumulation and depletion of electrons. It is defined as: $\Delta\rho = \rho_{total} - \rho_A - \rho_B$, where ρ_{total}, ρ_A and ρ_B represent the total charge density of the $Ga_{14}N_{16-n}Gd_2C_n$ monolayer, the pure $Ga_{16}N_{16}$ monolayer and the free standing Gd and C atoms, respectively. The yellow region represents the accumulation of electrons, whereas the cyan region represents the depletion of electrons. The yellow region is mainly located in N and C atoms, whereas the cyan region is mainly shown in Gd atoms. The above phenomenon is noticeable in the doped atoms and adjacent atomic positions. It is suggested that electrons in Gd and Ga atoms are depleted and transferred to N and C atoms.

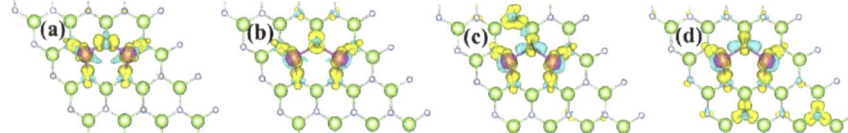

Figure 2. The charge density difference of (**a**) $Ga_{14}N_{16}Gd_2$, (**b**) $Ga_{14}N_{15}Gd_2C_1$, (**c**) $Ga_{14}N_{14}Gd_2C_2$ and (**d**) $Ga_{14}N_{13}Gd_2C_3$ monolayers. The isosurface level is 0.009 e·$^{-3}$. The yellow and cyan areas indicate the positive and negative electrons, respectively.

To clearly and quantitatively describe the electron transfer characters, the Bader analysis is established in Table 1. As in the above analysis, electrons depleted on Ga and Gd atoms and accumulated on N and C atoms. In $Ga_{14}N_{16}Gd_2$ monolayers, it is found that the Ga atoms lose about 1.35 |e|, the N atoms bonded without Gd atoms gain about 1.37 |e|

and the N atoms bonded with Gd atoms gain 1.47 |e|, which is 7.30% more than the former. Each Gd atom loses about 1.85 |e| and the introduction of C atoms has a negligible effect on this. The M site is special and is located between the two Gd atoms. The electron of N atoms at this position is 1.58 |e|, 15.33% more than that of other N atoms. When the C atom replaces this position, the obtained electron is 1.35 |e|, which becomes the same as the N atom bonded without Gd atoms, and no longer maintains specificity. The Ga atom bonded with C loses 1.24 |e|, which is reduced 7.46%.

Table 1. Calculated Bader analysis for Ga, N, Gd and C (except for M site) atoms, respectively. The M site is located between the two Gd atoms. Ga_1 and Ga_2 represent the Ga atom bonded with and without C, respectively, whereas N_1 and N_2 display the N atoms bonded with and without Gd atoms, respectively. The accumulation and depletion of electrons are indicated by + and −, respectively. The unit of all data is |e|.

	Ga_1	Ga_2	N_1	N_2	Gd	M	C
$Ga_{14}N_{16}Gd_2$	-	−1.34	1.47	1.37	−1.85	+1.58	-
$Ga_{14}N_{15}Gd_2C_1$	−1.24	−1.34	1.46	1.37	−1.79	+1.36	-
$Ga_{14}N_{14}Gd_2C_2$	−1.25	−1.35	1.46	1.37	−1.80	+1.35	1.06
$Ga_{14}N_{13}Gd_2C_3$	−1.23	−1.34	1.46	1.35	−1.79	+1.35	1.09

Figure 3a–f depicts the energy band structures of $Ga_{14}N_{16-n}Gd_2C_n$ monolayers without considering spin–orbital coupling (SOC). In $Ga_{14}N_{16}Gd_2$ monolayers (shown in Figure S1b), the conduction band minimum (CBM) is located at the Γ point along the high symmetry in the first Brillouin zone (BZ), whereas the valence band maximum (VBM) is located at K points. It can be seen from the energy band diagram, where the CBM maintains its original state, whereas the VBM rises slightly due to the orange energy level contributed by the C atom. The band gaps of the spin-up channels are 1.99 eV, 1.83 eV, 1.93 eV and 1.93 eV, respectively, whereas the spin-down channels are 2.05 eV, 0.56 eV, 0.74 eV and 0.46 eV, respectively. The band gap sharply decreased in the spin-down channel. This phenomenon is a consequence of the introduction of impurity energy levels near the Fermi energy level, which originates from the C atom. It can be seen that the C atom can transform the GaN:Gd monolayer into a spin-polarized semi-metal-like unique property with the spin-up channel maintaining a wide band gap and the spin-down channel having a small band gap.

The total magnetic moments of $Ga_{14}N_{16}Gd_2$, $Ga_{14}N_{15}Gd_2C_1$, $Ga_{14}N_{14}Gd_2C_2$, $Ga_{14}N_{13}Gd_2C_3$ are 14.00 μ_B, 15.00 μ_B, 16.00 μ_B and 17.00 μ_B, respectively. In Table 2, the spin details are depicted. For the $Ga_{14}N_{16}Gd_2$ monolayer, the magnetic moment is entirely contributed to by the Gd atoms. With an increasing number of C atoms, the ratios are reduced to 94.53%, 89.50% and 84.24%, respectively. Each C atom substitutes an N atom resulting in a hole, which is the reason for the increase in the magnetic moment by 1 μ_B. It is also shown that the T_C^{FMA} increases significantly under the influence of C atoms, with values of 58.22 K, 428.63 K and 319.32 K, respectively. A stable FM phase with a high magnetic moment and the highest T_C^{FMA} is obtained in the $Ga_{14}N_{13}Gd_2C_3$ monolayer. It is amended as 35.51 K, 264.46 K and 194.78 K, respectively. By comparing previous studies (shown in Table S2), Gd atoms in $Ga_{14}N_{16}Gd_2$ monolayers can introduce large magnetic moments, and the introduction of C atoms can further increase the magnetic moments.

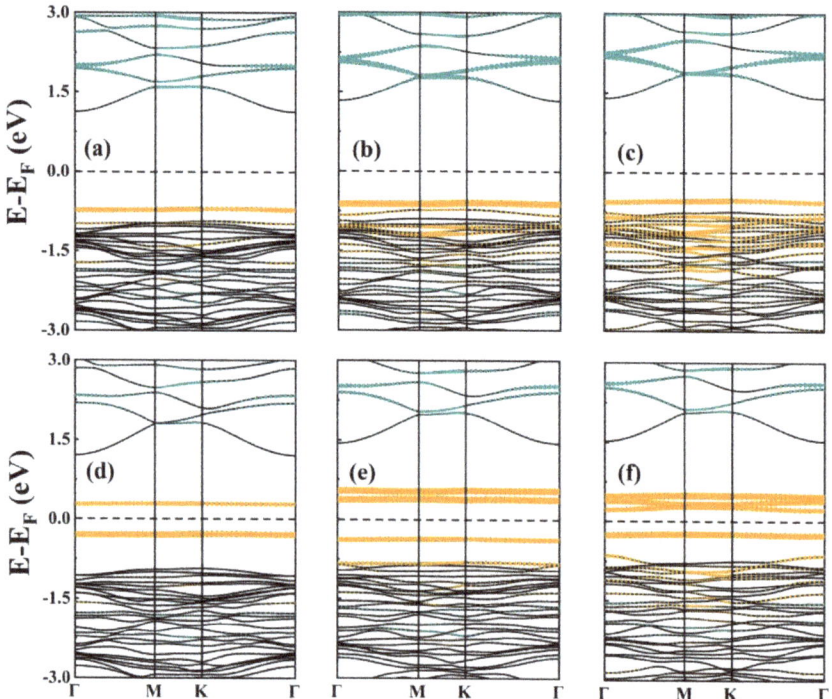

Figure 3. The spin-up band structures of (**a**) $Ga_{14}N_{15}Gd_2C_1$, (**b**) $Ga_{14}N_{14}Gd_2C_2$ and (**c**) $Ga_{14}N_{13}Gd_2C_3$ monolayers. The spin-down channels of (**d**) $Ga_{14}N_{15}Gd_2C_1$, (**e**) $Ga_{14}N_{14}Gd_2C_2$ and (**f**) $Ga_{14}N_{13}Gd_2C_3$ monolayers. The green and orange dotted lines represent the energy levels of Gd and C atoms, respectively. The dashed lines represent the Fermi level, which is taken to be 0.

Table 2. Magnetic moments (μ_B), energy difference ($\Delta E = E_{AFM} - E_{FM}$, meV), nearest-neighboring exchange coupling (J/eV) and estimated Curie temperature (T_C/K) of $Ga_{14}N_{16}Gd_2$, $Ga_{14}N_{15}Gd_2C_1$, $Ga_{14}N_{14}Gd_2C_2$, $Ga_{14}N_{13}Gd_2C_3$, $Ga_{14}N_{12}Gd_2C_4$ monolayers, respectively.

	M_{total} (μ_B)	M_{Gd} (μ_B)	ΔE (meV)	J (eV)	T_C^{FMA} (K)	T_C (K)
$Ga_{14}N_{16}Gd_2$	14.00	7.05	−4.08	−0.08	-	-
$Ga_{14}N_{15}Gd_2C_1$	15.00	7.09	15.05	0.31	58.22	35.51
$Ga_{14}N_{14}Gd_2C_2$	16.00	7.16	110.81	2.26	428.63	261.46
$Ga_{14}N_{13}Gd_2C_3$	17.00	7.16	82.55	1.68	319.32	194.78

To further investigate the magnetic mechanism, the PDOSs of $Ga_{14}N_{16-n}Gd_2C_n$ monolayers are plotted in Figure 4. It is further identified that the large magnetic moment stems mainly from the large exchange splitting of the Gd-4f state. The spin-up Gd-4f orbitals (below the VBM) are fully occupied while the spin-down orbitals (above the CBM) are fully unoccupied. The CBM is pushed to the Γ-point (shown in Figure 3) in the spin-down channel and decreases the electron effective mass which agrees with the previous report [39]. The Gd-4f orbitals are separated from the VBM (in the spin-up channel) and the CBM (in the spin-down channel) by about 4 and 7 eV, respectively. As a result, the coupling between Gd-4f and p-type orbitals should be weak [31]. In the spin-down channel, in agreement with the energy band analysis above, the C atom impurity energy level appears near the Fermi energy level and the CBM is reduced, leading to a small band gap. The Gd-5d orbitals and C-2p orbitals overlap near the Fermi energy level, forming p-d hybrid orbitals, and the C-2p near the Fermi energy level is significantly enhanced as the concentration of C atoms increases. Thus, we suggest that the stable FM phase originates from p-d hybridized orbitals.

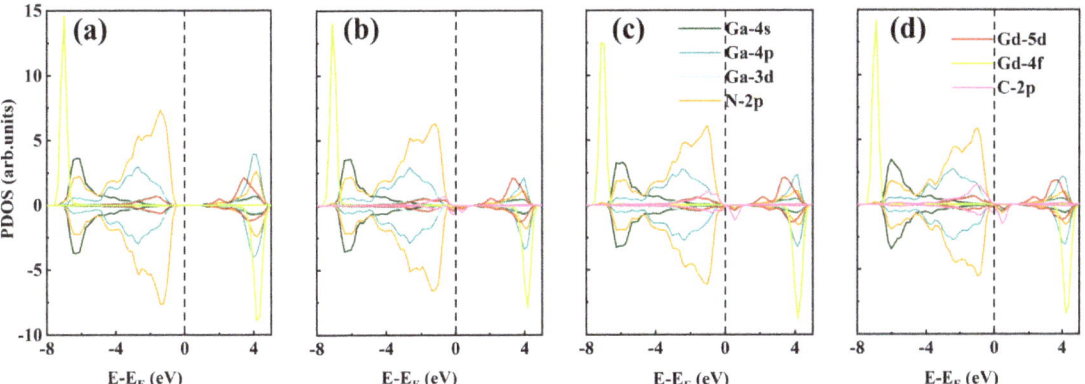

Figure 4. Spin-polarized partial density of states (PDOS) of (**a**) $Ga_{14}N_{16}Gd_2$, (**b**) $Ga_{14}N_{15}Gd_2C_1$, (**c**) $Ga_{14}N_{14}Gd_2C_2$ and (**d**) $Ga_{14}N_{13}Gd_2C_3$ monolayers, respectively. The dashed lines represent the Fermi level which is taken to be 0.

3.2. The Modulation by Strain Effect

The biaxial strains ($\varepsilon = \frac{a_0 - a}{a_0} \times 100$) ranging from -6% to 15% are performed on $Ga_{14}N_{16-n}Gd_2C_n$ monolayers. The total energy (E_{total}) and E_b of FM states are shown in Figure 5a,b, which is used to review the stability. The negative E_b indicates an exothermic reaction, and the larger $|E_b|$ means a more stable structure. The structural stability decreases as the curve decreases, whereas a rising one means that it is more stable. It can be clearly seen that structural stability decreases with increasing compressive strain. As the tensile strain increases, the stability of the system first increases, reaching a maximum at $\varepsilon = 2$, and then exhibits a significant decrease. Figure 5c shows the impact of strain effects on the band gap of the system. It can be seen that the compression strain has a nominal impact on the spin-up channel indicated by the solid line. For the spin-up band gap, similar trends exist for the total energy and E_b. Although the spin-down band gap shows slight fluctuations at $-4 \leq \varepsilon \leq 12$, the value decreases significantly when $\varepsilon \leq -5$ or $\varepsilon \geq 13$.

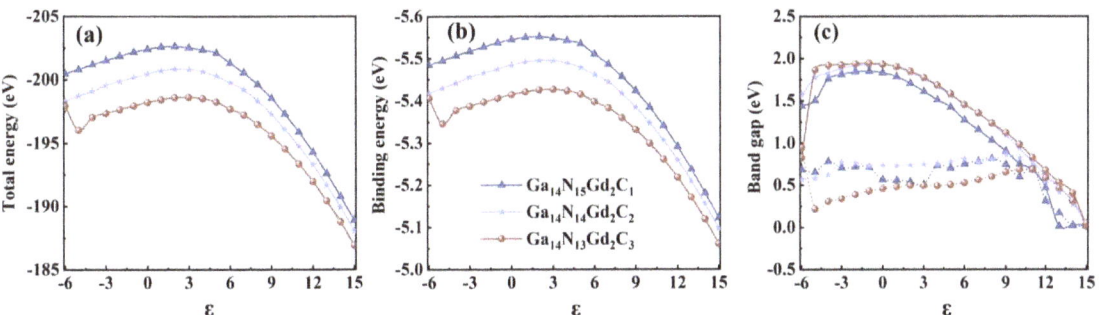

Figure 5. (**a**) E_{total} of FM phase, (**b**) E_b and (**c**) band gap of $Ga_{14}N_{16-n}Gd_2C_n$ monolayers. The dark blue (triangle), light blue (pentagram) and brown (circle) lines in the diagram represent $Ga_{14}N_{15}Gd_2C_1$, $Ga_{14}N_{14}Gd_2C_2$ and $Ga_{14}N_{13}Gd_2C_3$ monolayers, respectively.

ΔE and J are shown in Figure 6a,b and are used as criteria for the magnetic characters and predicting the evolution of the T_C. Positive and increasing values represent a stronger stable FM state. At $-4 < \varepsilon < 15$, the strain effect does not change the magnetic ground state of $Ga_{14}N_{16}Gd_2$, and the ground states of all structures are the FM states. When $\varepsilon \leq -5$, $Ga_{14}N_{13}Gd_2C_3$ possesses negative values, implying that the ground state is AFM. Additionally, the maximum values of ΔE for $Ga_{14}N_{15}Gd_2C_1$, $Ga_{14}N_{14}Gd_2C_2$ and $Ga_{14}N_{13}Gd_2C_3$ are obtained as 125.4 eV ($\varepsilon = -4$), 242.87 eV ($\varepsilon = 7$) and 125.91 eV ($\varepsilon = -3$), respectively. The T_C can be predicted to reach a maximum. Figure 6c,d demonstrates the T_C^{FMA} and T_C whose value increased to above T_R with strain effects. The highest T_C (T_C^{FMA}) is found for $Ga_{14}N_{15}Gd_2C_1$, $Ga_{14}N_{14}Gd_2C_2$ and $Ga_{14}N_{13}Gd_2C_3$ as 295.89 K (485.07 K, $\varepsilon = -4$), 564.57 K (925.53 K, $\varepsilon = 7$) and 297.09 K (487.04 K, $\varepsilon = -3$), respectively. It is clearly noticeable from the graph that the curve is most stable when $n = 2$, with T_C fluctuating around 300 K under strain effects.

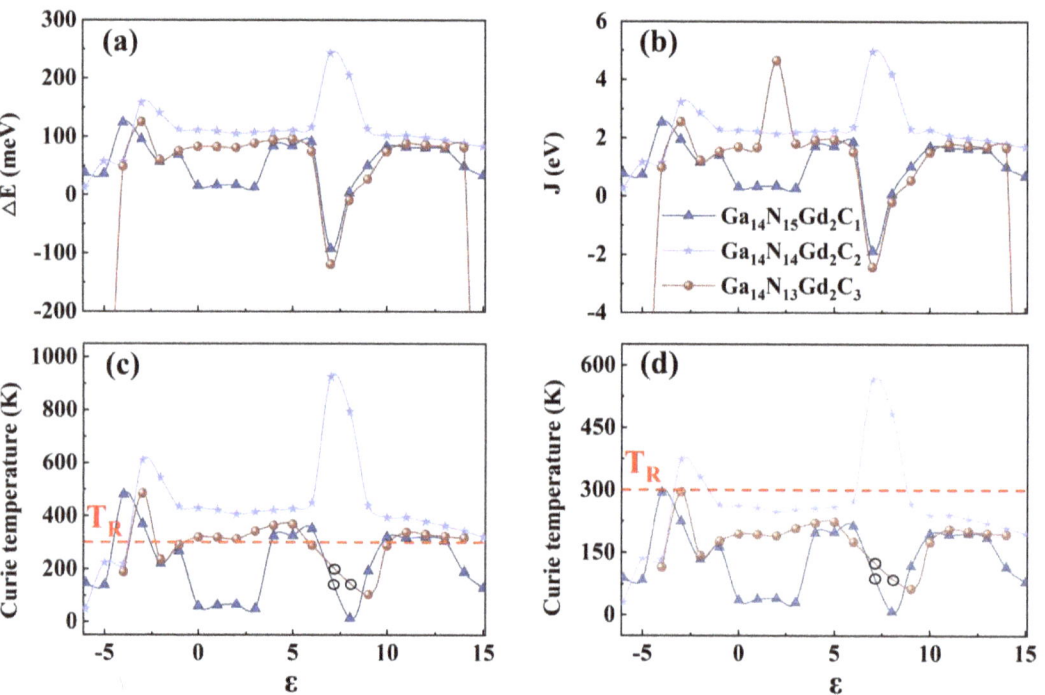

Figure 6. (**a**) Energy difference ($\triangle E$), (**b**) the magnetic coupling parameter (J), (**c**) T_C^{FMA} and (**d**) T_C. The dark blue (triangle), light blue (pentagram) and brown (circle) lines in the diagram represent the $Ga_{14}N_{15}Gd_2C_1$, $Ga_{14}N_{14}Gd_2C_2$ and $Ga_{14}N_{13}Gd_2C_3$ monolayers, respectively. The red dashed lines in the last two graphs indicate T_R. The black circles in the last two figures indicate this position as an AFM state without T_C.

The strains acting in the $Ga_{14}N_{14}Gd_2C_2$ monolayer of $\varepsilon = -6, 7$ and 15 are used as an example to understand the mechanism of orbital hybridization that causes the T_C change, whose PDOS is plotted in Figure 7. As can be seen in Figure 7b, the Gd-5d states and C-2p states overlap a lot near the Fermi energy, which indicates the coupling between them and the formatting of p-d orbitals. This is consistent with the results discussed above. A weakening of this coupling is found in Figure 7a,c, which is due to a reduction in the Gd-5d electronic state near the Fermi energy level, resulting in a reduction in T_C. In addition, the results obtained from the above analysis are further verified by the SOC method (shown in Figure S4). In conclusion, the replacement of N atoms with C atoms effectively transforms the GaN:Gd monolayer into a stable FM phase, which is further processed by biaxial strain to obtain GaN-based materials close to or even well above T_R.

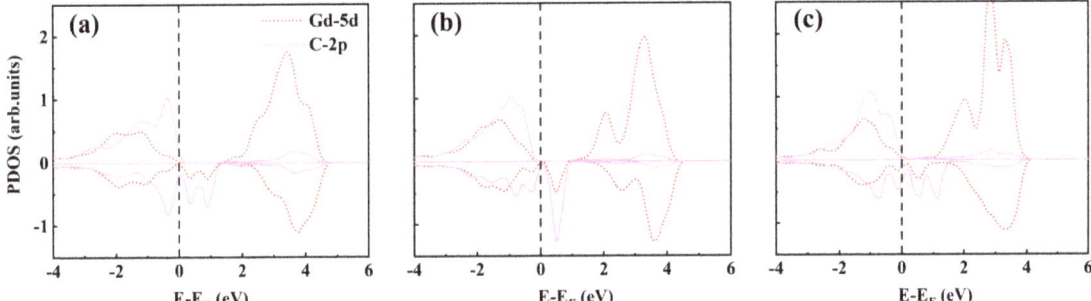

Figure 7. The PDOS of $Ga_{14}N_{14}Gd_2C_2$ monolayers at ε = (**a**) 6, (**b**) 7 and (**c**) 15, respectively.

4. Conclusions

In conclusion, the electronic and magnetic properties of $Ga_{14}N_{16-n}Gd_2C_n$ monolayers are studied by means of the DFT method. The total magnetic moment of the $Ga_{14}N_{16}Gd_2$ monolayer is 14 μ_B with a weak AFM phase. The magnetic moment stems mainly from the large exchange splitting of the Gd-4f state. It is suggested that the magnetic moment of the $Ga_{14}N_{16}Gd_2$ monolayer, in which the N atom has been substituted with a C atom, is increased with a stable FM phase. The main contributing factor is the hybridized p-d orbital between the Gd-5d and C-2p orbitals. A relatively high T_C of 261.46 K is observed in the $Ga_{14}N_{14}Gd_2C_2$ monolayers. In addition, the FM coupling and T_C can be further enhanced by suitable strain effects. The T_C of the monolayer can be significantly increased to 564.57 K, well above T_R.

Supplementary Materials: The following supporting information can be downloaded at: https://www.mdpi.com/article/10.3390/cryst13030531/s1, Figure S1: Band structures of (a) $Ga_{16}N_{16}$ (b) $Ga_{14}N_{16}Gd_2$, (c) $Ga_{23}N_{25}Gd_2$ and (d) $Ga_{34}N_{36}Gd_2$ monolayers. The blue and pink solid lines in the band structure represent spin up and spin down channels, respectively. PDOS of (e) $Ga_{16}N_{16}$ (f) $Ga_{14}N_{16}Gd_2$, (g) $Ga_{23}N_{25}Gd_2$ and (h) $Ga_{34}N_{36}Gd_2$ monolayers. The dash lines present the fermi-level which is taken to be 0.; Table S1: The E_b of FM phase, total magnetic moments (M_{total}), ΔE ($E_{AFM} - E_{FM}$), nearest-neighboring exchange (J), estimated T_C^{FMA} and T_C of $Ga_{14}N_{16}Gd_2$, $Ga_{14}N_{15}Gd_2C_1$, $Ga_{14}N_{14}Gd_2C_2$, $Ga_{14}N_{13}Gd_2C_3$, $Ga_{14}N_{12}Gd_2C_4$, respectively. Figure S2: PDOS with FM states of different substitution site: [1], [4], [M, 2], [5, 7], [2, M, 5] and [1, M, 2]. Figure S3: PDOS with AFM states of different substitution site: [1], [4], [M, 2], [5, 7], [2, M, 5] and [1, M, 2]. Figure S4: (a) E_{total} of FM phase, (b) E_b, (c) ΔE (d) J, (e) T_C^{FMA} and (f) T_C of $Ga_{14}N_{16-n}Gd_2C_n$ monolayers computed by the PBE method with SOC. The dark blue (triangle), light blue (pentagram) and brown (circle) lines in the diagram represent $Ga_{14}N_{15}Gd_2C_1$, $Ga_{14}N_{14}Gd_2C_2$ and $Ga_{14}N_{13}Gd_2C_3$ monolayers, respectively. The black circles in the last two figures indicate this position as an AFM state without T_C. Table S2: A brief summary of the magnetic properties for doped GaN materials. References [40–44] are cited in Supplementary Materials.

Author Contributions: Conceptualization, S.T. and H.X.; methodology, S.L., L.H. and S.T.; software, Y.H., H.X. and X.C.; validation, L.Z., Y.L. and H.X.; formal analysis, S.T.; investigation, S.T.; resources, S.T.; data curation, S.T.; writing—original draft prep-aration, S.T.; writing—review and editing, L.Z., H.X. and X.C.; visualization, Y.L., A.L. and H.X.; supervision, Y.L., A.L. and H.X.; project administration, H.X.; Supervision, R.X. All authors have read and agreed to the published version of the manuscript.

Funding: This research received no external funding.

Data Availability Statement: All data that support the findings of this study are included within the article (and Supplementary Files).

Acknowledgments: The authors would like to acknowledge the financial support from the Fundamental Research Funds for the Central Universities (2232022A-11), the Fundamental Research Funds for the Central Universities and Graduate Student Innovation Fund of Donghua University

(CUSF-DH-D-2022074) and the computational support from the Shanghai Supercomputer Center of the National Natural Science Foundation of Shanghai (21ZR1402200).

Conflicts of Interest: The authors declare no conflict of interest.

References

1. Wolf, S.A.; Awschalom, D.D.; Buhrman, R.A.; Daughton, J.M.; von Molnár, S.; Roukes, M.L.; Chtchelkanova, A.Y.; Treger, D.M. Spintronics: A Spin-Based Electronics Vision for the Future. *Science* **2001**, *294*, 1488–1495. [CrossRef]
2. Sato, K.; Katayama, Y. First principles materials design for semiconductor spintronics. *Semicond. Sci. Technol.* **2002**, *17*, 367–376. [CrossRef]
3. Zhao, Q.; Xiong, Z.; Luo, L.; Sun, Z.H.; Qin, Z.Z.; Chen, L.L.; Wung, N. Design of a new two-dimensional diluted magnetic semiconductor: Mn-doped GaN monolayer. *Appl. Surf. Sci.* **2017**, *396*, 480. [CrossRef]
4. Dietl, T.; Ohno, H.; Matsukura, F.; Cibert, J.; Ferrand, D. Zener Model Description of Ferromagnetism in Zinc-Blende Magnetic Semiconductors. *Science* **2000**, *287*, 1019–1022. [CrossRef]
5. Esch, A.V.; Bockstal, L.V.; Boeck, J.D.; Verbanck, G.; Steenbergen, A.S.V.; Wellmann, P.J.; Grietens, B.; Bogaerts, R.; Herlach, F.; Borghs, G. Interplay between the magnetic and transport properties in the III–V diluted magnetic semiconductor $Ga_{1-x}Mn_xAs$. *Phys. Rev. B* **1997**, *56*, 7392–7394.
6. Munekata, H.; Ohno, H.; Molnar, S.V.; Segmüller, A.; Chang, L.L.; Esaki, L. Diluted magnetic III-V semiconductors. *Phys. Rev. Lett.* **1989**, *63*, 1849–1852. [CrossRef]
7. Matsumoto, Y.J.; Murakami, M.; Shono, T.; Hasegawa, T.; Fukumura, T.; Kasawaki, M.; Ahmet, P.; Chiyow, T.; Koshihara, S.Y.; Koinuma, H. Room temperature ferromagnetism in transparent transition metal-doped titanium dioxide. *Science* **2001**, *291*, 854–856. [CrossRef]
8. Neal, J.R.; Behan, A.J.; Ibrahim, R.M.; Blythe, H.J.; Ziese, M.; Fox, A.M.; Gehring, G.A. Room temperature magneto-optics of ferromagnetic transition-metal-doped ZnO thin films. *Phys. Rev. Lett.* **2006**, *96*, 197208. [CrossRef] [PubMed]
9. Lee, J.S.; Lim, J.D.; Khim, Z.G.; Park, Y.D. Magnetic and structural properties of Co, Cr, V ion-implanted GaN. *J. Appl. Phys.* **2003**, *93*, 4512–4516. [CrossRef]
10. Du, K.; Xiong, Z.H.; Ao, L.; Chen, L.L. Tuning the electronic and optical properties of two-dimensional gallium nitride by chemical functionalization. *Vauum* **2021**, *185*, 110008. [CrossRef]
11. Zheng, F.F.; Xiao, X.; Xie, J.; Zhou, L.J.; Li, Y.Y.; Dong, H.L. Structures, properties and applications of two-dimensional metal nitrides: From nitride MXene to other metal nitrides. *2D Mater.* **2022**, *9*, 022001. [CrossRef]
12. Alaal, N.; Roqan, I.S. Tuning the electronic properties of hexagonal teo-demensional GaN monolayers via doping for enhanced optoelectronic applications. *ACS Appl. Nano Mater.* **2019**, *2*, 202–213. [CrossRef]
13. Chen, G.X.; Li, H.F.; Yang, X.; Wen, J.Q.; Pang, Q.; Zhang, J.M. Adsorption of 3d transition metal atoms on graphene-like gallium nitride monolayer: A first-principles study. *Superlattices Microstruct.* **2018**, *115*, 108–115. [CrossRef]
14. Hussain, F.; Cai, Y.Q.; Khan, M.J.I.; Imran, M.; Rashid, M.; Ullah, H.; Ahmad, E.; Kousar, F.; Ahmad, S.A. Enhanced ferromagnetic properties of Cu doped two-dimensional GaN mono-layer. *Int. J. Mod. Phys. C* **2015**, *26*, 1550009. [CrossRef]
15. Li, J.; Liu, H. Magnetism investigation of GaN monolayer doped with group VIII B transition metals. *J. Mater. Sci.* **2018**, *53*, 15986–15994. [CrossRef]
16. Chen, G.X.; Fan, X.B.; Li, S.Q.; Zhang, J.M. First-principles study of magnetic properties of alkali metals and alkaline earth metals doped two-dimensional GaN materials. *Acta Phys. Sin.* **2019**, *68*, 237303. [CrossRef]
17. Gutiérrez, C.A.H.; Moreno, Y.L.C.; Kuoppa, V.T.R.; Cardona, D.; Hu, Y.Q.; Kudriatsev, Y.; Serrano, M.A.Z.; Hernandez, S.G.; Lopez, M.L. Study of the heavily p-type doping of hexagonal cubic GaN with Mg. *Sci. Rep.* **2020**, *10*, 16858. [CrossRef] [PubMed]
18. Yeoh, K.H.; Yoon, T.L.; Lim, T.L.; Rusi; Ong, D.S. Monolayer GaN functionalized with alkali metal and alkaline earth metal atoms: A first-principles study. *Superlattices Microstruct.* **2019**, *130*, 428–436. [CrossRef]
19. Li, Y.; Xie, X.J.; Liu, H.; Wang, S.; Hao, Q.Y.; Liang, L.M.; Liu, C.C. Effect of carbon on the magnetic properties of Dy-implanted GaN films. *J. Alloys Compd.* **2018**, *762*, 887–891. [CrossRef]
20. Maskar, E.; Lamrani, A.F.; Belaiche, M.; Smairi, A.E.; Vu, T.V.; Rai, D.P. A DFT study of electronic, magnetic, optical and transport properties of rare earth element (Gd, Sm)-doped GaN material. *Mater. Sci. Semicond. Process.* **2022**, *139*, 106326. [CrossRef]
21. Maekawa, M.; Miyashita, A.; Sakai, S.; Kawasuso, A. Gadolinium-implanted GaN studied by spin-polarized position annihilation spectroscopy. *Phys. Rev. B* **2020**, *102*, 05442. [CrossRef]
22. Nobuaki, T.; Suzuki, A.; Yasushi, N.; Kawasuso, A. Room-temperature observation of ferromagnetism in diluted magnetic semiconductor GaGdN grown by RF-molecular beam epitaxy. *Solid State Commun.* **2002**, *122*, 651–653.
23. Asahi, H.; Zhou, Y.K.; Hashimoto, M.; Kim, M.S.; Li, X.J.; Emura, S.; Hasegawa, S. GaN-based magnetic semiconductors for nanospintronics. *J. Phys. Condens. Matter* **2004**, *16*, S555. [CrossRef]
24. Dhar, S.; Brandt, O.; Ramsteiner, M.; Kim, M.S.; Emura, S.; Hasegawa, S. Colossal magnetic moment of Gd in GaN. *Phys. Rev. Lett.* **2005**, *94*, 037205. [CrossRef]
25. Dhar, S.; Kammermeier, T.; Ney, A. Ferromagnetism and colossal magnetic moment in Gd-focused ion-beam-implanted GaN. *Appl. Phys. Lett.* **2006**, *89*, 06250. [CrossRef]

26. Liu, Z.Q.; Yi, X.Y.; Wang, J.W.; Kang, J.; Melton, A.G.; Shi, Y.; Lu, N.; Wang, J.X.; Li, J.M.; Ferguson, L. Ferromagnetism and its stability in n-type Gd-doped GaN: First-principles calculation. *Appl. Phys. Lett.* **2012**, *100*, 232408. [CrossRef]
27. Mitra, C.; Lambrecht, W.R.L. Interstitial-nitrogen- and oxygen-induced magnetism in Gd-doped GaN. *Phys. Rev. B* **2009**, *80*, 081202. [CrossRef]
28. Xie, R.K.; Xing, H.Z.; Zeng, Y.J.; Huang, Y.; Lu, A.J.; Chen, X.S. Room temperature ferromagnetism in Cu–Gd co-doped GaN nanowires: A first-principles study. *Phys. Lett. A* **2019**, *383*, 54–57. [CrossRef]
29. Thiess, A.; Blügel, S.; Dederichs, P.H.; Zeller, R.; Lambrecht, W.R.L. Systematic study of the exchange interactions in Gd-doped GaN containing N interstitials, O interstitials, or Ga vacancies. *Phys. Rev. B* **2015**, *92*, 100418. [CrossRef]
30. Dalpian, G.M.; Wei, S.H. Electron-induced stabilization of ferromagnetism in $Ga_{1-x}Gd_xN$. *Phys. Rev. B* **2005**, *72*, 115201. [CrossRef]
31. Liu, L.; Yu, P.Y.; Ma, Z.X.; Mao, S.S. Ferromagnetism in GaN:Gd: A density functional theory study. *Phys. Rev. Lett.* **2008**, *100*, 127203. [CrossRef]
32. Xie, R.K.; Xing, H.Z.; Zeng, Y.J.; Liang, Y.; Chen, S.X. First-principles calculations of GaN:Gd nanowires: Carbon-dopants-induced room-temperature ferromagnetism. *AIP Adv.* **2017**, *7*, 115003. [CrossRef]
33. Kresse, G.; Furthmuller, J. Efficiency of ab-initio total energy calculations for metals and semiconductors using a plane-wave basis set. *Comp. Mater. Sci.* **1996**, *6*, 15–50. [CrossRef]
34. Perdew, J.P.; Burke, K.; Ernzerho, M. Generalized Gradient Approximation Made Simple. *Phys. Rev. Lett.* **1996**, *77*, 3865–3868. [CrossRef] [PubMed]
35. Li, S.; Lu, A.J.; Xie, R.K.; Xing, H.Z.; Zeng, Y.J.; Huang, Y.; Chen, X.S. Tunable electronic and magnetic properties of functionalized (H, Cl, OH) germanium carbide Sheet. *J. Nanosci. Nanotechnol.* **2017**, *17*, 3927–3933. [CrossRef]
36. Lin, X.; Mao, Z.; Dong, S.G.; Jian, X.D.; Han, R.; Wu, P. First-principles study on the electronic structures and magnetic properties of TM-doped (TM = V, Cr, Mn, and Fe) tetragonal ScN monolayer. *J. Magn. Magn. Mater.* **2021**, *527*, 167764. [CrossRef]
37. Şahin, H.; Cahangirov, S.; Topsakal, M.; Bekaroglu, E.; Akturk, E.; Senger, R.T.; Ciraci, S. Monolayer honeycomb structures of group-IV elements and III-V binary compounds: First-principles calculations. *Phys. Rev. B* **2009**, *80*, 155453. [CrossRef]
38. Ranchal, R.; Yadav, B.S.; Trampert, A. Ferromagnetism at room temperature of c- and m-plane GaN: Gd films grown on different substrates by reactive molecular beam epitaxy. *J. Phys. D Appl. Phys.* **2013**, *46*, 075003. [CrossRef]
39. Liechtenstein, A.I.; Katsnelson, M.I.; Antropov, V.P.; Gubanov, V.A. Local spin density functional approach to the theory of exchange interactions inferromagnetic metals and allys. *J. Magn. Magn. Mater.* **1987**, *67*, 65–74. [CrossRef]
40. Zhang, Z.W.; Shang, J.Z.; Jiang, C.Y.; Rasmita, A.; Gao, W.B.; Wu, T. Direct photoluminescence probing of ferromagnetism in monolayer two-dimensional $CrBr_3$. *Nano Lett.* **2019**, *19*, 3138–3142. [CrossRef]
41. Huang, B.; Clark, G.; Moratalla, E.N.; Klein, D.R.; Cheng, R.; Seyler, K.L.; Zhong, D.; Schmidgall, E.; Mcguire, M.A.; Cobden, D.H.; et al. Layer-dependent ferromagnetism in a Van der Waals crystal down to the monolayer limit. *Nature* **2017**, *546*, 270–273. [CrossRef]
42. Deng, Y.J.; Yu, Y.J.; Song, Y.C.; Zhang, J.Z.; Wang, N.Z.; Sun, Z.Y.; Yi, Y.F.; Wu, Y.Z.; Wu, S.W.; Zhu, J.Y.; et al. Gata-tunable room-temperature ferromagnetism in two-dimensional Fe_3GeTe_2. *Nature* **2018**, *563*, 94–99. [CrossRef]
43. Li, B.; Wan, Z.; Wang, C.; Chen, P.; Huang, B.; Cheng, X.; Qian, Q.; Li, J.; Zhang, Z.W.; Sun, G.Z.; et al. Van der Waals epitaxial growth of air-stable $CrSe_2$ nanosheets with thickness-tunable magnetic order. *Nat. Mater.* **2021**, *20*, 818–825. [CrossRef]
44. Gong, C.; Li, L.; Li, Z.L.; Ji, H.W.; Stern, A.; Xia, Y.; Cao, T.; Bao, W.; Wang, C.Z.; Wang, Y.; et al. Discovery of intrinsic ferromagnetism in two-dimensional Van der Waals crystals. *Nature* **2017**, *546*, 265–269. [CrossRef]

Disclaimer/Publisher's Note: The statements, opinions and data contained in all publications are solely those of the individual author(s) and contributor(s) and not of MDPI and/or the editor(s). MDPI and/or the editor(s) disclaim responsibility for any injury to people or property resulting from any ideas, methods, instructions or products referred to in the content.

Article

Nanostructured Mn–Ni Powders Produced by High-Energy Ball-Milling for Water Decontamination from RB5 Dye

Wael Ben Mbarek [1], Mohammed Al Harbi [2], Bechir Hammami [2], Mohamed Khitouni [2,3,*], Luisa Escoda [1] and Joan-Josep Suñol [1]

[1] Department of Physics, Campus Montilivi s/n, University of Girona, 17003 Girona, Spain; benmbarek.wael@hotmail.fr (W.B.M.); joanjosep.sunyol@udg.edu (J.-J.S.)
[2] Department of Chemistry, College of Science, Qassim University, Buraydah 51452, Saudi Arabia; 421100416@qu.edu.sa (M.A.H.); b.hammami@qu.edu.sa (B.H.)
[3] Laboratory Inorganic Chemistry, UR-11-ES-73, Faculty of Science of de Sfax, University of Sfax, B.P. 1171, Sfax 3000, Tunisia
* Correspondence: kh.mohamed@qu.edu.sa; Tel.: +966-55-343-3072

Abstract: In this study, the degradation efficiency of Mn-20at%Ni and Mn-30at%Ni particle powders made by melt-spinning and high-energy ball-milling techniques is investigated in relation to the degradation of the azo dye Reactive Black 5. SEM, EDS, and XRD were used to analyze the powders' morphology, surface elemental composition, and phase structure. An ultraviolet-visible absorption spectrophotometer was used to measure the ball-milled powder's capacity to degrade, and the collected powders were examined using the FTIR spectroscopy method to identify the substituents in the extract. The impact of MnNi alloy on the azo dye Reactive Black 5's degradation and its effectiveness as a decolorizing agent were examined as functions of different parameters such as chemical composition, specific surface, and temperature. In comparison to the Mn-30at%Ni alloy, the powdered Mn-20at%Ni particles show better degrading efficiency and a faster rate of reaction. This remarkable efficiency is explained by the configuration of the valence electrons, which promotes more responding sites in the d-band when the Ni content is reduced. Therefore, increased electron transport and a hastened decolorization process are achieved by reducing the Ni concentration of RB5 solution with Mn80 particle powder. Additionally, this difference in their decolorization efficiency is explained by the fact that Mn-20at%Ni has the highest specific surface area of 0.45 m^2 g^{-1}. As the main result, the functional uses of nanostructured metallic powder particles as organic pollution decolorizers in the textile industry are greatly expanded by our study.

Keywords: nanomaterials; ball-milling; melt-spinning; decolorization; RB 5; XRD; UV-visible

1. Introduction

Due to their high toxicity and slow biodegradation rate, colored effluents have attracted a lot of attention [1–4]. Significant sources of contamination come from the textile industries' dye effluents [5]. Additionally, the textile industry consumes a significant amount of water and generates a lot of wastewater that contains dyes. Because it poses a major problem for sewage treatment stations, the appropriate regeneration of wastewater containing dyes in high quantities is necessary and critical work. Researchers have suggested a number of physicochemical and biological procedures and materials for the treatment of wastewater, including adsorption methods, biodegradation, the coagulation–flocculation method, advanced oxidation, and hypochlorite treatment, ozonation, and hypochlorite treatment [6–9]. The bulk of these processes, however, have limitations. For instance, Fenton, photocatalysts, and other oxidation procedures are highly expensive; biological processes require a long time; and occultation and adsorption may not be successful. Furthermore, it is beneficial to proceed with the cheapest, most accessible, most efficient techniques and materials when degrading azo dyes. One such practical method is reduction with zero-valent metals (ZVM) such as Mg, Ni, Fe, Zn, Co, or Al, and alloys, which have been investigated as attractive and advantageous routes due to their inexpensive cost merits, rapid degradation performance, and easy operation in the removal of azo dyes [10–18]. Specifically, the degradation reaction was determined by a redox reaction, wherein the surface metal loses electrons to break the active bonds (–N = N–) of the reactive dyes. In contrast to the bulk material, the fine powder form of metals and alloys offers more active surface sites for the reactive degradation of organic compounds. In particular, nanomaterials made up of inorganic components such as semiconductors, metal oxides, and nanocatalysts have attracted a lot of interest as wastewater treatment materials. A range of nanocatalysts are utilized in wastewater treatment, including photocatalysts [19,20], electrocatalysts [19], heterojunction photocatalytic materials [21], and Fenton-based catalysts [22] for analysis of their oxidation of organic contaminants [23] and of their antimicrobial effects [24]. The importance of nanoparticle photocatalytic processes, which are based on the interaction of light energy with metallic nanoparticles, may be seen in their extensive and effective photocatalytic activity on a variety of pollutants. These photocatalysts, which are often composed of semiconductor metals, can break down a range of persistent organic pollutants found in wastewater, such as dyes, detergents, pesticides, and volatile organic compounds [25]. Furthermore, halogenated and non-halogenated organic molecules, as well as many pharmaceuticals, personal care products, and heavy metals, can all be broken down very effectively by semiconductor nanocatalysts [26]. Semiconductor nanomaterials operate under reasonably benign conditions, and perform well even at low concentrations. The photoexcitation of an electron inside the catalyst is the fundamental idea behind how photocatalysis works. When TiO_2 is exposed to light (in this case, UV), holes ($h+$) and ejected electrons ($e-$) are generated in the conduction band. Water molecules (H_2O) absorb holes ($h+$) in an aqueous solution to create hydroxyl radicals (OH) [27]. The radicals have a potent and indiscriminate oxidizing effect. These hydroxyl radicals convert the organic pollutants into water and gaseous breakdown products [28]. On the other hand, it was discovered that the size, appropriate surface area, shapes, and roughness of nanocrystalline powders had a significant impact on their catalytic activity [29–33]. The use of mechanical alloying in high-energy ball mills is a simple and less complicated technique for producing fine metallic powder particles for decolorizing applications [29,30,34–37]. This is because these ball-milled particles frequently exhibit fine particles with rough surfaces and highly metastable nanocrystalline or amorphous morphologies. Indeed, milled powders undergo severe plastic deformation during this procedure to cause fracture and cold welding, which allows the grains to be refined to a steady state size (50 nm) for the majority of metals and alloys [38–40]. Repetitive fractures and cold welding of the powder particles cause interactions between the initial mixture's solid constituents. As a result of internal strains driven by a high density of dislocations and an important proportion of grain boundaries, mechanically stored enthalpy can act as a catalyst for the creation of nanocrystalline and/or amorphous structures [34,41], helping to enhance the material's mechanical, physical, and magnetic properties [42]. Recently, our research [29,30,35–37] reported on the remarkable effectiveness of mechanically alloyed Mn(Al), Ca(Al), Mn(Al,Co), and Mn(Al,Fe) powder particles in the degradation of azo dyes. In this study, new findings examining various Mn–Ni particle properties will be presented and evaluated against current studies, including the impact of composition and structure on the decomposition rate of RB5. To achieve this, we employ the method of melt spinning succeeded by ball-milling to produce Mn–Ni powder particles with a high surface area, which will increase the efficiency of dye molecule degradation.

2. Materials and Methods

The alloy ingots were made by arc melting a combination of Mn (99.9 wt%) and Ni (99.99 wt%) in a Ti-gettered argon environment to yield alloys with nominal compositions of Mn-30at%Ni and Mn-20at%Ni (at%) alloys. This bulk was heated by induction, melted, and then loaded into a spinning copper wheel through a nozzle that is 0.8 mm wide, producing

40 mm thick, rapidly quenched ribbons. Then, under an Ar atmosphere, the ribbon samples were placed in a ball-milling jar. The samples were milled in an inverse rotational direction, and the ball-milling speed for the jar was 500 rpm. In order to prevent the heat of the sample, powder agglomeration, and adhering to the jar walls and balls, a time of 5 min waiting period was used after each 10 min of milling. There were 15 h spent milling in total. The shape of the ball-milled (BM) powder was examined using scanning electron microscopy (SEM) in a DSM960A ZEISS microscope in secondary electron mode at a voltage of 15 kV. The SEM is equipped with a Vega Tescan energy dispersive X-ray spectrometry (EDS) analyzer. EDX was used to examine the semi-quantitative elemental composition.

Using Micromeritics ASAP 2020 M equipment, the Brunauer, Emmett, and Teller (BET) theory-based gas multilayer adsorption process was employed to determine the specific surface area of the MnNi powder. The observations were made in a nitrogen atmosphere after the powder had been degassed for 24 h at 300 °C.

By employing CuKα radiation with a Siemens D500 powder diffractometer at room temperature, the structural alterations of the milled powders were identified. A full pattern XRD Rietveld fitting approach was used to extract the microstructural properties.

A solution of RB5 with a concentration of 40 mg L^{-1} was employed to assess the colorant degradation process. 100 mL of solution was mixed with 0.25 g of ball-milled powder for each degradation experiment. On a Rotanta 460 r centrifuge, the solution samples were taken at periodic intervals and centrifuged for 15 min at 3000 rpm. An ultraviolet-visible absorption spectrophotometer (UV-Vis) was used to separate the supernatants and measure the color at the RB5 dye's maximum absorption wavelength (Shimadzu 2600 UV-visible). To identify the presence of the substituents in the extract, the powder was examined using a Fourier transform infrared spectroscopy (FTIR) spectrum. The same amount of ground sample was used to make the pellets in each example, and the spectra were calibrated to enable comparisons across various samples.

3. Results

The surface appearance of the BM Mn-30at%Ni (Figure 1(a1,a2)) and Mn-20at%Ni (Figure 1(b1,b2)) powder particles is shown in Figure 1, before they are used to degrade Reactive Black 5. As demonstrated, the severe plastic deformation imparted to the ribbons during the ball-milling process caused the nanostructured powder particles to be discovered to be irregular and/or spherical in shape, with corrugated surfaces. The average particle sizes of Mn-30at%Ni and Mn-20at%Ni alloys, respectively, are determined to be around 14 and 22 μm, showing a very uniform distribution (Figure 2(a2,b2)). The Mn-30at%Ni alloy was revealed to have an average particle size of 14 μm, which is somewhat smaller than the industry average. This small discrepancy can be attributed to the manual selection method of particle powder used for the SEM examination, and the standard error is 8 μm.

The EDS microanalysis shown in Figure 2(a1,b1) demonstrates that the BM powder is mostly made up of the basic elements Mn and Ni, and that there is no contamination from the milling tools or the oxygen in the air. The sputtering procedure that was utilized to prepare the samples for SEM observation is what caused the carbon that was discovered. Mn and Ni, respectively, have contents of 69.68:30.32 and 81.08:18.92 for Mn-30at%Ni and Mn-20at%Ni. These ratios, 70:30 and 80:20, are extremely similar to the nominal composition (Figure 2(a1,b1)).

Figure 1. Particle morphologies of the mechanical alloying (MA) Mn-30at%Ni (**a1**,**a2**) and Mn-20at%Ni (**b1**,**b2**) powders (Before degradation).

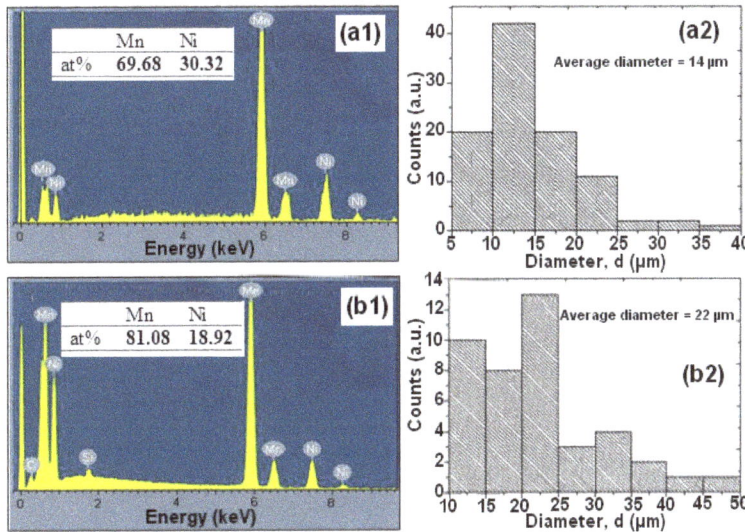

Figure 2. Nominal compositions as examined by EDX and the distribution of particle sizes of the mechanical alloying (MA) Mn-30at%Ni (**a1**,**a2**) and Mn-20at%Ni (**b1**,**b2**) powders (before degradation).

The two binary BM powders, Mn-30at%Ni and Mn-20at%Ni, were used to treat RB5 solutions, and Figure 3a shows representative photographs of these treatments. As shown on the test tubes, the supernatants of both aqueous solutions were separated at 0, 2, 5, 10, 13, 16, 17, and 18 h, as well as 0, 2, 5, and 15 h. It is obvious that both of the RB5 aqueous solutions have undergone decolorization. Figure 3b,c show the matching UV-Vis spectrum variations for both alloy compositions as a function of specified reaction time. The "–N = N–" azo produces the maximum absorbance at max = 597 nm in the visible area for the time t = 0 h (before treatment by BM powders) [29,35,36,43]. According to previous studies, the relationship between the intensity of this peak and the solution's azo dye content is linear [43,44]. The dye's benzene and naphthalene rings, respectively, are associated with the other two bands in the UV area, which are located at 310 and 230 nm [45,46]. The degradation and evolution of the RB5 chromophores are indicated by the decrease in intensity at 597 nm. In other words, this reduction provides details regarding the azo band cleavage, the creation of (–NH_2) groups, and consequently, the breakdown of RB5 in solution. The intensification of the absorbance peak at 267 nm served as additional evidence of the azo band cleavage. The same outcomes have been reported in our most recent research [29,36] when Black 5 dye solutions were reductively degraded by Ca-35at%Al and Mn-15at%Al nanostructured powders. Three hours before the binary Mn-30at%Ni effect on RB5 discoloration, the binary Mn-20at%Ni causes the solution to become completely discolored in 15 h (Figure 3d). However, it appears that when employing the two Mn–Ni powder compositions, the efficiencies of the RB5 decolorization at the same dosage and temperature are different. However, when the Ni content rises and the Mn content falls in the alloy's composition, the question of why there is a 3 h delay in the kinetics of decolorization arises. As was discovered in this work, the same dye solution and Reactive Black 5 dye were decolored using Mn–Ni particle powders with various Mn and Ni contents under the same conditions (milling time, temperature, etc.). The mass transfer effects on the degradation method can be removed as a result. Mn and Ni are both transition metals that make up the alloy's component and have nearly full or partially filled 3d bands. Therefore, it is reasonable to assume that the relative contents of Mn and Ni are responsible for the variances in their decolorization capabilities. In other words, the varied decolorization qualities of Mn–Ni with varying Ni levels depend on the variable Mn and Ni characteristics. The adsorbate in the solution during the decolorization process is the same for both types of nanostructured powders; therefore, the Mn-Ni alloy's electronic structure mostly affects the adsorption of dyes. The van der Waals forces, charge density effect, and transfer forces between the alloys and dye molecules are comparable due to the identical physical characteristics of these nanostructured powders. Therefore, the primary topic of our discussion is covalent or localized bonding. All atoms share the valence electrons of metallic atoms to create an electron cloud with a significantly distributed status, based on the energy band theory [47,48]. The valence electron configurations of Mn and Ni are $4s^23d^5$ and $4s^23d^8$, respectively. According to magnetic measurements, each atom of Mn and Ni had an average of 1.22 and 0.6 holes in the d band, respectively [48,49]. More unpaired electrons are present when there are more holes. More unpaired electrons correlate to a higher adsorption capacity because the latter can establish a localized adsorption bond with the adsorbate molecules [50,51]. Because Mn80 particle powders have more d band holes per unit of an atom than Mn70 ribbons, they exhibit the highest adsorption efficiency during the first 15 h of the decolorization procedure. However, it is important to keep in mind that these variations could be explained by the varied reduction potentials of Mn and Ni, as well as the various solubilities of respective hydroxides. Given that Mn and Ni are the same components in both alloys, it appears that these differences are related to their elemental composition. The standard oxidation potential of Mn/Mn^{2+} (1.18 V) is more negative than that of Ni/Ni^{2+} (0.23 V), yet it is less negative than that of H^+/H_2 (0.00 V). As a result, the transfer of electrons from Ni to H^+ takes longer than it does from Mn to H^+. According to these hypotheses, decreasing the Ni concentration of RB5 solution with Mn80 particles powder results in improved electron transport and a faster decolorization process. Similar results were found in our most recent work [29], where it was shown that the Mn–Al–Fe powder decolorization reaction occurred more rapidly than the MnAlCo mixture. The different oxidation potentials of Fe/Fe^{2+} (0.44 V) and Co/Co^{2+} (0.34 V) were used to explain this result. Additionally, the rapid decolorization of acid orange II solution is primarily dependent on the amount of Fe present in the Fe–Co–Si–B ribbons [50]. Due to the fact that Fe has a lower oxidation potential than Co, there is a greater electron transfer between Fe and H^+ compared to Co and H^+.

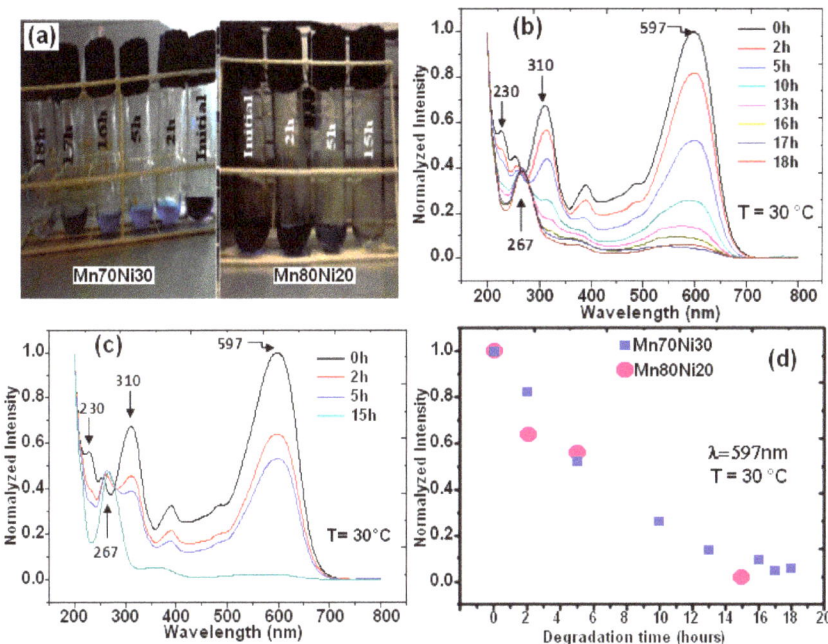

Figure 3. Image of Black 5 solutions processed by MA powders before and after degradation (**a**), UV absorption spectra at different times for Mn-30at%Ni (**b**) and Mn-20at%Ni (**c**) alloys, and the decolorization % from UV absorption intensity at 597 nm versus reaction time for both alloys (**d**).

Four experiments were conducted at temperatures of 25, 30, 40, and 50 °C to track the influence of temperature on the efficiency of the decolorization effect on the degradation of RB5. The exponential function fits the degradation behavior, as seen in Equation (1) [49]:

$$I = I_0 + I_1 e^{-t/t_0} \qquad (1)$$

where I is the standardized intensity of the absorption peak, I_0 and I_1 are fitting constants, t is the decolorization time, and t_0 is the moment when the intensity drops to e^{-1} of the original condition, this was calculated by fitting the data points.

The Arrhenius-type Equation (2) can be used to determine the thermal activation energy limit ΔE if we assume that the process is thermally activated:

$$t_0 = \tau_0 - e^{-E/RT} \qquad (2)$$

where R is the gas constant and τ_0 is a time pre-factor. Using BM $Mn_{80}Ni_{20}$ powder and a temperature range of 25–50 °C, Figure 4b shows the Arrhenius curve of $\ln(t_0)$ as a function of $1/T$ for the decomposition processes of azo dye.

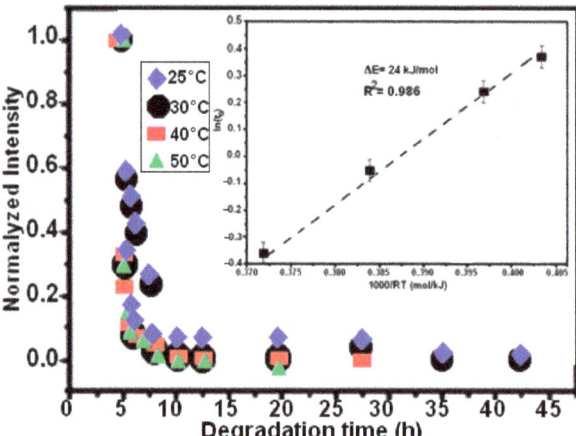

Figure 4. The plot of the decay time (t_0) against temperature and the standardized UV-vis absorption intensity at 597 nm for the MA Mn-20at%Al powder at various temperatures. The fitting of the Arrhenius-type equation to generate the activation energy is represented by the solid lines.

As a result, 24 ± 5 kJ mol^{-1} is the expected ΔE value. The complete degrading reaction took place in 7.5 h, maintaining the extremely high efficiency observed at a temperature of 50 °C. It has previously been noted that activation energy values range between 30 and 80 kJ mol^{-1} for the decolorizing processes of several azo-colorants using metallic particles [31,32,51]. The activation energy for binary $Mn_{85}Al_{15}$ was determined to be 14 kJ mol^{-1} in our recent work on the degradation of RB5 [35]. According to Aboli-Ghasemabadi et al., the activation energy for the MnAl-based degradation of Orange II was 49 ± 5 kJ mol^{-1}. For the heterogeneous decomposition of Orange II with peroxymonosulfate accelerated by mesoporous $MnFe_2O_4$, Deng et al. obtained a value of $\Delta E = 27.7$ kJ mol^{-1} [52]. The activation energy for typical thermal processes is commonly between 60 and 250 kJ mol^{-1}, according to Cheng et al. [53]. In practice, it is impossible to compare the reaction rate to those reported using other materials because of the various utilized conditions, particle weight/solution volume fractions, and type forms of dyes or starting concentrations. The specific surface area of powdered particle components is another concept that can affect the effectiveness of a reaction. The small specific surface area of the powder particles can be used to explain the low reaction efficiency of the ball-milled $Mn_{80}Ni_{20}$ (15 h/30 °C and 7.5 h/50 °C) powder in the current investigation. It is well known that all catalysts preserve a significant amount of surface area [54,55]. The BET method yielded specific surface areas of 0.32 and 0.45 m^2 g^{-1} for Mn-30at%Ni (b) and Mn-20at%Ni, respectively. During the experiment process, the pH values of the solution vary from acid (pH~6 at the beginning of the reaction) to alkaline (pH~9 at the end of the reaction) character. The initial acid behavior around a pH of 6 is due to the solution of the dye. When the reduction reaction takes place, aromatic amines are generated with an increase in pH of around 9. This indicates that the process generates bases that increase the pH. These findings lead us to the conclusion that the redox processes in acidic and alkaline solutions may be related to the phenomenon of degradation. One possible explanation is that as a consequence of the redox process, the protons of the aqueous medium together with the hydrogen atoms cause the generation of hydrogen gas adhered to the metallic surface, and it is the latter that reacts with the azo group, generating aromatic amines. The proposed mechanisms in basic and acidic media, however, were recapped in Table 1. For acid or neutral conditions, the presence of H_3O^+ on the metal surface can facilitate the reduction process, acting as an intermediary in the overall process. The active H2 will reduce the adsorbed azo dye molecules, thereby increasing the decomposition efficiency [35]. Using metallic compositions or ZVM as oxidants, several authors have investigated the effects of an acid medium on the reduction of "–N =

N–" bonds [31,56,57]. They claimed that in an acid media, 100% of the azo dye solution was degraded (an expected result, because in acidic conditions, insoluble hydroxides that prevent the closure of the circuit generated for the electron mobility are not formed).

Table 1. The proposed mechanisms of the degradation process of RB5 in acid or neutral mediums.

The Mechanism in Acid or Neutral Medium	Micropile
$Mn \rightarrow Mn^{2+} + 2e^-$ $H_2O \rightarrow H_3O^+ + OH^-$ $2H_3O^+ + 2e^- \rightarrow H_2 + 2H_2O$ $Mn^{2+} + 2OH^- \rightarrow Mn(OH)_2 \downarrow$ $R-N = N-R' + H_2 \rightarrow R-NH-NH-R'$ $R-NH-NH-R' + H_2 \rightarrow R-NH_2 + R'-NH_2$	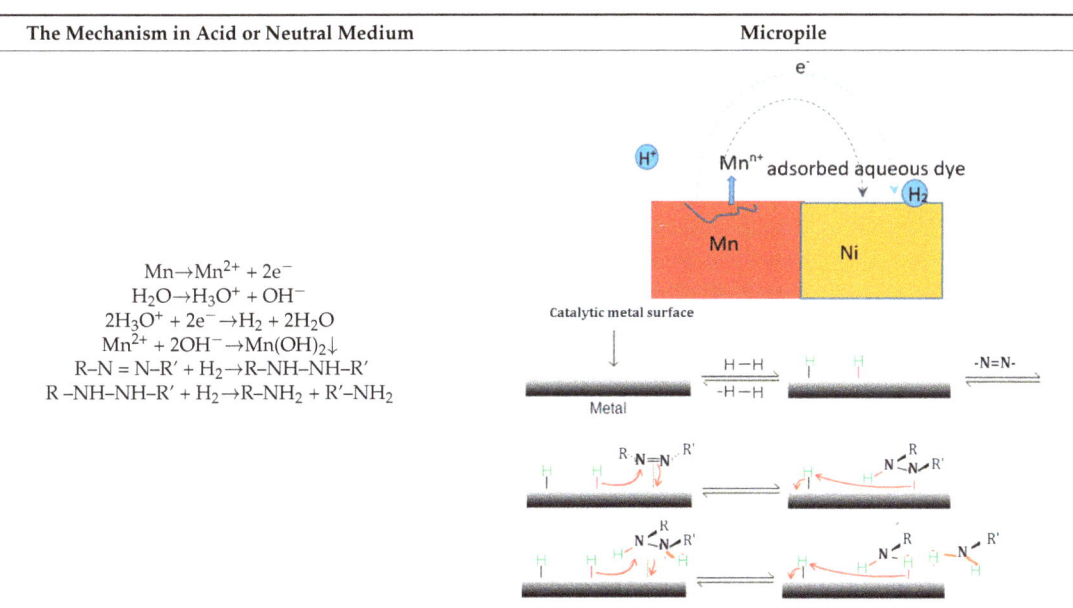

SEM analysis is utilized to describe the powder particle surfaces that were produced after the degradation process. In Figure 5(a1,a2), micrographs of ball-milled $Mn_{70}Al_{30}$ particles show that some reaction products, such as flowered and nanobristles forms, are uniformly adsorbed over the entire surface of the particles. In contrast, in Figure 5(b1,b2), micrographs of $Mn_{80}Ni_{20}$ particles show that nanorods are uniformly distributed on all of the particles' surfaces. Mn_2NiO_4 oxide makes up the majority of these nanorods. Numerous corrosion pits were seen on the surface of the alloy for both particle powders, proving that pitting corrosion on the alloy particles occurred throughout the RB5 degradation process [9,35,58,59]. The related EDS studies shown in Figure 5(a3,b3) reveal the approximate elemental compositions as follows: Mn: 56.63/Ni: 10.58/O: 33.06 for the $Mn_{80}Ni_{20}$ alloy, and Mn: 40.51/Ni: 30.08/O: 29.41 for the $Mn_{80}Ni_{20}$ alloy.

On the other hand, XRD was used to analyze changes in the microstructure of the $Mn_{70}Ni_{30}$ and $Mn_{80}Ni_{20}$ alloys that were noticed after the degradation of RB5. Figure 6 displays the results that were attained. A closer look at the XRD diffractograms of the BM $Mn_{70}Ni_{30}$ and $Mn_{80}Ni_{20}$ powders, which were obtained before the degradation of RB5, reveals that the MnNi phase is primarily present. Ball-milled $Mn_{70}Ni_{30}$ and $Mn_{80}Ni_{20}$ powders have average crystallite diameters of 18 ± 5 nm and 12 ± 5 nm, respectively. By increasing the number of atoms in the boundary regions, this nanocrystalline structure may also contribute to an increase in the reactivity of the powdered $Mn_{80}Ni_{20}$ particles within the aqueous-colored solutions. After degradation, for both powders, we note the appearance of newly identified $Mn_2Ni_1O_4$ oxide (JCPD 98-017-4001), and the $Mn(OH)_2$ (JCPD 00-008-0171), and $Ni(OH)_2$ (JCPD 00-014-0117) hydroxides beside the Mn(Ni) phase (JCPD 98-010-4918) (Figure 6a,b). The new products resemble micro-precipitates covering the surface of the particle powders.

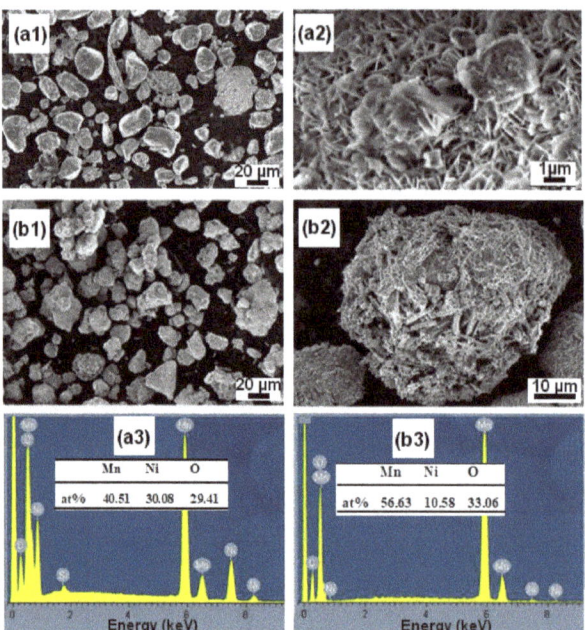

Figure 5. Surface morphologies of the Mn-30at%Ni (**a1**,**a2**) and Mn-20at%Ni (**b1**,**b2**) particle powders, respectively, collected after RB5 decomposition, and their nominal contents as determined by EDX (**a3**,**b3**).

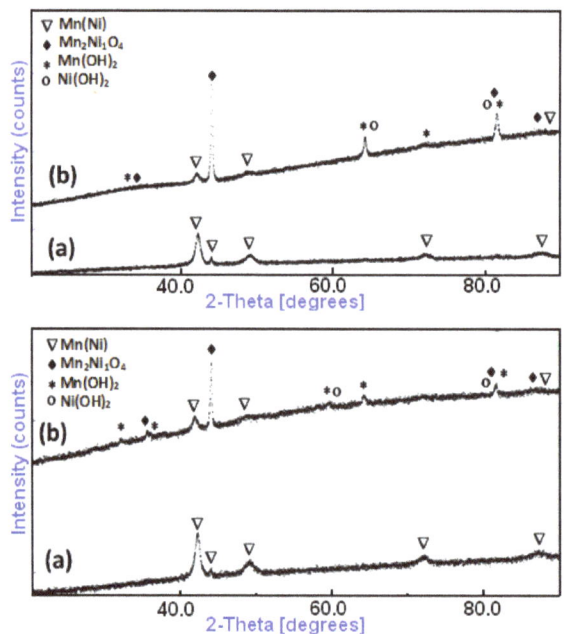

Figure 6. XRD diffraction patterns of MA Mn-30at%Ni (**a**) and Mn-20at%Ni (**b**) powders before and after degradation, respectively.

The FTIR spectrums of $Mn_{70}Ni_{30}$ and $Mn_{80}Ni_{20}$ powders obtained after the degradation reaction of RB5 compared to the initial RB5 powder are given in Figure 7. The FTIR spectrum of RB5 dye before degradation (Figure 7a) showed bands at 3411 cm^{-1}: O-H stretching; 2961 cm^{-1}: skeletal vibration of the benzene ring; 2354 cm^{-1}: aldehyde C-H vibration; 1745 cm^{-1}: C = C stretching vibration; 1635 cm^{-1}: azo bond (–N = N–); 1492 cm^{-1}: C = C aromatic skeletal vibrations; 1186 cm^{-1}: C–OH stretching vibration; 1045 cm^{-1}: C–OH stretching vibration; 1025 cm^{-1}: benzene mode coupling with stretching vibration of –SO_3; 805 cm^{-1}: –CH_3 skeletal vibration; 616 cm^{-1}: Sulfonic group. All bands are identified on the basis of pieces of information from previous works that studied the degradation of RB5 molecule [60–63] given in Table 2. The availability of active functional groups and the surface characteristics of the biosorbents after interaction with the dye are changes that are evident in the FTIR spectra of both powders after degradation. Some classic RB5 bonds degrade during the degradation, while new bonds arise. Bonds of the naphthalene ring's C = C aromatic skeletal vibration at 1400–1600 cm^{-1} and the azo bond's –N = N– peak at 1635 cm^{-1} are both clearly diminished. All FTIR spectra still contain the 2961 cm^{-1} peak, which is a small peak attributed to the skeletal vibration of the benzene ring. After degradation, a few additional peaks at 1750, 1520, and 1480 cm^{-1} that represent the stretching vibrations of the C = C, N–H, and C–N bonds, respectively, appear (Figure 7a,b). This proves that the existence of amines is promoted by the reductive dissociation of the –N = N bond [29,35,64]. Since amines are recognized to be water pollutants, using Mn-Ni powders in industrial processes must be carried out in combination with an additional amine adsorption procedure. In their recent work, Ben Mbarek and al. [65] were able to effectively remove unwanted intermediate chemicals from reduction processes, mainly aromatic amines. A variety of adsorbents were used, including wood, graphene oxide, activated carbon, and fine particles. These new results indicated that graphene oxide and activated carbon are the best secondary product adsorbents. Indeed, the regeneration of metal powders, synthesized by high-energy mechanosynthesis, and wastewater from discoloration, can serve as a reference for environmental applications in the purification of industrial effluents. On the other hand, the surface of the metals is covered by remnants of adhered dye and by the layer of oxides and/or hydroxides preventing the reaction between the active metal element of the alloy and the H+ of the aqueous solution. The hydrogen produced would be responsible for the attack of the azo group and its reduction in accordance with the mechanism proposed in the work. However, the reduction of the azo group could occur through the direct reaction of the attached dye and Mn, which is the active reducing agent. The oxide layers of both Mn and Ni are more soluble in acidic conditions. A system for removing these oxides would be washing with an acid solution and subsequent removal by rinsing with water. Then, the dry product is rubbed and the action is repeated. Once finished, the powders are dried at room temperature, or ventilation is used to remove the water from the washing. Therefore, maximum regeneration and use of metal powder particles may be implemented to reduce the costs associated with their use, or as possible solutions to minimize material modifications.

The effectiveness of materials used in the removal of organic pollutants depends on the chemical composition in addition to other factors. The first influences both the adsorption and reduction mechanism as well as the kinetics of the reaction. It is difficult to compare bimetallic reducing agents given that having the same composition but a different morphology or structure can produce different efficiencies. Another factor is the pH; the pH directly influences the redox process and the formation of precipitates. Table 3 gives an approximate comparison of the efficiency of the present Mn-Ni powders used in the removal of RB5 organic pollutants to some nanoreductives applied and reported in the previous literature.

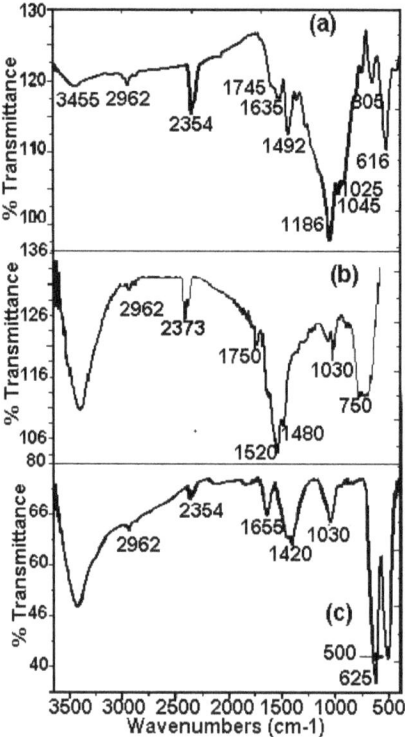

Figure 7. FTIR spectrum of (**a**) RB5 powder before decomposition, and (**b**,**c**) powders Mn-30at%Ni and Mn-20at%Ni, respectively, after decomposition.

Table 2. FTIR bonds identification for RB5 [45–48].

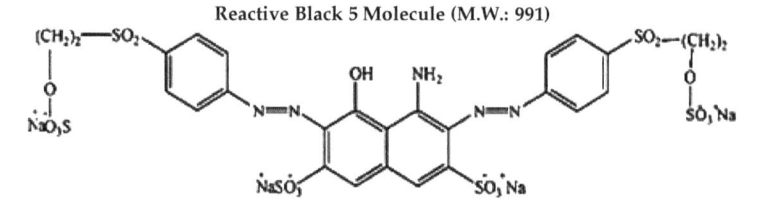

Peak/Band	Identification
3411 cm^{-1}	O–H stretching vibration
2936 cm^{-1}	Benzene ring skeletal vibration
2354 cm^{-1}	C–H aldehyde vibration
1745 cm^{-1}	C = C stretching vibration
1635 cm^{-1}	azo bond (–N = N–)
1528 cm^{-1}	N–H stretching vibration
1492 cm^{-1}	C = C aromatic skeletal vibrations
1260 cm^{-1}	C–N stretching vibration
1186 cm^{-1}	C–OH stretching vibration
1045 cm^{-1}	C–OH stretching vibration
1028 cm^{-1}	Benzene mode coupling with stretching vibration of –SO$_3$
804 cm^{-1}	–CH$_3$ skeletal vibration
616 cm^{-1}	Sulfonic group

Table 3. List of some nanoreductives used in the removal of RB5 organic pollutants compared with the Mn–Ni powders used in the present work.

Alloy Name	Route of Synthesis	RB5 Concentration	Alloy Dose	Time	Removal	Ref.
Zero-valent iron powder	Commercial	100 mg/L	0.5g/L	120 min	100%	[56]
Mn–Al	Ball-milled	40 mg/L	2.5 g/L	20 min	100%	[35]
Ca–Al	Ball-milled	40 mg/L	1g/L	1 min	100%	[36]
FeSiB	Ball-milled	40mg/L	2.5 g/L	3 min	100%	[65]
MnAlFe	Ball-milled	40mg/L	2.5 g/L	5 min	100%	[29]
MnAlCo	Ball-milled	40mg/L	2.5 g/L	3 min	100%	[29]
Mn-30%Ni	Ball-milled	40mg/L	2.5 g/L	17 h	100%	[Present work]
Mn-20%Ni	Ball-milled	40mg/L	2.5 g/L	15 h	100%	[Present work]

4. Conclusions

In the current work, melt-spinning followed by high-energy ball-milling was successfully used to synthesize Mn-30at%Ni and Mn-20at%Ni powders. Through the use of several analysis techniques, including XRD, SEM-EDX, UV-visible, and FTIR, the efficiency and kinetics of the degrading response of these two alloys in the decolorization reaction of an aqueous solution of Reactive Black 5 were studied. Both particles displayed exceptional performance as decolorizing powders of Reactive Black 5 azo dye aqueous solution, despite the differences in chemical composition and crystalline size. The powdered Mn-20at%Ni particles exhibit good degrading efficiency and demonstrate a faster rate of reaction than the Mn-30at%Ni alloy. The valence electron arrangement explains this remarkable efficiency. When reducing Ni content, it promotes a higher number of reacting sites in the d-band. Later, this enhances electron transport between Mn and H^+, hastening the decolorization of RB5 solution containing powdered Mn-20at%Ni particles. Additionally, Mn-20at%Ni has the highest specific surface area (0.45 $m^2\ g^{-1}$), which helps to account for the disparity in their decolorization effectiveness. The FTIR spectrum investigation shows an oxidative dissociation of the $-N = N-$ bond, which promotes the presence of amines. The usage of Mn-Ni particle powders in industrial operations must be supplemented with an additional amine adsorption procedure, because these amines are also water pollutants. Chemical processes of decolorization in acidic and basic media are effectively proposed on the basis of all experimental findings linking the degradation process to redox reactions in acidic and alkaline solutions.

Author Contributions: Conceptualization, W.B.M. and M.A.H.; formal analysis, L.E. and B.H.; data curation, W.B.M., M.A.H. and M.K.; writing—original draft preparation, M.K., L.E. and J.-J.S.; writing—review and editing, M.K., B.H. and J.-J.S.; supervision. All authors have read and agreed to the published version of the manuscript.

Funding: This research received no external funding.

Institutional Review Board Statement: It is not necessary. This research does not involve humans or animals.

Informed Consent Statement: It is not mandatory/necessary in this research.

Data Availability Statement: Data may be requested from the authors.

Conflicts of Interest: The authors declare no conflict of interest.

References

1. Amin, N.K. Removal of direct blue-106 dye from aqueous solution using new activated carbons developed from pomegranate peel: Adsorption equilibrium and kinetics. *J. Hazard. Mater.* **2009**, *165*, 52–62. [CrossRef]
2. Yahagi, T.; Degawa, M.; Seino, Y.; Matsushima, T.; Nagao, M.; Sugimura, T.; Hashimoto, Y. Mutagenicity of carcinogenic azo dyes and their derivatives. *Cancer Lett.* **1975**, *1*, 91–96. [CrossRef] [PubMed]
3. Saratale, R.G.; Saratale, G.D.; Chang, J.S.; Govindwar, S.P. Bacterial decolorization and degradation of azo dyes: A review. *J. Taiwan Inst. Chem. Eng.* **2011**, *42*, 138–157. [CrossRef]

4. Çatalkaya, E.Ç.; Bali, U.; Şengül, F. Photochemical degradation and mineralization of 4-chlorophenol. *Environ. Sci. Pollut. Res.* **2003**, *10*, 113–120. [CrossRef] [PubMed]
5. Fu, W.; Yang, H.; Chang, L.; Hari-Bala; Li, M.; Zou, G. Anatase TiO_2 nanolayer coating on strontium ferrite nanoparticles for magnetic photocatalyst. *Colloids Surf. A Physico-Chem. Eng. Asp.* **2006**, *289*, 47–52. [CrossRef]
6. Asghar, A.; Raman, A.A.A.; Daud, W.M.A.W. Advanced oxidation processes for in-situ production of hydrogen peroxide/hydroxyl radical for textile wastewater treatment: A review. *J. Clean. Prod.* **2015**, *87*, 826–838. [CrossRef]
7. Khan, T.A.; Dahiya, S.; Ali, I. Use of kaolinite as adsorbent: Equilibrium, dynamics and thermodynamic studies on the adsorption of Rhodamine B from aqueous solution. *Appl. Clay Sci.* **2012**, *69*, 58–66. [CrossRef]
8. Tseng, W.J.; Lin, R.D. $BiFeO_3/\alpha$-Fe_2O_3 core/shell composite particles for fast and selective removal of methyl orange dye in water. *J. Colloid Interface Sci.* **2014**, *428*, 95–100. [CrossRef]
9. Qin, X.D.; Zhu, Z.W.; Liu, G.; Fu, H.M.; Zhang, H.W.; Wang, A.M.; Li, H.; Zhang, H.F. Ultrafast degradation of azo dyes catalyzed by cobalt-based metallic glass. *Sci. Rep.* **2016**, *5*, 18226. [CrossRef]
10. Chen, S.S.; Hsu, H.D.; Li, C.W. A new method to produce nanoscale iron for nitrate removal. *J. Nanopart. Res.* **2004**, *6*, 639–647. [CrossRef]
11. Fan, J.; Guo, Y.H.; Wang, J.J.; Fan, M.H. Rapid Decolorization of azo Dye Methyl Orange in Aqueous Solution by Nanoscale zerovalent Iron Particles. *J. Hazard. Mater.* **2009**, *166*, 904–910. [CrossRef]
12. Kumar, M.; Chakraborty, S. Chemical denitrification of water by zero-valent magnesium powder. *J. Hazard. Mater.* **2006**, *135*, 112–121. [CrossRef]
13. Noubactep, C. Elemental metals for environmental remediation: Learning from cementation process. *J. Hazard. Mater.* **2010**, *181*, 1170–1174. [CrossRef]
14. Kanel, S.R.; Manning, B.; Charlet, L.; Choi, H. Removal of arsenic(III) from groundwater by nanoscale zero-valent iron. *Environ. Sci. Technol.* **2005**, *39*, 1291–1298. [CrossRef]
15. Schrick, B.; Blough, J.L.; Jones, A.D.; Mallouk, T.E. Hydrodechlorination of trichloroethylene to hydrocarbons using bimetallic nickel-iron nanoparticles. *Chem. Mater.* **2002**, *14*, 5140–5147. [CrossRef]
16. Chang, J.H.; Cheng, S.F. The remediation performance of a specific electrokinetics integrated with zero-valent metals for perchloroethylene contaminated soils. *J. Hazard. Mater.* **2006**, *131*, 153–162. [CrossRef] [PubMed]
17. Xiong, Z.; Zhao, D.; Pan, G. Rapid and complete destruction of perchlorate in water and ion-exchange brine using stabilized zero-valent iron nanoparticles. *Water Res.* **2007**, *41*, 3497–3505. [CrossRef] [PubMed]
18. Hu, J.; Lo, I.M.C.; Chen, G. Fast removal and recovery of Cr(VI) using surface-modified jacobsite ($MnFe_2O_4$) nanoparticles. *Langmuir* **2005**, *21*, 11173–11179. [CrossRef]
19. Dutta, A.K.; Maji, S.K.; Adhikary, B. γ-Fe_2O_3 Nanoparticles: An Easily Recoverable Effective Photo-Catalyst for the Degradation of Rose Bengal and Methylene Blue Dyes in the Waste-Water Treatment Plant. *Mater. Res. Bull.* **2014**, *49*, 28–34. [CrossRef]
20. Huang, Y.; Fan, W.; Long, B.; Li, H.; Zhao, F.; Liu, Z.; Tong, Y.; Ji, H. Visible Light $Bi_2S_3/Bi_2O_3/Bi_2O_2CO_3$ Photocatalyst for Effective Degradation of Organic Pollutions. *Appl. Catal. B Environ.* **2016**, *185*, 68–76. [CrossRef]
21. Low, J.; Yu, J.; Jaroniec, M.; Wageh, S.; Al-Ghamdi, A.A. Heterojunction Photocatalysts. *Adv. Mater.* **2017**, *29*, 1601694. [CrossRef] [PubMed]
22. Kurian, M.; Nair, D.S. Heterogeneous Fenton Behavior of Nano Nickel Zinc Ferrite Catalysts in the Degradation of 4-Chlorophenol from Water under Neutral Conditions. *J. Water Process Eng.* **2015**, *8*, e37–e49. [CrossRef]
23. Ma, H.; Wang, H.; Na, C. Microwave-Assisted Optimization of Platinum-Nickel Nanoalloys for Catalytic Water Treatment. *Appl. Catal. B Environ.* **2015**, *163*, 198–204. [CrossRef]
24. Chaturvedi, S.; Dave, P.N.; Shah, N.K. Applications of Nano-Catalyst in New Era. *J. Saudi Chem. Soc.* **2012**, *16*, 307–325. [CrossRef]
25. Lin, S.T.; Thirumavalavan, M.; Jiang, T.Y.; Lee, J.F. Synthesis of ZnO/Zn Nano Photocatalyst Using Modified Polysaccharides for Photodegradation of Dyes. *Carbohydr. Polym.* **2014**, *105*, 1–9. [CrossRef]
26. Adeleye, A.S.; Conway, J.R.; Garner, K.; Huang, Y.; Su, Y.; Keller, A.A. Engineered Nanomaterials for Water Treatment and Remediation: Costs, Benefits, and Applicability. *Chem. Eng. J.* **2016**, *286*, 640–662. [CrossRef]
27. Anjum, M.; Miandad, R.; Waqas, M.; Gehany, F.; Barakat, M.A. Remediation of Wastewater Using Various Nano-Materials. *Arab. J. Chem.* **2019**, *12*, 4897–4919. [CrossRef]
28. Akhavan, O. Lasting Antibacterial Activities of Ag-TiO_2/Ag/a-TiO_2 Nanocomposite Thin Film Photocatalysts under Solar Light Irradiation. *J. Colloid Interface Sci.* **2009**, *336*, 117–124. [CrossRef]
29. Mbarek, W.; Saurina, J.; Escoda, L.; Pineda, E.; Khitouni, M.; Suñol, J.J. Effects of the Addition of Fe, Co on the Azo Dye Degradation Ability of Mn-Al Mechanically Alloyed Powders. *Metals* **2020**, *10*, 1578. [CrossRef]
30. AboliGhasemabadi, M.; Mbarek, W.B.; Cerrillo-Gil, A.; Roca-Bisbe, H.; Casabella, O.; Blanquez, P.; Pineda, E.; Escoda, L.; Sunol, J.J. Azo-dye degradation by Mn–Al powders. *J. Environ. Manag.* **2020**, *258*, 110012. [CrossRef]
31. Wang, J.Q.; Liu, Y.H.; Chen, M.W.; Xie, G.Q.; Louzguine-Luzgin, D.V.; Inoue, A.; Perepezko, J.H. Rapid Degradation of Azo Dye by Fe-Based Metallic Glass Powder. *Adv. Funct. Mater.* **2012**, *22*, 2567–2570. [CrossRef]
32. Wang, J.Q.; Liu, Y.H.; Chen, M.W.; Louzguine-Luzgin, D.V.; Inoue, A.; Perepezko, J.H. Excellent capability in degrading azo dyes by MgZn-based metallic glass powders. *Sci. Rep.* **2012**, *2*, 418. [CrossRef] [PubMed]
33. Sapkota, B.B.; Mishra, S.R. A Simple Ball Milling Method for the Preparation of p-CuO/n-ZnO Nanocomposite Photocatalysts with High Photocatalytic Activity. *J. Nanosci. Nanotechnol.* **2013**, *13*, 6588–6596. [CrossRef]

34. Suryanarayana, C. Mechanical alloying and milling. *Prog. Mater. Sci.* **2001**, *46*, 1–184. [CrossRef]
35. Ben Mbarek, W.; Azabou, M.; Pineda, E.; Fiol, N.; Escoda, L.; Sunol, J.J.; Khitouni, M. Rapid degradation of azo-dye using Mn–Al powders produced by ball-milling. *RSC Adv.* **2017**, *7*, 12620–12628. [CrossRef]
36. Ben Mbarek, W.; Pineda, E.; Fiol, N.; Escoda, L.; Sunol, J.J.; Khitouni, M. High efficiency decolorization of azo dye Reactive Black 5 by Ca-Al particles. *J. Environ. Chem. Eng.* **2017**, *5*, 6107–6113. [CrossRef]
37. Ghasemabadi, M.A.; Mbarek WBen Casabella, O.; Roca-Bisbe, H.; Pineda, E.; Escoda, L.; Sunol, J.J. Application of mechanically alloyed MnAl particles to de-colorization of azo dyes. *J. Alloys Comp.* **2018**, *741*, 240–245. [CrossRef]
38. Kuyama, J.; Inui, H.; Imaoka, S.; Ishihara, K.N.; Shinhu, P. Nanometer-sized crystals formed by the mechanical alloying in the Ag-Fe system. *Jpn. J. Appl. Phys.* **1991**, *30*, L854. [CrossRef]
39. Kuschke, W.M.; Keller, R.M.; Grahle, P.; Mason, R.; Arzt, E. Mechanisms of powder milling investigated by X-ray diffraction and quantitative metallography. *Int. J. Mater. Res.* **1995**, *86*, 804–813. [CrossRef]
40. Mhadhbi, M.; Khitouni, M.; Azabou, M.; Kolsi, A. Characterization of Al and Fe nanosized powders synthesized by high energy mechanical milling. *J. Mater. Charact.* **2008**, *59*, 944–950. [CrossRef]
41. El-Eskandarany, M.S. *Mechanical Alloying for Fabrication of Advanced Engineering Materials*; Noyes Publications/William Andrew Publishing: Norwich, NY, USA, 2001.
42. Suryanarayana, C.; Koch, C.C. *Non-Equilibrium Processing of Materials*; Suryanarayana, C., Ed.; Pergamon: New York, NY, USA, 1999; pp. 313–344.
43. Cao, J.; Wei, L.; Huang, Q.; Wang, L.; Han, S. Reducing degradation of azo dye by zero-valent iron in aqueous solution. *Chemosphere* **1999**, *38*, 565–571. [CrossRef]
44. Nam, S.; Tratnyek, P.G. Reduction of azo dyes with zero-valent iron. *Water Res.* **2000**, *34*, 1837–1845. [CrossRef]
45. Wu, F.; Deng, N.; Hua, H. Degradation mechanism of azo dye C. I. reactive red 2 by iron powder reduction and photooxidation in aqueous solutions. *Chemosphere* **2000**, *4*, 1233–1238. [CrossRef]
46. Stylidi, M.; Kondarides, D.I.; Verykios, X.E. Pathways of solar light-induced photocatalytic degradation of azo dyes in aqueous TiO_2 suspensions. *Appl. Catal. B* **2003**, *40*, 271–286. [CrossRef]
47. Pauling, L. The nature of the interatomic forces in metals. *Phys. Rev.* **1938**, *54*, 899–904. [CrossRef]
48. Pauling, L. A resonating-valence-bond theory of metals and intermetallic compounds. *Proc. R. Soc. London. Ser. A Math. Phys. Sci.* **1949**, *196*, 343–362. [CrossRef]
49. Dowden, D.A. Heterogeneous catalysis. Part I. Theoretical basis. *J. Chem. Soc.* **1950**, *56*, 242–265. [CrossRef]
50. Zhang CZhu ZZhang, H.; Sun, Q.; Liu, K. Effects of cobalt content on the decolorization properties of Fe-Si-B amorphous alloys. *Results Phys.* **2018**, *10*, 1–4. [CrossRef]
51. Ponder, S.M.; Darab, J.G.; Mallouk, T.E. Remediation of Cr(VI) and Pb(II) aqueous solutions using supported, nanoscale zero-valent iron. *Environ. Sci. Technol.* **2000**, *34*, 2564–2569. [CrossRef]
52. Deng, J.; Feng, S.; Ma, X.; Tan, C.; Wang, H.; Zhou, S.; Zhang, T.; Li, J. Heterogeneous degradation of Orange II with peroxymonosulfate activated by ordered mesoporous $MnFe_2O_4$. *Separ. Purif. Technol.* **2016**, *167*, 181–189. [CrossRef]
53. Chen, J.X.; Zhu, L.Z. Heterogeneous UV-Fenton catalytic degradation of dyestuff in water with hydroxyl-Fe pillared bentonite. *Catal. Today* **2007**, *126*, 463–470. [CrossRef]
54. Luo, X.K.; Li, R.; Huang, L.; Zhang, T. Nucleation and growth of nanoporous copper ligaments during electrochemical dealloying of Mg-based metallic glasses. *Corros. Sci.* **2013**, *67*, 100–108. [CrossRef]
55. Yang, Y.Y.; Li, Z.L.; Wang, G.; Zhao, X.P.; Crowley, D.E.; Zhao, Y.H. Computational identification and analysis of the key biosorbent characteristics for the biosorption process of Reactive Black 5 onto fungal biomass. *PLoS ONE* **2012**, *7*, e33551. [CrossRef] [PubMed]
56. Satapanajaru, T.; Chompuchan, C.; Suntornchot, P.; Pengthamkeerati, P. Enhancing Decolorization of Reactive Black 5 and Reactive Red 198 during Nano Zerovalent Iron Treatment. *Desalination* **2011**, *266*, 218–230. [CrossRef]
57. Zhang, C.Q.; Zhu, Z.W.; Zhang, H.F.; Hu, Z.Q. Rapid decolorization of Acid Orange II aqueous solution by amorphous zero-valent iron. *J. Environ. Sci.* **2012**, *24*, 1021–1026. [CrossRef] [PubMed]
58. Zhang, L.; Gao, X.; Zhang, Z.; Zhang, M.; Cheng, Y.; Su, J. A doping lattice of aluminum and copper with accelerated electron transfer process and enhanced reductive degradation performance. *Sci. Rep.* **2016**, *6*, 31797. [CrossRef]
59. Mbarek, W.B.; Escoda, L.; Saurina, J.; Pineda, E.; Alminderej, F.M.; Khitouni, M.; Suñol, J.J. Nanomaterials as a Sustainable Choice for Treating Wastewater: A Review. *Materials* **2022**, *15*, 8576. [CrossRef]
60. Méndez-Martínez, A.J.; Dávila-Jiménez, M.M.; Ornelas-Dávila, O.; Elizalde-González, M.P.; Arroyo-Abad, U.; Sirés, I.; Brillas, E. Electrochemical reduction and oxidation pathways for Reactive Black 5 dye using nickel electrodes in divided and undivided cells. *Electrochim. Acta* **2012**, *59*, 140–149. [CrossRef]
61. Almeida, E.J.R.; Corso, C.R. Comparative study of toxicity of azo dye Procion Red MX-5B following biosorption and biodegradation treatments with the fungi *Aspergillus niger* and *Aspergillus terreus*. *Chemosphere* **2014**, *112*, 317–322. [CrossRef]
62. Shilpa, S.; Shikha, R. Biodegradation of Dye Reactive Black-5 by a Novel Bacterial Endophyte. *Int. Res. J. Environ. Sci.* **2015**, *4*, 44–53.
63. El Bouraie, M.; SalahEdin, W. Biodegradation of Reactive Black 5 by *Aeromonas hydrophila* strain isolated from dye contaminated textile wastewater. *Sustain. Environ. Res.* **2016**, *26*, 209–216. [CrossRef]

64. Elías, V.R.; Sabre, E.V.; Winkler, E.L.; Satuf, M.L.; Rodriguez-Castellón, E.; Casuscelli, S.G.; Eimer, G.A. Chromium and titanium/chromium-containing MCM-41 mesoporous silicates as promising catalysts for the photobleaching of azo dyes in aqueous suspensions. A multitechnique investigation. *Microporous Mesoporous Mater.* **2012**, *163*, 85–95. [CrossRef]
65. Ben Mbarek, W.; Daza, J.; Escoda, L.; Fiol, N.; Pineda, P.; Khitouni, M.; Suñol, J.J. Removal of Reactive Black 5 Azo Dye from Aqueous Solutions by a Combination of Reduction and Natural Adsorbents Processes. *Metals* **2023**, *13*, 474. [CrossRef]

Disclaimer/Publisher's Note: The statements, opinions and data contained in all publications are solely those of the individual author(s) and contributor(s) and not of MDPI and/or the editor(s). MDPI and/or the editor(s) disclaim responsibility for any injury to people or property resulting from any ideas, methods, instructions or products referred to in the content.

Article

Synthesis, Structural and Magnetic Characterization of Superparamagnetic $Ni_{0.3}Zn_{0.7}Cr_{2-x}Fe_xO_4$ Oxides Obtained by Sol-Gel Method

Abdulrahman Mallah [1,*], Fatimah Al-Thuwayb [1], Mohamed Khitouni [1], Abdulrahman Alsawi [2], Joan-Josep Suñol [3], Jean-Marc Greneche [4] and Maha M. Almoneef [5]

[1] Department of Chemistry, College of Science, Qassim University, Buraydah 51452, Saudi Arabia; 411207168@qu.edu.sa (F.A.-T.); kh.mohamed@qu.edu.sa (M.K.)
[2] Department of Physics, College of Science, Qassim University, Buraydah 51452, Saudi Arabia; ansaoy@qu.edu.sa
[3] Department of Physics, University of Girona, Campus Montilivi, 17071 Girona, Spain; joanjosep.sunyol@udg.edu
[4] Institut des Molécules et Matériaux du Mans (IMMM), UMR CNRS 6283, Université du Maine, Avenue Olivier Messiaen, CEDEX 9, 72085 Le Mans, France; jean-marc.greneche@univ-lemans.fr
[5] Physics Department, Faculty of Science, Princess Nourah bint Abdulrahman University, Riyadh 11564, Saudi Arabia
* Correspondence: a.mallah@qu.edu.sa; Tel.: +966-163-012-507

Abstract: The sol-gel process was used to produce ferrite $Ni_{0.3}Zn_{0.7}Cr_{2-x}Fe_xO_4$ compounds with x = 0, 0.4, and 1.6, which were then subsequently calcined at several temperatures up to 1448 K for 48 h in an air atmosphere. X-ray diffraction (XRD), scanning electron microscopy (SEM), vibrating sample magnetometer (VSM), and ^{57}Fe Mössbauer spectrometry were used to examine the structure and magnetic characteristics of the produced nanoparticles. A single-phase pure $Ni_{0.3}Zn_{0.7}Cr_{2-x}Fe_xO_4$ nanoparticle had formed. The cubic $Fd3m$ spinel structure contained indexes for all diffraction peaks. The crystallite size is a perfect fit for a value of 165 ± 8 nm. Based on the Rietveld analysis and the VSM measurements, the low magnetization Ms of $Ni_{0.3}Zn_{0.7}Cr_{2-x}Fe_xO_4$ samples was explained by the absence of ferromagnetic Ni^{2+} ions and the occupancy of Zn^{2+} ions with no magnetic moments in all tetrahedral locations. Moreover, because of the weak interactions between Fe^{3+} ions in the octahedral locations, the magnetization of the current nanocrystals is low or nonexistent. According to Mössbauer analyses, the complicated hyperfine structures are consistent with a number of different chemical atomic neighbors, such as Ni^{2+}, Zn^{2+}, Cr^{3+}, and Fe^{3+} species that have various magnetic moments. A Fe-rich neighbor is known to have the highest values of the hyperfine field at Fe sites, while Ni- and Cr-rich neighbors are responsible for the intermediate values and Zn-rich neighbors are responsible for the quadrupolar component.

Keywords: sol-gel method; ferrite oxides; superparamagnetic

Citation: Mallah, A.; Al-Thuwayb, F.; Khitouni, M.; Alsawi, A.; Suñol, J.-J.; Greneche, J.-M.; Almoneef, M.M. Synthesis, Structural and Magnetic Characterization of Superparamagnetic $Ni_{0.3}Zn_{0.7}Cr_{2-x}Fe_xO_4$ Oxides Obtained by Sol-Gel Method. Crystals 2023, 13, 894. https://doi.org/10.3390/cryst13060894

Academic Editor: Andrey Prokofiev

Received: 12 April 2023
Revised: 7 May 2023
Accepted: 25 May 2023
Published: 30 May 2023

Copyright: © 2023 by the authors. Licensee MDPI, Basel, Switzerland. This article is an open access article distributed under the terms and conditions of the Creative Commons Attribution (CC BY) license (https:// creativecommons.org/licenses/by/ 4.0/).

1. Introduction

Due to a remarkable combination of their outstanding physical, chemical, and structural features, as well as a variety of promising applications, mixed ferrite systems have attracted increasing attention in recent years. Due to their significant magnetic properties, particularly in the radio frequency region, physical flexibility, high electrical resistivity, mechanical hardness, and chemical stability, spinel ferrites are among an important class of magnetic materials [1–8]. The majority of mixed oxides, except the lithium-nickel oxide and the lithium-cobalt oxide systems, possess a spinel structure of the general formula $M_xM'_{3-x}O_4$ (M, M' = Ni, Co, Fe, Mn, Cr...). Thirty-two oxygen atoms are arranged between tetrahedral (A) and octahedral (B) sites in a cubic closed-packed arrangement that makes up the unit cell of spinel ferrites.

The compositions and synthesis methods of spinel ferrite nanostructures have an impact on their chemical and structural characteristics, and the cation substitutions determine the corresponding electric and magnetic properties. In ferrites, whose crystal chemical formula is $(M^{2+}_{1-x}Fe^{3+}_{x})_{tetra} (M^{2+}_{x} Fe^{3+}_{2-x})_{octa} O_4$ [9], a specific cationic distribution is allowed. The "degree of inversion", denoted by the letter "x", corresponds to the fraction of sites (A) that are occupied by Fe^{3+} cations. Two extreme examples associated with the change of the cation configuration can be distinguished. One is the standard spinel (x = 0), which has all Fe^{3+} cations in the (B) locations and all M^{2+} cations in the (A) locations. The other is an inverse spinel (x = 1), where the M^{3+} cations are distributed equally in the A and B sites and all M^{2+} ions occupy the (B) regions. It is also quite common to find spinels that have a somewhat inverse cation distribution, which falls in between normal and inverse spinels (0 < x < 1).

The features of spinel ferrites are greatly influenced by the cation distribution, which is connected to the process of elaboration [10]. Furthermore, when the particle size is in the nanometer range, it has a noticeable impact on the physical and chemical characteristics of ferrite and other magnetic materials. High surface-to-volume ratio nanoparticles show noticeable magnetic properties in comparison to their massive counterparts [11–13]. With the reduction in grain size, ferrite nanoparticles exhibit characteristics such as high field irreversibility [14], change in Neel temperature [15], higher coercivity values [16], lower saturation magnetization values, modified lowered, or increased magnetic moments [17–19], etc. Among the ferrites that are an important component of magnetic ceramic materials, nanosized nickel ferrite, $NiFe_2O_4$, has desirable characteristics for use as soft magnets due to its high electrical resistivity, low coercivity, and low saturation magnetization values. These qualities make it a suitable material for magnetic and magneto-optical applications. Nanostructured $NiFe_2O_4$ is a great core material for power transformers in electronics and communication systems, since it has nearly no hysteresis losses [20,21]. In addition, $NiFe_2O_4$ nanoparticles are well known for their use in electrical, electronic, and catalytic applications, as well as for their ability to sense gases and humidity [22]. It has been noted that the magnetic properties of $NiFe_2O_4$ are significantly influenced by the size of the crystallite. For bulk systems, the impact of the structure on the magnetic characteristics of $NiFe_2O_4$ has been demonstrated [23,24].

According to Brook and Kingery [25], depending on their structure, $NiFe_2O_4$ samples display ferrimagnetism, superparamagnetism, or paramagnetism. In actuality, superparamagnetism is associated with materials with grains smaller than 10 nm, and ferrimagnetism is associated with samples with grains larger than 15 nm, while paramagnetism is related to the non-crystalline (or amorphous) materials that might resemble a totally disordered state. In contrast to their bulk counterpart, $NiFe_2O_4$ nanoparticles exhibit higher coercivity and lower saturation magnetization values [26–29]. The magnetic properties of $NiFe_2O_4$ result in the super-exchange interactions between the tetrahedral and octahedral sublattices as the metallic ions are surrounded by oxygen atoms. In addition, the magnetic properties are also influenced by the magnetocrystalline anisotropy, the canting effect resulting from the presence of triangular cationic platelets and antiferromagnetic interactions, and the dipolar interactions between projected moments on the nanoparticle surface. To improve some of their electric or magnetic properties, ferrite nanocrystals can be doped with several metallic species, including chromium, copper, manganese, and zinc [30–32]. For instance, they noted that the Cu replacement in NiZn-ferrite increases permeability, saturation magnetization, and magnetic losses; tan δ (dielectric losses) at 1.3 GHz is <0.01 in NiCuZn, but it is substantially lower (6×10^{-3}) in NiZn [33,34]. On the other hand, cations such as Cr^{3+}, in particular, show a significant predilection for substituting into the B-site and a tendency for anti-ferromagnetic coupling with Fe. This causes magnetic fluctuations in the system, resulting in interesting magnetic characteristics when one of the cations is partially or completely substituted with Cr^{3+}. As a result, it is interesting to investigate its magnetic properties with Cr in place of Fe, which is expected to change the frustration in the system [35–37]. Sijo et al. [38] investigated the impact of the Cr substitution on the structural, magnetic, and dielectric properties in NiCrFeO$_4$ nanoparticles prepared by the solution combustion method. Because of the distribution of cations between the tetrahedral A and octahedral B-sites, NiCrFeO$_4$ exhibited significant changes in its atomic arrangement depending critically on preparation temperature. B-sites are where chromium ions chose to concentrate. Ni^{2+} and Fe^{3+} were distributed differently between octahedral and tetrahedral sites as a result of temperature, which had an impact on their characteristics and material properties. They noticed that substituting one Cr cation for one Fe cation in NiFe$_2$O$_4$ decreased the transition temperature. Furthermore, their Mössbauer investigations at room temperature showed the presence of small superparamagnetic particles and larger-sized ferrimagnetic particles. On the other hand, while spinel ferrites can be made using a variety of processes, the sol-gel method produces high-purity, nanocrystalline, and homogeneous ferrites, which have an important influence on the structural and magnetic properties.

In this work, we have studied the impact of Cr substitutions on nickel, zinc, and ferrite nanocrystalline structures produced by the sol-gel method. At ambient temperature, we specifically describe the structural and magnetic characteristics of these doped materials. In addition, we have examined in detail the Mössbauer spectra of the doped samples.

2. Materials and Methods

2.1. Preparation Method

Due to its rapid reaction rate, low preparation temperature, and generation of tiny particles, the sol-gel method is frequently utilized in the synthesis of ferrite nanocrystals. Mixed oxides were prepared by this method based on the Pechini procedure utilizing stoichiometric proportions of the precursors' Ni(NO$_3$)$_2$6H$_2$O, Zn(NO$_3$)$_2$6H$_2$O, Fe(NO$_3$)$_3$9H$_2$O, and Cr(NO$_3$)$_3$9H$_2$O. Thus, in our experiment, nanocrystalline Ni$_{0.3}$Zn$_{0.7}$Cr$_{2-x}$Fe$_x$O$_4$ with $x = 0$, 0.4, and 1.6 were synthesized by this method. To create a mixed solution, stoichiometric proportions of metal nitrates were first fully dissolved in distilled water. Then, as a chelating agent, measured quantities of citric acid were added and dissolved while stirring. The molar ratio of nitrates to citric acid was set to 1:1, and a little amount of ammonia solution was added to the mixture to change the pH to 7. Ethylene glycol was then added as a polymerization agent after the solution had been heated to 373 K while being stirred. Until the gel was obtained, the heating was continued. The produced wet gel was dried to obtain a dry foam, which was then milled in a mortar; then each sample was heated in air atmosphere at 473 K for 24 h, 773 K for 24 h, 973 K for 24 h, 1173 K for 24 h and finally at 1448 K for 48 h to achieve the desired crystalline phase. All samples were milled after each heat process.

2.2. Measurements and Characterizations

Using a Philips® PW1800 X-ray diffractometer with CuKα radiation ($\lambda = 1.54056$ Å) operating at 40 kV and 30 mA, X-ray diffraction patterns (XRD) of the produced nanocrystals were obtained. The Rietveld refinement method was applied with Fullprof software [39]. A Hitachi® S4160 Field Emission Scanning Electron Microscope (FE-SEM) was used to examine the samples' morphology and average particle size. Meghnatis Daghigh Kavir Co.® Vibrating Sample Magnetometer (VSM) was used to assess the samples' magnetic characteristics while they were at room temperature. A maximum magnetic field of 9 kOe was applied during these tests, it should be noted.

Following the synthesis of these nanocrystals from the precursors at the atomic level, a ^{57}Co source diffused into a rhodium matrix was used for conventional ^{57}Fe Mössbauer transmission spectrometry at 300 and 77 K. In order to describe the broadened quadrupolar and/or magnetic components, the Mössbauer spectra were fitted using the Mosfit program [40], which involved quadrupolar components and magnetic sextets with Lorentzian lines or discontinuous distribution of quadrupolar and/or magnetic sextets. The proportions of Fe species are obtained from the relative absorption area of the relevant component, assuming the same value of recoilless f-factor, while the values of isomer shift refer to that of α-Fe at room temperature.

3. Results

3.1. Structural Analyses

Figure 1 depicts the XRD diffractograms for the produced $Ni_{0.3}Zn_{0.7}Cr_{2-x}Fe_xO_4$ (x = 0; 0.4; 1.6) ferrites. In accordance with well-crystallized samples, all the XRD patterns show well-defined and relatively thin Bragg peaks. XRD patterns of all the ferrite particles present the spinel structure and are well indexed to (111), (220), (311), (222), (400), (422), (511), (440), (620), (533), (622), and (642) crystal planes of cubic spinel phase. The cubic $Fd3m$ spinel structure had indexes for all diffraction peaks. Knowing the distribution of cations between the tetrahedral (A) and octahedral (B) sites is important to refine the structure of $Ni_{0.3}Zn_{0.7}Cr_{2-x}Fe_xO_4$ samples using the Rietveld method. Using X-ray powder diffraction data, the Rietveld structural refinement was done using the FULLPROF-suite software; this is a well-established technique for extracting structural details from powder diffraction data. Measurements using the (in-field) Mössbauer spectroscopy method have been used to determine the cations distributions for ferrites with the general formula AB_2O_4. In earlier research, Mössbauer characterization of some Ni-Zn ferrites [41] and Cr-substituted ferrites [42,43] revealed that Zn^{2+} ions prefer to occupy the tetrahedral (A) regions, while Ni^{2+} ions are distributed on the octahedral (B) regions, and Fe^{3+} and Cr^{3+} ions are distributed on both sites. These cations' distribution has been verified in various investigations [44–47]. As a result, the cation distribution for the $Ni_{0.3}Zn_{0.7}Cr_{2-x}Fe_xO_4$ samples can be written as $\left(Zn^{2+}_{0.7}Cr^{3+}_{0.3}\right)_A \left[Ni^{2+}_{0.3}Cr^{3+}_{1.7}\right]_B O^{2-}_4$ (for x = 0), $\left(Zn^{2+}_{0.7}Fe^{3+}_{0.3}\right)_A \left[Ni^{2+}_{0.3}Cr^{3+}_{1.6}Fe^{3+}_{0.1}\right]_B O^{2-}_4$ (for x = 0.4), and $\left(Zn^{2+}_{0.7}Fe^{3+}_{0.3}\right)_A \left[Ni^{2+}_{0.3}Cr^{3+}_{0.4}Fe^{3+}_{1.3}\right]_B O^{2-}_4$ (for x = 1.6). The Rietveld refinement was carried out using this created formula. The atomic locations for (A) cations, (B) cations, and O were taken at positions 8a (1/8, 1/8, 1/8), 16d (1/2, 1/2, 1/2), and 32e (x, y, z).

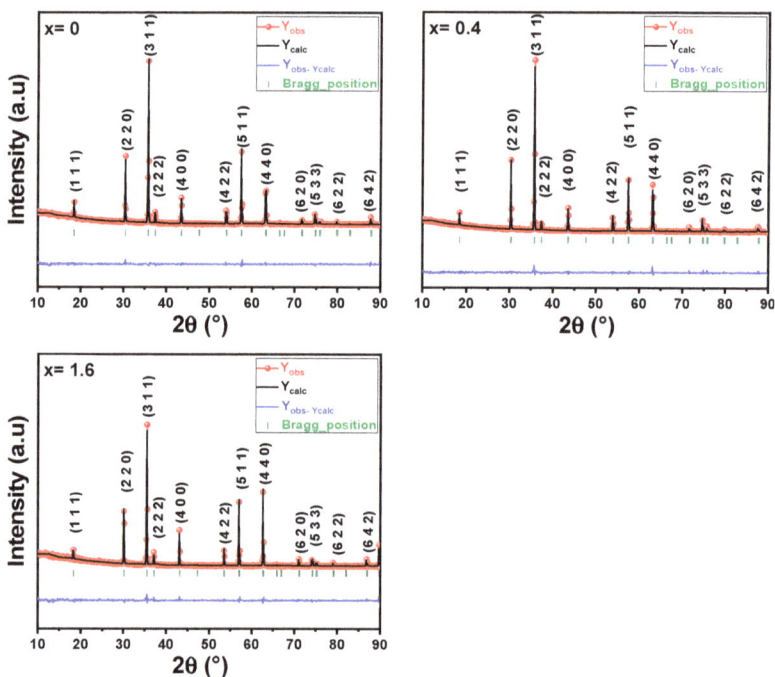

Figure 1. XRD patterns with Rietveld refinement for $Ni_{0.3}Zn_{0.7}Cr_{2-x}Fe_xO_4$ ferrites: the bottom line (green) represents the difference between the XRD data (red) and calculated fit (black), and the blue lines represent the Bragg positions.

In the structural refinement procedure of our samples, we have followed the standard steps of the Rietveld method, which consists of following the sequence: (i) refinement of overall Scale factor + background coefficients (all other parameters are kept fixed); (ii) the same + refinement of detector zero offset (or sample displacement in Bragg-Brentano geometry) + refinement of lattice parameters; (iii) the same + refinement of shape parameters + refinement of asymmetry parameters; (iv) the same + refinement of atomic positions + refinement of global Debye-Waller parameter or thermal agitation factors; and (v) the same + refinement of site occupancy rate. In our refinements, we have taken care to respect this sequence of steps to release the different parameters. This ensures a stability of the refinement with all the parameters released. Table 1 provides the refined values for the structural parameters. Small residual values in the goodness of fit (χ^2) confirm the good agreement between the computed and observed XRD data, as seen in Figure 1. The parameters presented in Table 1 with uncertainty in brackets (such as lattice constant, cell volume, isotropic thermal agitation parameter, bond length, and bond angle) are those which are refined. The parameters presented without uncertainty (such as atomic position and occupancy factor) are those fixed.

Table 1. Structural characteristics for $Ni_{0.3}Zn_{0.7}Cr_{2-x}Fe_xO_4$ ferrites obtained following the structural refinement by the Rietveld method. B_{iso}: isotropic thermal agitation parameter. Definitions of structural parameters are given in the text. R factors: R_P = profile factor, R_B = Bragg factor, and R_F = crystallographic factor). χ^2: The goodness of fit. The numbers in parentheses are estimated standard deviations to the last significant digit.

		Fe Content		x = 0	x = 0.4	x = 1.6
Space group					Fd – 3m	
Cell parameters		Lattice constant a (Å)		8.3242 (5)	8.3329 (5)	8.3906 (5)
		Cell volume V (Å^3)		476.80 (6)	578.62 (6)	590.71 (6)
Atoms	Zn/Cr/Fe	Wyckoff positions		4c	4c	4c
		Site symmetry		$-43\,m$	$-43\,m$	$-43\,m$
		Atomic positions	x = y = z	1/8	1/8	1/8
		Occupancy factors		$\left(Zn^{2+}_{0.7}Cr^{3+}_{0.3}\right)_A$	$\left(Zn^{2+}_{0.7}Fe^{3+}_{0.3}\right)_A$	$\left(Zn^{2+}_{0.7}Fe^{3+}_{0.3}\right)_A$
		Biso (Å^2)		2.2 (4)	1.9 (5)	1.98 (4)
	Ni/Fe/Cr	Wyckoff positions		16d	16d	16d
		Site symmetry		$-3\,m$	$-3\,m$	$-3\,m$
		Atomic positions	x = y = z	1/2	1/2	1/2
		Occupancy factors		$\left[Ni^{2+}_{0.3}Cr^{3+}_{1.7}\right]_B$	$\left[Ni^{2+}_{0.3}Cr^{3+}_{1.6}Fe^{3+}_{0.1}\right]_B$	$\left[Ni^{2+}_{0.3}Cr^{3+}_{0.4}Fe^{3+}_{1.3}\right]_B$
		Biso (Å^2)		2.4 (4)	1.5 (5)	1.8 (3)
	O	Wyckoff positions		32e	32e	32e
		Site symmetry		3 m	3 m	3 m
		Atomic positions	x = y = z	0.2582 (1)	0.2572 (2)	0.2558 (2)
		Occupancy factors		O^{2-}_4	O^{2-}_4	O^{2-}_4
		Biso (Å^2)		2.1 (5)	1.2 (8)	2.8 (7)
Structural parameters		RA (Å)		1.920 (1)	1.905 (2)	1.904 (2)
		RB (Å)		2.015 (1)	2.027 (2)	2.049 (2)
		θA-O-B (°)		122.5 (4)	122.9 (7)	123.3 (7)
		θB-O-B (°)		93.8 (4)	93.2 (7)	92.8 (7)
		Dc (nm)		156	163	172
Agreement factors		R_P (%)		6.29	6.73	7.45
		R_{wp} (%)		8.32	9.15	9.59
		R_F (%)		4.26	5.01	8.06
		χ^2 (%)		1.20	1.43	1.51

The lattice parameters rise linearly with Fe substitution, as seen in Table 1. This is explained by the Fe^{3+} ion's larger radius $r_{Fe}^{3+} = 0.67$ Å compared to the Cr^{3+} ion's smaller radius $r_{Cr}^{3+} = 0.63$ Å [48]. Additionally, the oxygen atomic locations that were determined are indicative of the spinel structure [49,50]. The computed and provided crystallite size (D_c) values in Table 1 rise from 156 nm for x = 0 to 173 nm for x = 1.6, respectively.

Additionally, as Fe^{3+} content increases in $Ni_{0.3}Zn_{0.7}Cr_{2-x}Fe_xO_4$ (x = 0; 0.4; 1.6) samples, the cation-oxygen bond length at the tetrahedral sites (R_A) decreases; however, the cation-oxygen bond length at the octahedral sites (R_B) increases. The values of the bond angles (θ_{A-O-B} and θ_{B-O-B}) were also presented in Table 1 for $Ni_{0.3}Zn_{0.7}Cr_{2-x}Fe_xO_4$ samples. The θ_{A-O-B} bond angle is concerned with the A-O-B interactions, while the θ_{B-O-B} bond angle is related to the B-O-B interactions. From the data presented in Table 1, the increase in bond angle (θ_{A-O-B}) with Fe^{3+} substitution indicates an increase in the strength of A-B exchange interactions, and the decrease in the bond angle (θ_{B-O-B}) indicates a decrease in the strength of B-B exchange interactions [51,52].

On the other hand, it is well known that due to the crystalline size effect and intrinsic strain effect, the X-ray diffraction peak broadens in nanocrystals, and this peak broadening typically consists of two parts: instrumental broadening and physical broadening [53,54]. The following connection can be used to fix this instrumental broadening:

$$\beta_D = \left[(\beta_m^2 - \beta_i^2)^{1/2} \times (\beta_m - \beta_i)\right]^{1/2} \quad (1)$$

where β_m is the measured broadening, β_i is the instrumental broadening, and β_D is the corrected broadening. Here, crystalline silicon has been used as a standard reference material for position calibration and instrumental broadening calculation. To estimate the crystallite size using XRD profiles, a number of models have been created, including the Halder-Wagner approach, the Williamson-Hall plot, and the Scherrer equation [55]. Using the Williamson-Hall (W-H) plot model [56,57], the average crystallite size for each studied oxide sample was determined in the current study. This model contains a distinct component for predicting peak broadening related to crystallite microstrain and uses diffraction peak broadening from at least four diffraction peaks as a foundation for calculating crystallite size. The interpretation of crystallite size from diffraction data was based on spherical particles with cubic symmetry. The Williamson-Hall plot was effective in determining the size of the crystallites and the inherent microstrain in the nanoparticle of the microcrystalline structure [58–60]. This approach assumes that the Scherrer equation is followed by the corrected line broadening (β_D) of a Bragg reflection (hkl) arising from the small crystallite size <D>:

$$<D> = \frac{k\lambda}{\beta_D \cos\theta} \quad (2)$$

where <D> is the effective crystallite size normal to the reflecting planes, k is the shape factor (~0.9), λ is the X-ray wavelength, and θ_{hkl} is the Bragg angle.

Additionally, the Williamsom-Hall model takes into account uniform lattice microstrain caused by crystal flaws in the nanocrystals along the crystallographic direction. In other words, it takes into account the isotropic lattice microstrain [61]. The physical broadening of the XRD profile is really impacted by this intrinsic microstrain, and this later-induced peak broadening is stated as:

$$\beta_s = 4\varepsilon \tan\theta \quad (3)$$

Therefore, it is possible to express the overall broadening caused by crystallite size and lattice strain in a specific peak with the (hkl) value as:

$$\beta_{hkl} = \beta_D + \beta_s \quad (4)$$

where β_{hkl} is the full width at half of the maximum intensity (FWHM) for the chosen diffraction planes.

$$\beta_{hkl} = \frac{k\lambda}{D}\frac{1}{\cos\theta} + 4\varepsilon\tan\theta \tag{5}$$

When this Equation (5) is re-arranged, we get:

$$\beta_{hkl}\cdot\cos\theta = \frac{k\lambda}{D} + 4\varepsilon\sin\theta \tag{6}$$

Equation (6) is a straight-line equation that takes the crystals' isotropic nature into account. One may determine the lattice microstrain (ε) from the slope straight line of the plot created with $4\sin\theta$ along the x-axis and $\beta_{hkl}\cdot\cos\theta$ along the y-axis for the three samples' oxides and the inverse of average crystallite size (D) from the intercept. For samples of oxides with x = 0, x = 0.4, and x = 1.6, the average particle sizes were determined to be roughly 110, 122, and 131 nm, respectively. The size of a coherently diffracting domain is considered to be the size of a crystallite, which is not always the same as particle size. Meanwhile, the estimated lattice strain values for the samples with x = 0, 0.4, and 1.6 are roughly 0.022, 0.015, and 0.01%, respectively. The atomic arrangement within the crystal lattice is primarily responsible for the lattice contraction or expansion in the crystallites, which is where the lattice microstrain originates. On the other hand, due to size refinement and internal-external stresses that cause lattice strain, numerous defects (point defects such vacancies, stacking faults, grains boundaries, dislocations, etc.) also get generated at the lattice structure. The estimated average crystallite size for the tree samples was approximately 20% less than the theoretical estimates of crystallite size (see Table 1) calculated from the Rietveld refinement.

3.2. Mössbauer Spectrometry

The Mössbauer spectra obtained at 300 and 77 K on the two samples containing Fe are shown in Figure 2. The 300 K spectrum of $Ni_{0.3}Zn_{0.7}Cr_{1.6}Fe_{0.4}O_4$ shows clearly a pure quadrupolar feature, while that of $Ni_{0.3}Zn_{0.7}Cr_{0.4}Fe_{1.6}O_4$ results from a complex hyperfine structure which consists of a quadrupolar component with narrow lines superimposed on a broadened single line feature. The first spectrum can be described by the summation of quadrupolar components, while the second spectrum at 300 K must be described by at least 2 or 3 components: a central quadrupolar doublet, a broadened single line, and possibly a magnetic sextet. At 77 K, the Mössbauer spectra are composed of a magnetic component with broadened lines and a quadrupolar component or a broadened single line. The mean values of isomer shift suggest the presence of exclusively Fe^{3+} species (0.30 and 0.35 for $Ni_{0.3}Zn_{0.7}Cr_{1.6}Fe_{0.4}O_4$ and $Ni_{0.3}Zn_{0.7}Cr_{0.4}Fe_{1.6}O_4$ at 300 K and 0.42 and 0.45, respectively, at 77 K). However, the lack of resolution of the hyperfine structures prevents estimating accurately the proportions of Fe located in the octahedral and tetrahedral sites: it should be necessary to use in-field Mössbauer spectrometry to get such proportions.

These complex hyperfine structures are consistent with a variety of different chemical atomic neighbors, including Ni^{2+}, Zn^{2+}, Cr^{3+}, and Fe^{3+} species which possess different magnetic moments. The highest values of the hyperfine field at Fe sites are a priori the result of a Fe-rich neighboring, the intermediate values of a Ni and Cr-rich neighboring, while the quadrupolar component is assigned to Zn-rich neighboring.

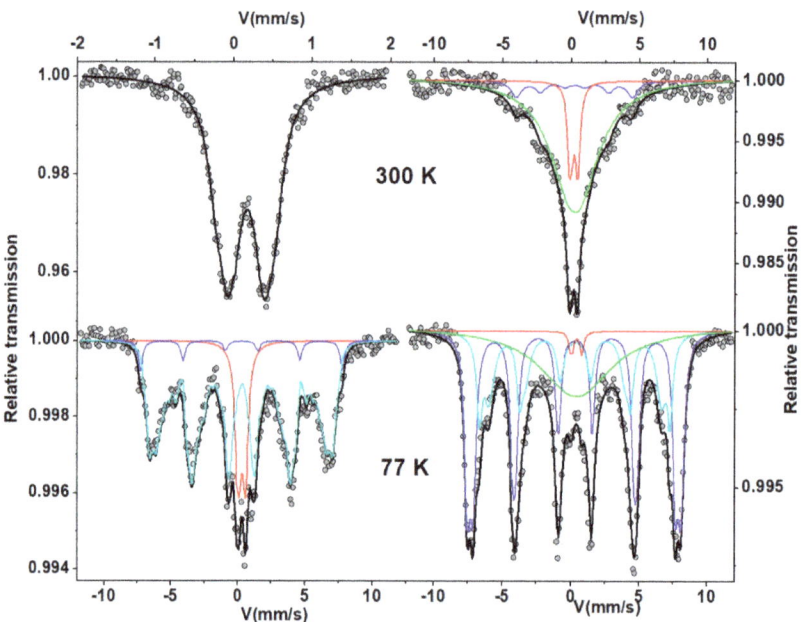

Figure 2. Mössbauer spectra obtained on Ni$_{0.3}$Zn$_{0.7}$Cr$_{1.6}$Fe$_{0.4}$O$_4$ (**left**) and Ni$_{0.3}$Zn$_{0.7}$Cr$_{0.4}$Fe$_{1.6}$O$_4$ (**right**) at 300 (**top**) and 77 K (**bottom**): the red, dark blue, cyan and green components correspond to the Zn-rich, Fe-rich, Ni-rich, and Cr-rich neighboring, respectively (see text).

3.3. Morphological Study

Figure 3a–c display the SEM images of Ni$_{0.3}$Zn$_{0.7}$Cr$_{2-x}$Fe$_x$O$_4$ (x = 0; 0.4; 1.6) ferrite samples at two different magnifications (a–c). As shown, the ferrite samples of Ni$_{0.3}$Zn$_{0.7}$Cr$_2$O$_4$, Ni$_{0.3}$Zn$_{0.7}$Cr$_{1.6}$Fe$_{0.4}$O$_4$, and Ni$_{0.3}$Zn$_{0.7}$Cr$_{0.4}$Fe$_{1.6}$O$_4$ have typical particle sizes between 0.5 and 2 µm. These microscopic particles were also dispersed in the material and were dense. It is clear that each particle results from the aggregation of many crystals because the particle sizes determined from SEM images are larger than those estimated using the XRD data. In addition, when the iron proportion increases from 0 to 1.6 at %, the shape of the particles changes from multifaceted to spherical and homogeneous. In addition, magnetic attraction can be the cause of particle adhesion.

3.4. Magnetic Characteristics

A magnetometer (VSM) is used to examine the synthesized materials' magnetic characteristics while they are at room temperature. The M-H curves of the Ni$_{0.3}$Zn$_{0.7}$Cr$_{2-x}$Fe$_x$O$_4$ (x = 0; 0.4; 1.6) ferrite samples are shown in Figure 4. Table 2 provides the matching saturation magnetization (Ms) and coercivity (Hc) values for various percentages of x. As shown, the maximum value of Ms is found around 0.46 emu/g in the Ni$_{0.3}$Zn$_{0.7}$Cr$_2$O$_4$ sample. While the ferrite samples Ni$_{0.3}$Zn$_{0.7}$Cr$_{1.6}$Fe$_{0.4}$O$_4$ and Ni$_{0.3}$Zn$_{0.7}$Cr$_{0.4}$Fe$_{1.6}$O$_4$ exhibit nearly unmeasurable remanence and coercivity with no discernible hysteresis. The magnetization of the samples has been shown to be related to various factors, such as sublattice interactions, spin tilt, the magnitude of the individual moment of the cations, grain size, etc. A strong A-O-B interaction also causes a ferrimagnetic order to form in the spinels. Meanwhile, the presence of chromium in the site (B) reduces the strength of the A-O-B interaction, which can eventually be destroyed [62]. On the other hand, superparamagnetism has this special characteristic [63]. The small value of Ms corresponds to spin canting in the surface layer of nanoparticles. In contrast to local symmetry for those atoms close to the surface layer, which resulted in a decreased Ms in these nanocrystals, the canting and disorder of the surface layer spins may be caused by broken super-exchange bonds. Additionally, because of the relatively low Zn content (around 0.7), the Zn^{2+} cations in the tetrahedral sites occupy fewer spins in the A sites, decreasing the A-B super-exchange interactions, and the spines of the octahedral sites are unable to maintain collinearity with the tiny tetrahedral spins. Consequently, the Ms looks to be relatively low for x = 0 or null for x = 0.4 and 1.6.

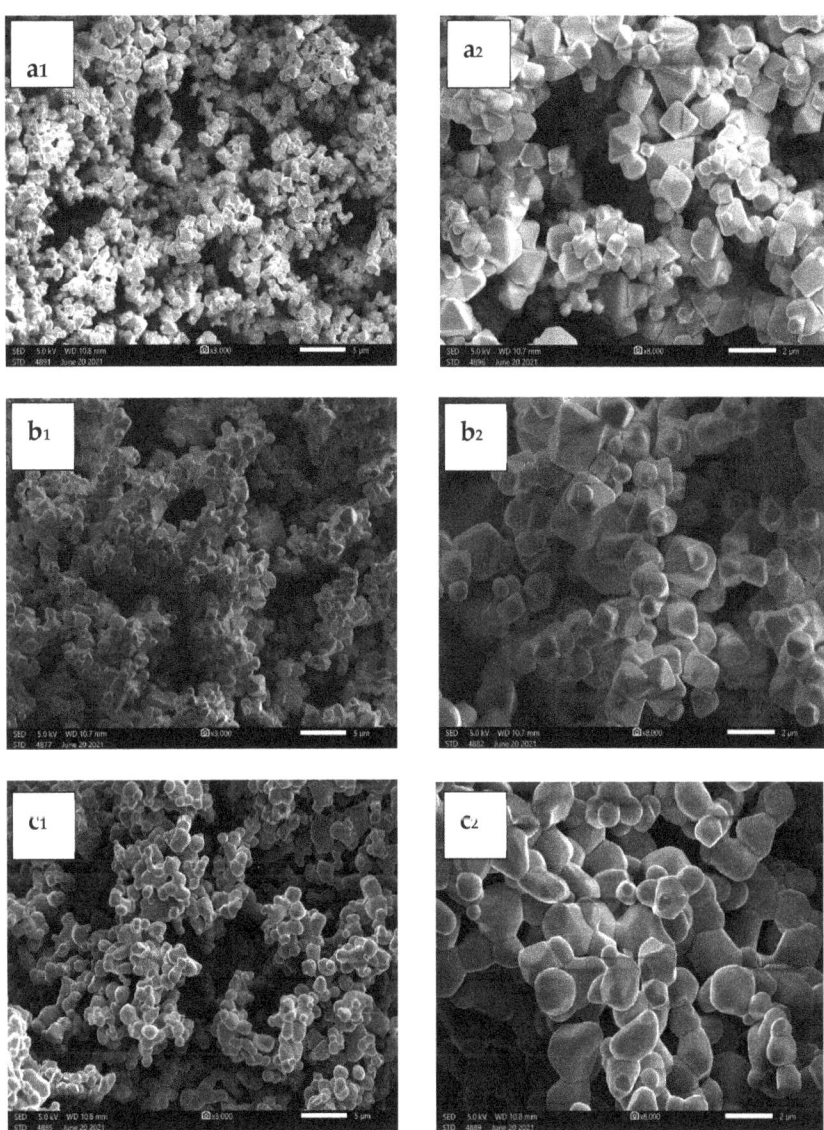

Figure 3. SEM images of the $Ni_{0.3}Zn_{0.7}Cr_{2-x}Fe_xO_4$ (x = 0 (**a**); 0.4 (**b**); 1.6(**c**)) ferrite samples (**left** scale bars: 5 μm, **right** scale bars: 2 μm).

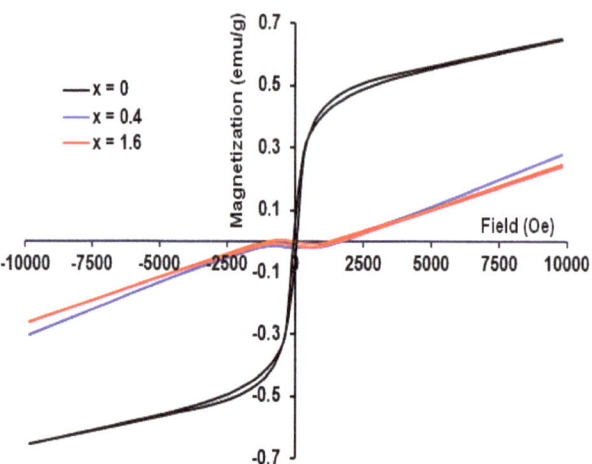

Figure 4. Hysteresis loops of $Ni_{0.3}Zn_{0.7}Cr_{2-x}Fe_xO_4$ (x = 0, 0.4 and 1.6) samples at room temperature.

Table 2. Cation distribution, crystalline size, lattice parameter, saturation magnetization, and coercivity of $Ni_{0.3}Zn_{0.7}Cr_{2-x}Fe_xO_4$ compounds.

x_{Fe}	Composition	Cation Distribution	D (nm) ±2	a (Å) ±0.0005	Ms (emu/g)	Hc (Oe) ±5
0	$Ni_{0.3}Zn_{0.7}Cr_2O_4$	$\left(Zn^{2+}_{0.7}Cr^{3+}_{0.3}\right)_A\left[Ni^{2+}_{0.3}Cr^{3+}_{1.7}\right]_B O^{2-}_4$	156	8.3242	0.46	316
0.4	$Ni_{0.3}Zn_{0.7}Cr_{1.6}Fe_{0.4}O_4$	$\left(Zn^{2+}_{0.7}Fe^{3+}_{0.3}\right)_A\left[Ni^{2+}_{0.3}Cr^{3+}_{1.6}Fe^{3+}_{0.1}\right]_B O^{2-}_4$	163	8.3329	--	...
1.6	$Ni_{0.3}Zn_{0.7}Cr_{0.4}Fe_{1.6}O_4$	$\left(Zn^{2+}_{0.7}Fe^{3+}_{0.3}\right)_A\left[Ni^{2+}_{0.3}Cr^{3+}_{0.4}Fe^{3+}_{1.3}\right]_B O^{2-}_4$	172	8.3906	-----	...

The low coercivity value of 316 Oe (Table 2) implies that the $Ni_{0.3}Zn_{0.7}Cr_2O_4$ sample is soft magnetic. This is due to the presence of grain boundaries in the sample, which need less energy to align along the external magnetic field in comparison to domain movement. Previous studies have shown that the presence of the paramagnetic ion Cr^{3+} introduced magnetic dilution into ferrites similar to that produced by non-magnetic substitution, which can induce interesting properties in ferrites. Metal cations can be used to adjust the magnetic characteristics of a system for a specific application.

Based on the Rietveld analysis (Table 1) and the VSM measurements (Figure 3), the absence of the Ni^{2+} ions as ferromagnetic ions and the occupancy of Zn^{2+} ions with zero magnetic moments in all tetrahedral sites—i.e., replacing the entire Fe^{3+} ions in the octahedral sites—are responsible for the small Ms of $Ni_{0.3}Zn_{0.7}Cr_{2-x}Fe_xO_4$. Due to the weakened B-B connection in ferrites and the weak interactions between Fe^{3+} ions in the octahedral sites, the magnetization of the current nanocrystals is low or nonexistent. Additionally, Kodama [64] and Priyadharsini et al. [65] reported that the core-shell model, which explains how the finite size effects of the nanoparticles lead to a canting of spins on their surface and subsequently reduce their magnetizability, can be used to understand the low value of Ms (compared to that of bulk Ni-ferrite (56 emu/g) [66]).

4. Conclusions

In the current study, the sol-gel synthesis process was used to create $Ni_{0.3}Zn_{0.7}Cr_{2-x}Fe_xO_4$ ferrite samples with x = 0, 0.4, and 1.6. The dry gel samples were subsequently heated at temperatures 473 (for 24 h), 773 (for 24 h), 1173 K (for 24 h), and 1448 K (for 48 h) in an air atmosphere. The structure, microstructure, and magnetic characteristics of the prepared ferrites particles were studied by the use of X-ray diffraction (XRD), scanning electron microscopy (SEM), vibrating sample magnetometer (VSM), and ^{57}Fe Mössbauer spectrometry. All diffraction peaks, as determined by the XRD investigation, were indexed in the cubic $Fd3m$ spinel structure with a crystallite size fitted for 140–160 nm.

The proposed cation distributions revealed that Fe^{3+} and Cr^{3+} ions are distributed on both sites, whereas Ni^{2+} ions are distributed on the octahedral B-sites, and Zn^{2+} ions prefer to reside on the tetrahedral A-sites. Based on this distribution the detailed formulas are described as $\left(Zn^{2+}_{0.7}Cr^{3+}_{0.3}\right)_A \left[Ni^{2+}_{0.3}Cr^{3+}_{1.7}\right]_B O^{2-}_4$, $\left(Zn^{2+}_{0.7}Fe^{3+}_{0.3}\right)_A \left[Ni^{2+}_{0.3}Cr^{3+}_{1.6}Fe^{3+}_{0.1}\right]_B O^{2-}_4$, and $\left(Zn^{2+}_{0.7}Fe^{3+}_{0.3}\right)_A \left[Ni^{2+}_{0.3}Cr^{3+}_{0.4}Fe^{3+}_{1.3}\right]_B O^{2-}_4$ for x = 0, 0.4, and 1.6, respectively.

The VSM analyses demonstrate that ferrite samples $Ni_{0.3}Zn_{0.7}Cr_{1.6}Fe_{0.4}O_4$ and $Ni_{0.3}Zn_{0.7}Cr_{0.4}Fe_{1.6}O_4$ possess superparamagnetic properties (with almost immeasurable remanence and coercivity). The phenomenon was justified by the fact that all tetrahedral positions were occupied by Zn^{2+} ions without magnetic moments and were devoid of ferromagnetic Ni^{2+} cations. Further, because of the weak interactions between Fe^{3+} ions in the octahedral locations, the magnetization of the current nanocrystals is low or nonexistent.

The complex hyperfine structures were consistent with a variety of different chemical atomic neighbors, including Ni^{2+}, Zn^{2+}, Cr^{3+}, and Fe^{3+} species, which possess different magnetic moments. At Fe sites, the quadrupolar component was attributed to the Zn-rich neighboring region, while the highest values of the hyperfine field are a priori the result of a Fe-rich neighboring. To overcome the lack of resolution of the hyperfine structures that prevent accurately estimating the proportions of Fe located in the octahedral and tetrahedral sites, it will be necessary to use in-field Mössbauer spectrometry in our future work.

Author Contributions: Conceptualization, A.M. and M.K.; formal analysis, F.A.-T. and J.-M.G.; data curation, A.M.; investigation, A.A. and M.M.A.; writing—original draft preparation, M.K. and F.A.-T.; writing—review and editing, M.K., J.-M.G and J.-J.S.; supervision, A.M. All authors have read and agreed to the published version of the manuscript.

Funding: This research received no external funding.

Data Availability Statement: Data will be requested from the authors.

Acknowledgments: The authors want to thank Qassim university.

Conflicts of Interest: The authors declare no conflict of interest.

References

1. Mandal, K.; Mandal, S.P.; Agudo, P.; Pal, M. A study of nanocrystalline (Mn–Zn) ferrite in SiO$_2$ matrix. *Appl. Surf. Sci.* **2001**, *182*, 386–389. [CrossRef]
2. Yang, H.; Shen, L.; Zhao, L. Magnetic properties of nanocrystalline Li$_{0.5}$Fe$_{2.1}$Cr$_{0.4}$O$_4$ ferrite. *Mater. Lett.* **2003**, *57*, 2455–2459. [CrossRef]
3. Yang, H.; Wang, D.; Wang, Z.; Zhao, M.; Li, M.; Wang, L. A study of the photovoltage properties of nanocrystalline LiFe$_5$O$_8$. *Mater. Chem. Phys.* **1997**, *48*, 212–215. [CrossRef]
4. Saad, M.M.H.E.; Alsobhi, B.O.; Almeshal, A. Structural, elastic, thermodynamic, electronic, magnetic, thermoelectric and optical investigation of chromate spinels TCr$_2$O$_4$ [T = V^{2+}, Mn^{2+}, Fe^{2+}] for optoelectronic applications. *Mater. Chem. Phys.* **2023**, *294*, 127041. [CrossRef]
5. Chen, Q.; Zhang, Z.J. Size-dependent superparamagnetic properties of MgFe$_2$O$_4$ spinel ferrite nanocrystallites. *J. Appl. Phys.* **1998**, *73*, 3156–3158. [CrossRef]
6. Li, F.; Wang, H.; Wang, L.; Wang, J. Magnetic properties of ZnFe$_2$O$_4$ nanoparticles produced by a low-temperature solid-state reaction method. *J. Magn. Magn. Mat.* **2007**, *309*, 295–299. [CrossRef]
7. Sun, S.; Zeng, H.; Robinson, D.B.; Raoux, S.; Rice, P.M.; Wang, S.X.; Li, G. Preparation and Reversible Phase Transfer of CoFe$_2$O$_4$ Nanoparticles. *J. Am. Chem. Soc.* **2004**, *126*, 2782. [CrossRef]
8. Hyeon, T.; Chung, Y.; Park, J.; Lee, S.S.; Kim, Y.W.; Park, B.H. Monodisperse MFe$_2$O$_4$ (M = Fe, Co, Mn) Nanoparticles. *J. Phys. Chem. B* **2002**, *106*, 6831. [CrossRef]
9. Pradhan, S.K.; Bid, S.; Gateshki, M.; Petkov, V. Microstructure characterization and cation distribution of nanocrystalline magnesium ferrite prepared by ball milling. *Mater. Chem. Phys.* **2005**, *93*, 224–230. [CrossRef]

10. Sepelak, V.; Becker, K.D. Mössbauer studies in the mechanochemistry of spinel ferrites. *J. Mater. Synth. Process* **2000**, *8*, 155–166. [CrossRef]
11. Roy, S.; Dubenko, I.; Edorh, D.D.; Ali, N. Size-induced variations in structural and magnetic properties of double exchange $La_{0.8}Sr_{0.2}MnO_3-\delta$ nano-ferromagnet. *J. Appl. Phys.* **2004**, *96*, 1202. [CrossRef]
12. McHenry, M.E.; Laughlin, D.E. Nano-scale materials development for future magnetic applications. *Acta Mater.* **2000**, *48*, 223. [CrossRef]
13. Jacob, J.; Abdul Khadar, M. VSM and Mössbauer study of nanostructured hematite. *J. Magn. Magn. Mater.* **2010**, *322*, 614–621. [CrossRef]
14. Kodama, R.H.; Berkowitz, A.E.; McNiff, E.J.; Foner, S. Surface Spin Disorder in $NiFe_2O_4$ Nanoparticles. *Phys. Rev. Lett.* **1996**, *77*, 394. [CrossRef] [PubMed]
15. Chinnasamy, C.N.; Narayanasamy, A.; Ponpandian, N.; Justin Joseyphus, R.; Jayadevan, B.; Tohji, K.; Chattopadhyay, K. Grain size effect on the Néel temperature and magnetic properties of nanocrystalline $NiFe_2O_4$ spinel. *J. Magn. Magn. Mater.* **2002**, *238*, 281. [CrossRef]
16. Shi, Y.; Ding, J. Strong unidirectional anisotropy in mechanically alloyed spinel ferrites. *J. Appl. Phys.* **2001**, *90*, 4078. [CrossRef]
17. Lin, D.; Nunes, A.C.; Majkrzak, C.F.; Berkowitz, A.E. Polarized neutron study of the magnetization density distribution within a $CoFe_2O_4$ colloidal particle II. *J. Magn. Magn. Mater.* **1995**, *145*, 343. [CrossRef]
18. Oliver, S.A.; Harris, V.G.; Hamdeh, H.H.; Ho, J.C. Large zinc cation occupancy of octahedral sites in mechanically activated zinc ferrite powders. *Appl. Phys. Lett.* **2000**, *76*, 2761. [CrossRef]
19. Jiang, J.Z.; Goya, G.F.; Rechenberg, H.R. Magnetic properties of nanostructured $CuFe_2O_4$. *J. Phys. Condens. Matter* **1999**, *11*, 4063. [CrossRef]
20. Abraham, T. Economics of ceramic magnet. *Am. Ceram. Soc. Bull.* **1994**, *73*, 62–65.
21. Seyyed Ebrahimi, S.A.; Azadmanjiri, J. Evaluation of $NiFe_2O_4$ ferrite nanocrystalline powder synthesized by a sol–gel auto-combustion method. *J. Non-Cryst. Solids* **2007**, *353*, 802. [CrossRef]
22. Rashad, M.M.; Fouad, O.A. Synthesis and characterization of nano-sized nickel ferrites from fly ash for catalytic oxidation of CO. *Mater. Chem. Phys.* **2005**, *94*, 365. [CrossRef]
23. Blum, S.L. Microstructure and properties of ferrites. *J. Am. Ceram. Soc.* **1958**, *41*, 489. [CrossRef]
24. Lin, C. Behavior of ferro-or ferrimagnetic very fine particles. *J. Appl. Phys.* **1961**, *32*, S233. [CrossRef]
25. Brook, R.J.; Kingery, W.D. Nickel ferrite thin films: Microstructures and magnetic properties. *J. Appl. Phys.* **1967**, *38*, 3589. [CrossRef]
26. Pradeep, A.; Priyadharsini, P.; Chandrasekaran, G. Production of single phase nano size $NiFe_2O_4$ particles using sol–gel auto combustion route by optimizing the preparation conditions. *Mater. Chem. Phys.* **2008**, *112*, 572. [CrossRef]
27. Nathani, H.; Misra, R.D.K. Surface effects on the magnetic behavior of nanocrystalline nickel ferrites and nickel ferrite-polymer nanocomposites. *Mater. Sci. Eng. B* **2004**, *113*, 228. [CrossRef]
28. George, M.; John, A.M.; Nair, S.S.; Joy, P.A.; Anantharaman, M.R. Finite size effects on the structural and magnetic properties of sol–gel synthesized $NiFe_2O_4$ powders. *J. Magn. Magn. Mater.* **2006**, *302*, 190. [CrossRef]
29. Xianghui, H.; Zhenhua, C. A study of nanocrystalline $NiFe_2O_4$ in a silica matrix. *Mater. Res. Bull.* **2005**, *40*, 105. [CrossRef]
30. Gubbala, S.; Nathani, H.; Koizol, K.; Misra, R.D.K. Magnetic properties of nanocrystalline Ni-Zn, Zn-Mn, Ni-Mn ferrites synthesized by reverse micelle technique. *J. Phys. B* **2004**, *348*, 317–328. [CrossRef]
31. Saafan, S.A.; Meaz, T.M.; El-Ghazzawy, E.H.; El Nimr, M.K.; Ayad, M.M.; Bakr, M. A.C. and D.C. conductivity of NiZn ferrite nanoparticles in wet and dry conditions. *J. Magn. Magn. Mater.* **2010**, *322*, 2369–2374. [CrossRef]
32. Singhal, S.; Chandra, K. Cation distribution and magnetic properties in chromium-substituted nickel ferrites prepared using aerosol route. *J. Solid State Chem.* **2007**, *180*, 296–300. [CrossRef]
33. Tsay, C.Y.; Liu, K.S.; Lin, T.F.; Lin, I.N. Microwave sintering of NiCuZn ferrites and multilayer chip inductors. *J. Magn. Magn. Mater.* **2000**, *209*, 189–192. [CrossRef]
34. Aphesteguy, J.C.; Damiani, A.; DiGiovanni, D.; Jacobo, S.E. Microwave-absorbing characteristics of epoxy resin composite containing nanoparticles of NiZn- and NiCuZn-ferrite. *J. Phys. B* **2009**, *404*, 2713–2716. [CrossRef]
35. Eustace, D.A.; Docherty, F.T.; McComb, D.W.; Craven, A.J. ELNES as a Probe of Magnetic Order in Mixed Oxides. *J. Phys. Conf. Ser.* **2006**, *26*, 165. [CrossRef]
36. Sijo, A.K. Magnetic and structural properties of $CoCr_xFe_{2-x}O_4$ spinels prepared by solution self-combustion method. *Ceram. Int.* **2017**, *43*, 2288–2290. [CrossRef]
37. Raghasudha, M.; Ravinder, D.; Veerasomaiah, P. FTIR Studies and Dielectric Properties of Cr Substituted Cobalt Nano Ferrites Synthesized by Citrate-Gel Method. *Nanosci. Nanotech.* **2013**, *3*, 105–114. [CrossRef]
38. Sijo, A.K.; Dimple, P.D.; Roy, M.; Sudheesh, V.D. Magnetic and dielectric properties of $NiCrFeO_4$ prepared by solution self-combustion method. *Mater. Res. Bull.* **2017**, *94*, 154–159. [CrossRef]
39. Rietveld, H.M. A profile refinement method for nuclear and magnetic structures. *J. Appl. Cryst.* **1969**, *2*, 65. [CrossRef]
40. Teillet, J.; Varret, F. MOSFIT Program, Université du Maine, Le Mans, France. *unpublished*.
41. Chakrabarti, M.; Sanyal, D.; Chakrabarti, A. Preparation of $Zn_{(1-x)}Cd_xFe_2O_4$ (x = 0.0, 0.1, 0.3, 0.5, 0.7 and 1.0) ferrite samples and their characterization by Mössbauer and positron annihilation techniques. *J. Phys. Condens. Matter* **2007**, *19*, 1. [CrossRef]

42. Patange, S.M.; Shirsath, S.E.; Jadhav, S.S.; Jadhav, K.M. Cation distribution study of nanocrystalline $NiFe_{2-x}Cr_xO_4$ ferrite by XRD, magnetization and Mössbauer spectroscopy. *Phys. Status Solidi A* **2012**, *209*, 347. [CrossRef]
43. Gismelseed, A.M.; Yousif, A.A. Mössbauer study of chromium-substituted nickel ferrites. *Phys. B* **2005**, *370*, 215. [CrossRef]
44. Anwar, M.S.; Ahmed, F.; Koo, B.H. Enhanced relative cooling power of $Ni_{1-x}Zn_xFe_2O_4$ ($0.0 \leq x \leq 0.7$) ferrites. *Acta Mater.* **2014**, *71*, 100. [CrossRef]
45. Hakim, M.A.; Nath, S.K.; Sikder, S.S.; Maria, K.H. Cation distribution and electromagnetic properties of spinel type Ni–Cd ferrites. *J. Phys. Chem. Solids* **2013**, *74*, 1316. [CrossRef]
46. Lohar, K.S.; Patange, S.M.; Mane, M.L.; Shirsath, S.E. Cation distribution investigation and characterizations of $Ni_{1-x}Cd_xFe_2O_4$ nanoparticles synthesized by citrate gel process. *J. Mol. Struct.* **2013**, *1032*, 105–110. [CrossRef]
47. Khalaf, K.A.; Al Rawas, A.D.; Gismelssed, A.M.; Al Jamel, A.; Al Ani, S.K.; Shongwe, M.S.; Al Riyami, K.O.; Al Alawi, S.R. Influence of Cr substitution on Debye-Waller factor and related structural parameters of $ZnFe_{2-x}Cr_xO_4$ spinels. *J. Alloys Compd.* **2017**, *701*, 474–486. [CrossRef]
48. Shannon, R.D. Revised effective ionic radii and systematic studies of interatomic distances in halides and chalcogenides. Acta crystallographica section A: Crystal physics, diffraction, theoretical and general crystallography. *Acta Cryst.* **1976**, *32*, 751–767. [CrossRef]
49. Hcini, S.; Kouki, N.; Omri, A.; Dhahri, A.; Bouazizi, M.L. Effect of sintering temperature on structural, magnetic, magnetocaloric and critical behaviors of Ni-Cd-Zn ferrites prepared using sol-gel method. *J. Magn. Magn. Mater.* **2018**, *464*, 91–102. [CrossRef]
50. Kouki, N.; Hcini, S.; Boudard, M.; Aldawas, R.; Dhahri, A. Microstructural analysis, magnetic properties, magnetocaloric effect, and critical behaviors of $Ni_{0.6}Cd_{0.2}Cu_{0.2}Fe_2O_4$ ferrites prepared using the sol–gel method under different sintering temperatures. *RSC Adv.* **2019**, *9*, 1990–2001. [CrossRef]
51. Kumar, G.; Kotnala, R.K.; Shah, J.; Kumar, V.; Kumar, A.; Dhiman, P.; Singh, M. Cation distribution: A key to ascertain the magnetic interactions in a cobalt substituted Mg–Mn nanoferrite matrix. *Phys. Chem. Chem. Phys.* **2017**, *19*, 16669. [CrossRef]
52. Sharma, R.; Thakur, P.; Kumar, M.; Thakur, J.N.; Negi, N.S.; Sharma, P.; Sharma, V. Improvement in magnetic behaviour of cobalt doped magnesium zinc nano-ferrites via co-precipitation route. *J. Alloys Compd.* **2016**, *684*, 569–581. [CrossRef]
53. Mhadhbi, M.; Khitouni, M.; Azabou, M.; Kolsi, A. Characterization of Al and Fe nanosized powders synthesized by high energy mechanical milling. *Mater. Charact.* **2008**, *59*, 944–950. [CrossRef]
54. Khitouni, M.; Kolsi, A.W.; Njah, N. The effect of boron additions on the disordering and crystallite refinement of Ni3Al powders during mechanical milling. *Ann. Chim. Sci. Mat.* **2003**, *28*, 17–29. [CrossRef]
55. Nath, D.; Singh, F.; Das, R. X-ray diffraction analysis by Williamson-Hall, Halder-Wagner and size-strain plot methods of CdSe nanoparticles- a comparative study. *Mater. Chem. Phys.* **2020**, *239*, 122021. [CrossRef]
56. Williamson, G.K.; Hall, W.H. X-ray line broadening from filed aluminum and wolfram. *Acta Metall.* **1953**, *1*, 22–31. [CrossRef]
57. Klug, H.P.; Alexander, L.E. *X-ray Diffraction Procedure for Polycrystalline and Amorphous Materials*; Wiley: New York, NY, USA, 1974.
58. Bindu, P.; Thomas, S. Estimation of lattice strain in ZnO nanoparticles: X-ray peak profile analysis. *J. Theor. Appl. Phys.* **2014**, *8*, 123–134. [CrossRef]
59. Chenari, H.M.; Seibel, C.; Hauschild, D.; Reinert, F.; Abdollahian, H. Titanium dioxide nanoparticles: Synthesis, X-ray line analysis, and chemical composition study. *Mater. Res.* **2016**, *19*, 1319–1323. [CrossRef]
60. Kumar, S.; Mote, V.; Prakash, R.; Kumar, V. X-ray analysis of α-Al_2O_3 particles by Williamson–Hall methods. *Mater. Focus* **2016**, *5*, 545–549. [CrossRef]
61. Hall, W.H. X-ray line broadening in metals. *Proc. Phys. Soc. Sect.* **1949**, *62*, 741–743. [CrossRef]
62. Hcini, S.; Boudard, M.; Zemni, S.; Oumezzine, M. Critical behavior of $Nd_{0.67}Ba_{0.33}Mn_{1-x}Fe_xO_3$ (x = 0 and 0.02) manganites. *Ceram. Int.* **2015**, *41*, 2042–2049. [CrossRef]
63. Leslie-Pelecky, D.L.; Rieke, R.D. Magnetic properties of nanostructured materials. *Chem. Mater.* **1996**, *8*, 1770–1783. [CrossRef]
64. Kodama, R.H. Magnetic Nanoparticle. *J. Magn. Magn. Mater.* **1999**, *200*, 359–372. [CrossRef]
65. Priyadharsini, P.; Pradeep, A.; Rao, P.S.; Chandrasekaran, G. Structural, spectroscopic and magnetic study of nanocrystalline Ni–Zn ferrites. *Mater. Chem. Phys.* **2009**, *116*, 207–213. [CrossRef]
66. Maaz, K.; Karim, S.; Mumtaz, A.; Hasanain, S.K.; Liu, J.; Duan, J.L. Synthesis and magnetic characterization of nickel ferrite nanoparticles prepared by co-precipitation route. *J. Magn. Magn. Mater.* **2009**, *321*, 1838–1842. [CrossRef]

Disclaimer/Publisher's Note: The statements, opinions and data contained in all publications are solely those of the individual author(s) and contributor(s) and not of MDPI and/or the editor(s). MDPI and/or the editor(s) disclaim responsibility for any injury to people or property resulting from any ideas, methods, instructions or products referred to in the content.

Article

The Effect of Transition Metals Co-Doped ZnO Nanotubes Based-Diluted Magnetic Semiconductor for Spintronic Applications

Muhammad Adil Mahmood [1], Rajwali Khan [1,2,*], Sattam Al Otaibi [3], Khaled Althubeiti [4], Sherzod Shukhratovich Abdullaev [5,6], Nasir Rahman [1], Mohammad Sohail [1] and Shahid Iqbal [7]

[1] Department of Physics, University of Lakki Marwat, Lakki Marwat 28420, KP, Pakistan
[2] Department of Physics, Zhejiang University, Hangzhou 310000, China
[3] Department of Electrical Engineering, College of Engineering Taif University, P.O. Box 11099, Taif 21944, Saudi Arabia; srotaibi@tu.edu.sa
[4] Department of Chemistry, College of Science, Taif University, P.O. Box 11099, Taif 21944, Saudi Arabia; k.althubeiti@tu.edu.sa
[5] Faculty of Chemical Engineering, New Uzbekistan University, Tashkent 10000, Uzbekistan
[6] Scientific Department, Tashkent State Pedagogical University Named after Nizami, Tashkent 100183, Uzbekistan
[7] Department of Physics, University of Wisconsin-La Crosse, La Crosse, WI 54601, USA
* Correspondence: rajwalipak@zju.edu.cn

Citation: Mahmood, M.A.; Khan, R.; Al Otaibi, S.; Althubeiti, K.; Abdullaev, S.S.; Rahman, N.; Sohail, M.; Iqbal, S. The Effect of Transition Metals Co-Doped ZnO Nanotubes Based-Diluted Magnetic Semiconductor for Spintronic Applications. *Crystals* 2023, *13*, 984. https://doi.org/10.3390/cryst13070984

Academic Editor: Ikai Lo

Received: 9 May 2023
Revised: 6 June 2023
Accepted: 17 June 2023
Published: 21 June 2023

Copyright: © 2023 by the authors. Licensee MDPI, Basel, Switzerland. This article is an open access article distributed under the terms and conditions of the Creative Commons Attribution (CC BY) license (https://creativecommons.org/licenses/by/4.0/).

Abstract: The Impact of Co and Gd on the structural, magnetic and dielectric properties of ZnO nanotubes synthesized by co-precipitation is reported. The results demonstrate that incorporating Co and Gd into ZnO diminished crystallinity while retaining the optimum orientation. The outcomes of transmission electron microscopy and scanning electron microscopy examined that the Co and Gd dopants had no effect on the morphology of the produced nanotubes. It was also discovered that as the frequency and concentration of Gd co-dopant decreased, the dielectric constant and loss values increased. When doping was present, the dielectric constant and ac electrical conductivity response was found to be inversely related. Ultimately, at 300K, Co and Gd co-doped ZnO nanotubes exhibited ferromagnetic properties. When Gd doping was increased to 3%, the ferromagnetic response increased. Since then, increasing the Gd co-doping, the ferromagnetic response decreased. For the same sample ($Zn_{0.96-x}Co_{0.04}Gd_{0.03}O$ nanotubes), the electrical conductivity exhibited also superior to pure and low Gd doped ZnO. Its high ferromagnetism is usually caused by magnetic impurities replaced on the ZnO side. Therefore, considering the behaviour of these nanotubes, it can be sued spin-based electronics.

Keywords: diluted magnetic semiconductor; ferromagnetism; dielectric properties; oxygen vacancies; spintronic devices

1. Introduction

Diluted magnetic semiconductor (DMS) obtained by doping a dilute amount of transition metals [1–8] with semiconductors, has fascinated enormous interest among researchers and is being intensively studied for spintronics-related applications. Significant improvements in DMS for spintronic electrical devices, including the single electronic charge, quantum hall effect, semiconductor laser, and resistive switching, have been made as a result of the discovery of $Ga_{1-x}Mn_xAs$ ferromagnet (FM) material [9,10]. These devices are intriguing due to the potential for obtaining DMS Curie temperatures (T_C) above room temperature ferromagnetism (RTFM). A number of earlier studies [11–14] showed that by doping of TMs doped to host II–VI and III–V semiconductors created an FM. Many practical applications favor this characteristic, however, it raised some debate regarding FM origin. A recent study for spin-based electronics has the potential to learn more about the RTFM of these structures because many materials containing nonmagnetic ions have displayed unexpected FM behavior. [14–20].

The insertion of 3d electron TMs into the lattice of II–VI semiconductor materials results in the creation of the DMS [12]. Zinc oxide (ZnO) is a typical substance used as a host for the introduction of TM ions. It has a wurtzite lattice, a huge direct band gap, and a high exciton binding energy, and it is an optically transparent II–VI semiconductor. Furthermore, room temperature ferromagnetism (RTFM) is observed in doped ZnO as a result of the free charge carriers produced by the doping of specific transition metal ions [21]. Fe, Mn, Ni and Co doping elements are discovered to be of considerable attention among the transition metal ions due to the capacity to introduce ferromagnetic behavior by replacing out the non-magnetic element Zn ion within the ZnO crystal lattice with a 3d transition metal ion. The choice of co-dopant ions depends on their valence states, coordination number, ionic radius and distinctive magnetic contribution. When contrasted with other three-dimensional metals, Co is preferred specifically because of its similar ionic radius (0.58 Å) to that of Zn (0.60 Å), as well as its large magnetic moment (μ_{Co} = 1.8 μ_B) (d_7 low spin configuration). Recently, Hao et al. published their findings on the discovery of the fundamental RTFM in Mn-doped ZnO nanoparticles. [22]. Later, using this material, Oshio et al. [23] investigated the ZnO epitaxial film's leakage current. The performance of Co-ZnO nanoparticles RTFM and semiconductor quantum dots was recently examined in order to demonstrate that Co doping produced defects and oxygen vacancies [24]. These results indicate that samples of (Co, Zn) co-doped SnO_2 have tunable RTFM and that O_2 annealing vacancies may be major method to improve the RTFM. Optoelectronic and electronic devices, ceramic industries, optical components, sensors, catalysis, biomedical, lighting, etc. are only a few of the technological applications that the ZnO material demonstrates due to its many defect-related features. According to Liu, Changzhen, et al., the optical band gap of Fe-doped ZnO nanoparticles increased with increasing Fe concentration, and they concluded that the observed RTFM is not caused by secondary phase $ZnFe_2O_4$ or the metallic Fe clusters. [25]. According to Saleh et al., A ferromagnetic secondary phase known as $\gamma-Fe_2O_3$ is responsible for RTFM in Fe-doped nanocrystalline ZnO particles, and the energy gap reduced with boosting Fe concentration [26]. The cause of ferromagnetism in ZnO-based DMSs is still up for debate despite several experimental and theoretical publications on ZnO. High-frequency electronic devices also require the superior dielectric, magnetic, and electrical conductivity of TMs-ZnO. [27–29].

The co-precipitation method was used in this study to synthesis ZnO, $Zn_{0.96}Co_{0.04}O$ and $Zn_{0.96-x}Co_{0.04}Gd_xO$ (x = 0, 0.01, 0.03, 0.04) nanotubes (NTs) with varying Gd contents while keeping a fixed Co content. The results suggest that the crystallinity was affected by the changing Co concentration, and this revealed a decrease in the rate at which photoinduced electron-hole pairs in ZnO recombined. Zinc NTs' magnetic and dielectric responses were improved as a result of this phenomena. The outcomes were evaluated in relation to magnetic and optical properties in respect to structure and microstructure.

2. Materials and Methods

2.1. Synthesis

The compound, Cobalt chloride [$CoCl_2.4H_2O$], Zinc chloride [$ZnCl_2.6H_2O$] and Gadolinium cloride [$GdCl_3.6H_2O$], were collected from Alfa-Aesar. The ZnO, $Zn_{0.96}Co_{0.04}O$ and $Zn_{0.96-x}Co_{0.04}Gd_xO$ (x = 0, 0.01, 0.03, 0.04) nanotubes (NTs) were created by adopting the co-precipitation method. The process involved dissolving zinc chloride ($ZnCl_2.6H_2O$) in 50 milliliter of pure water (aqua), adding 20 milliliter of aqueous ammonia solution (1.5 M) dropwise, and stirring vigorously while the pH was maintained between 1.0 and 10.2 [30,31]. After the reaction, washing and centrifugation were used to get brownish-type powder. Afterwards, they were dried at 60 °C for 24 h before being annealed in a furnace for 5 h at 400 °C. Before annealing, the powder was washed 6 times to remove the chlorine from the samples. Zinc, Co, and Gd acetate were gently added while stirring, then the precipitates were separated for the Co and Gd co-doped specimens, as shown in Figure 1.

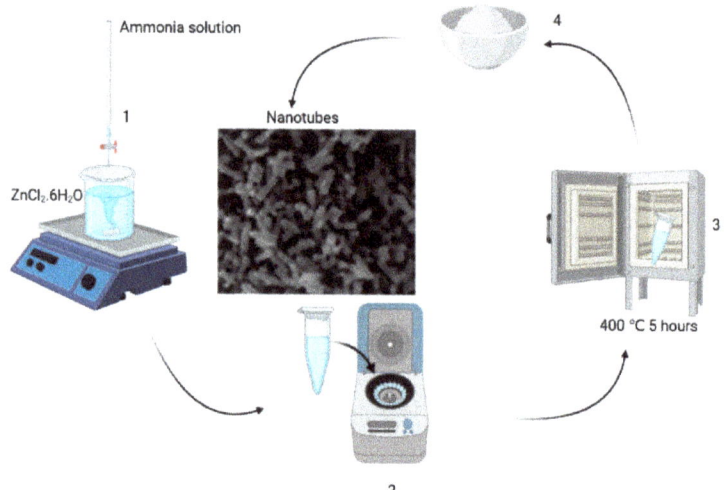

Figure 1. Co-precipitation method used in this work.

2.2. Instruments

Cu Kα radiation (λ = 1.5406 Å) was utilized from XRD in order to perform structural characterization. The sample's lattice parameters and volumes were calculated via Rietveld refinements using the High Score Plus application. A field emission scanning electron microscope, also known as a FE-SEM, was utilized so that the particle shapes could be analyzed. The particles' nature was determined using transmission electron microscopy. An impedance analyzer was used to determine the dielectric properties as well as the electrical conductivity from a frequency range of 40 Hz to 7 MHz and a frequency range of 50 Hz to 7 MHz. For the purpose of calculating the parameters of diluted magnetic semiconductors, a Quantum Design superconducting Quantum interface Device was utilized.

3. Results and Discussions

3.1. Structural Properties

Figure 2 shows the XRD results for ZnO, $Zn_{0.96}Co_{0.04}O$ and $Zn_{0.96-x}Co_{0.04}Gd_xO$ (x = 0, 0.01, 0.03, 0.04) nanotubes (NTs). All of the specimens crystallized with the hexagonal structure (wurtzite-type $P6_3mc$ space group) and exhibited no evidence of impurities. All the Lattice constants of a = 3.244 Å, c = 5.186 for pure ZnO, Å, a = 3.243 Å, c = 5.190 Å for only Co doped ZnO sample, a = 3.241 Å, c = 5.189 for 1% Gd doped sample, a = 3.237 Å, c = 5.187 for 3% Gd doped sample and a = 3.235 Å, c = 5.178 Rietveld refinement software was used to obtain a weighted profile factor R_{WP}= 9.34% for a 4% Gd doped sample, 9.27 and the goodness-of-fit χ^2 = 2.571, 2.581) for the all as shown in Figure 2a,d. The unit cell volume (V) for pure ZnO is 47.19 Å^3 and it is increased to V = 47.35 Å^3 for $Zn_{0.96}Co_{0.04}Gd_{0.04}O$ specimen. For $Zn_{0.93}Co_{0.04}Gd_{0.03}O$, the unit cell volume raised to V = 47.35Å^3, which is more significant than pure and $Zn_{0.96}Gd_{0.04}O$. It is predicted since the ionic radius of Co^{2+} (0.65Å) and Gd^{3+} (94Å) are larger than that of Zn^{2+} (0.60 Å). The replacement of Zn^{2+} ions by Co^{2+} and Gd^{3+} ions in the structure of wurtzite explains this phenomenon. This is due to the fact that zinc ions are less electronegative and have smaller ionic radii than Co and Gd ions [1,24,26]. Also, the XRD findings are consistent with the theory that Co^{2+} and Gd^{3+} ions were added to the wurtzite structure, most likely in place of Zn ions. The geometry of the wurtzite unit cell changed as a result. According to the XRD patterns, Co (fixed) and Gd co-dopants in ZnO enhances in concentration, causing the peaks to shift at a smaller angle, as shown in schematic 2(b). Due to Gd and Co doping, it is advised that samples be changed to a disordered state.

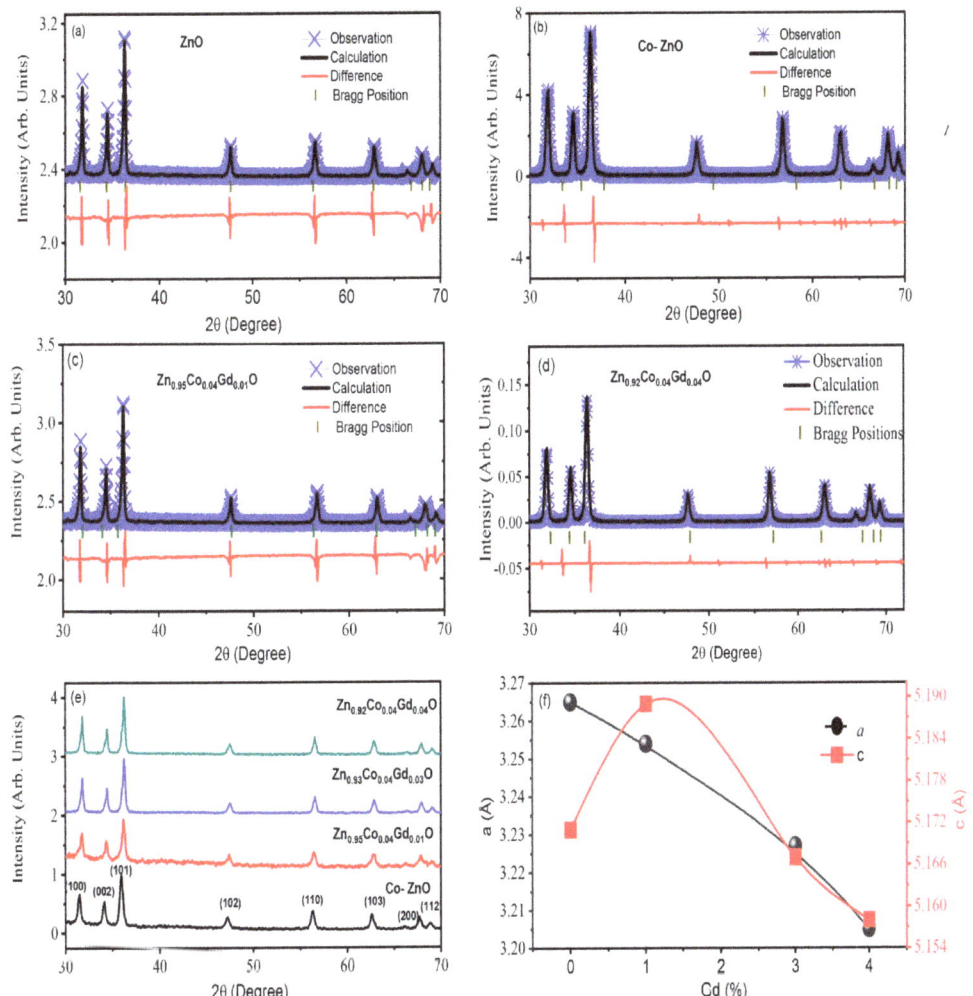

Figure 2. (**a**–**d**) The ZnO Rietveld refinement and $Zn_{0.96}Co_{0.04}O$ and $Zn_{0.96-x}Co_{0.04}Gd_xO$ (Gd = 0, 0.01, 0.03). (**e**) XRD for all the samples, (**f**) Lattice parameters versus Gd doping determined from single doped XRDs & $Zn_{0.96-x}Co_{0.04}Gd_xO$ (Co = 0, 0.01, 0.03, 0.04) specimens.

For the purpose of using the Scherrer formula to evaluate the crystalline size of the samples, the XRD peaks broadening was used [29,32], and it was discovered that as the Co content rose from 0% to 4%, size shrank. This decline is explained by the limitation of grain expansion brought on by the appearance of Co and Gd in ZnO. This demonstrates that the high levels of Co both prevent grain formation and do not generate O_2 vacancies to enable densification. Particle size is reduced in strongly doped specimens due to grain boundary segregation, though. The smaller ionic radius of Co (70 pm) ions contrasted to Zn (74 pm) ions is the source of the drop in lattice parameter with a greater Co concentration. The lattice parameters, X-ray density, unit cell volume and grain size calculations are summarized in Table 1.

Table 1. The fluctuation of structural values was estimated using ZnO, $Zn_{0.96}Co_{0.04}O$ and $Zn_{0.96-x}Co_{0.04}Gd_xO$ (x = 0, 0.01, 0.03, 0.04) nanotubes (NTs).

Sample	hkl	2θ (°)	d-Spacing (Å)	Grain Size (nm)	Lattice Constant		Unite Cell Volume (Å3)
					a (Å)	c (Å)	
ZnO	100	31.522	2.7773	18.34	3.244	5.186	47.19
	002	34.442	2.5972	16.11	-	-	-
	101	36.352	2.4704	17.45	-	-	-
Gd (0%)	100	31.847	2.8079	16.54	3.243	5.190	47.20
	002	34.645	2.5973	15.61	-	-	-
	101	36.465	2.4715	19.34	-	-	-
Gd (1%)	100	31.902	2.8082	19.32	3.241	5.189	47.27
	002	34.670	2.5976	15.35	-	-	-
	101	36.490	2.4719	19.01	-	-	-
Gd (3%)	100	36.498	2.8085	19.75	3.237	5.187	47.30
	002	34.689	2.5980	19.54	-	-	-
	101	36.512	2.4721	19.4	-	-	-
Gd (4%)	100	36.459	2.8088	20.25	3.235	5.178	47.35
	002	34.692	2.5984	19.54	-	-	-
	101	36.534	2.4727	20.46	-	-	-

The obtained compositions of Zn, O, Co, and Gd (in weight percent) are shown by the EDX spectra in Figure 3a–c. As was to be expected, the majority of the chemical elements in pure ZnO are composed of zinc and oxygen, however, specimens that have been co-doped with Co and Gd also show Co and Gd peaks. As illustrated in the Figure 3a–c, the amount required to produce the sample was found to be very near to the percent by weight of the doped TMs. Figure 3a–c exhibits the morphological and elemental examination of pure, 3 and 4 weight percent Gd and Co co-doped ZnO that was subjected to scanning electron microscopy examination. These findings suggest that the hexagonal wurtzite structure of the NTs has the same morphology [33]. The samples have a tube-like structure and become more aggregated as the Co and Gd concentration rises.

The transmission electron microscopy (TEM) images of ZnO, $Zn_{0.93}Co_{0.04}Gd_{0.03}O$, and $Zn_{0.92}Co_{0.04}Gd_{0.04}O$ Nanotubes are demonstrated in Figure 4a–c. The transmission electron microscopy pictures of ZnO, $Zn_{0.93}Co_{0.04}Gd_{0.03}O$, and $Zn_{0.92}Co_{0.04}Gd_{0.04}O$ Nanotubes showed a tube-like structure with a homogeneous dissemination. According to the transmission electron microscopy investigation, the mean tube size for pure ZnO is approximately 15 nm, 17 nm for $Zn_{0.93}Co_{0.04}Gd_{0.03}O$, and 19 nm for $Zn_{0.92}Co_{0.04}Gd_{0.04}O$. Co and Gd co-doping increases tube size, which is reliable with the X-rays Diffraction observations. The middle panels of Figure 3a–c show selected area electron diffraction (SEAD) pictures of the generated Nanotubes. The selected area electron diffraction image of the ZnO sample's fringes reveals that ZnO formed into its polycrystalline tetragonal structure. Furthermore, it is significant to observe that the broad X-rays Diffraction peak of Co and Gd-doped ZnO indicates microscopic crystallite size and, thus, a substantial surface area. The transmission electron microscopy micrograph makes it clear that Co-Gd co-doped ZnO has a larger size than pure ZnO, which suggests that it has a larger surface area. For improved photo degradation efficiency, the sample needs to have a bigger surface area.

3.2. Dielectric Properties

3.2.1. Dielectric Constant

For all of the samples that were annealed at 400 °C, Figure 5a displays the frequency (*f*) dependency of the dielectric constant (ε_r). The samples were created by using a presser to form them into round pellets with an 8mm diameter, and then their dielectric loss, capacitance and a.c conductivity was calculated. The capacitance was calculated using gold glue electrodes in accordance with the methodology described in Refs. [3,34]. The following relationship (1) was utilized to evaluate the ε_r

$$\varepsilon_r = Cd/\varepsilon_0 A \tag{1}$$

where C denotes capacitance, d denotes cylinder height, and A denotes pellet cross-sectional area. At frequencies over 1.5×10^4 Hz, all samples ε_r values decline equally with rising f and remain constant. Large values of reported dielectric constants are often induced by either space charge polarization (SCP) or rotation dielectric polarization. These two types of polarization can be distinguished by their respective abbreviations: SCP and RDP. The interfacial area contains oxygen vacancies (OVS), which can be ionized to generate single or double ionized vacancies, vacancy clusters, dangling bonds, and other defects [31,32].

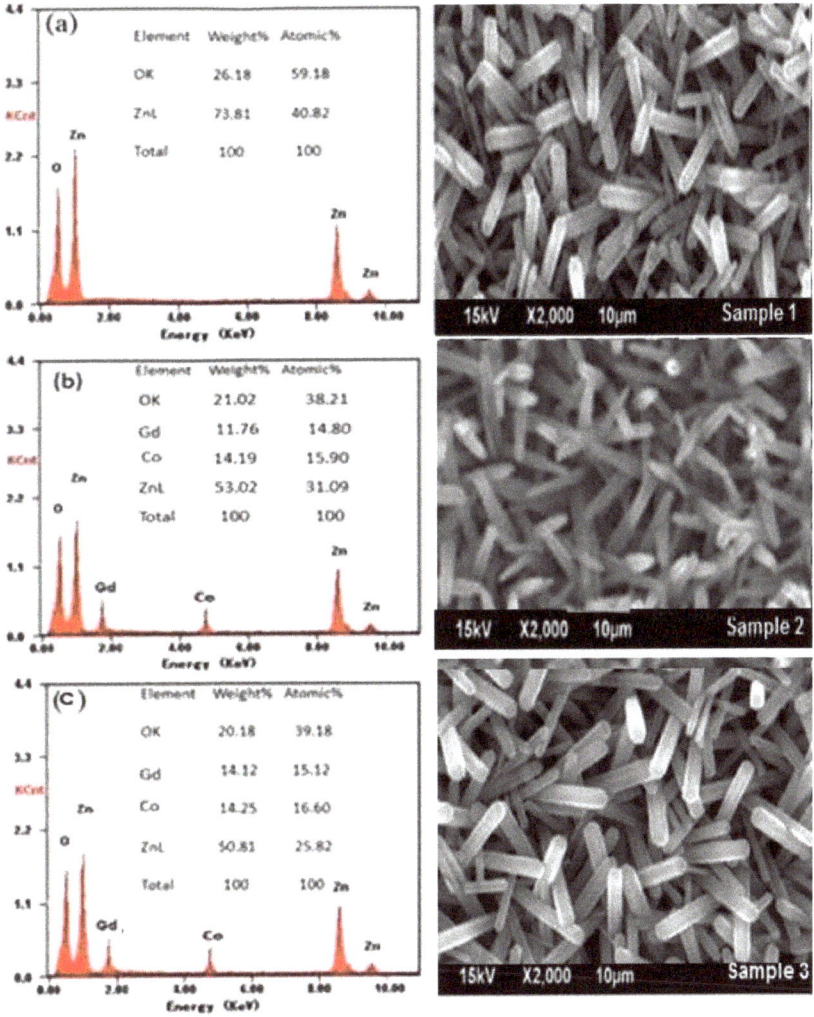

Figure 3. EDX pictures of (**a**) ZnO, (**b**) $Zn_{0.93}Co_{0.04}Gd_{0.03}O$, (**c**) $Zn_{0.92}Co_{0.04}Gd_{0.04}O$ along with the associated SEM pictures.

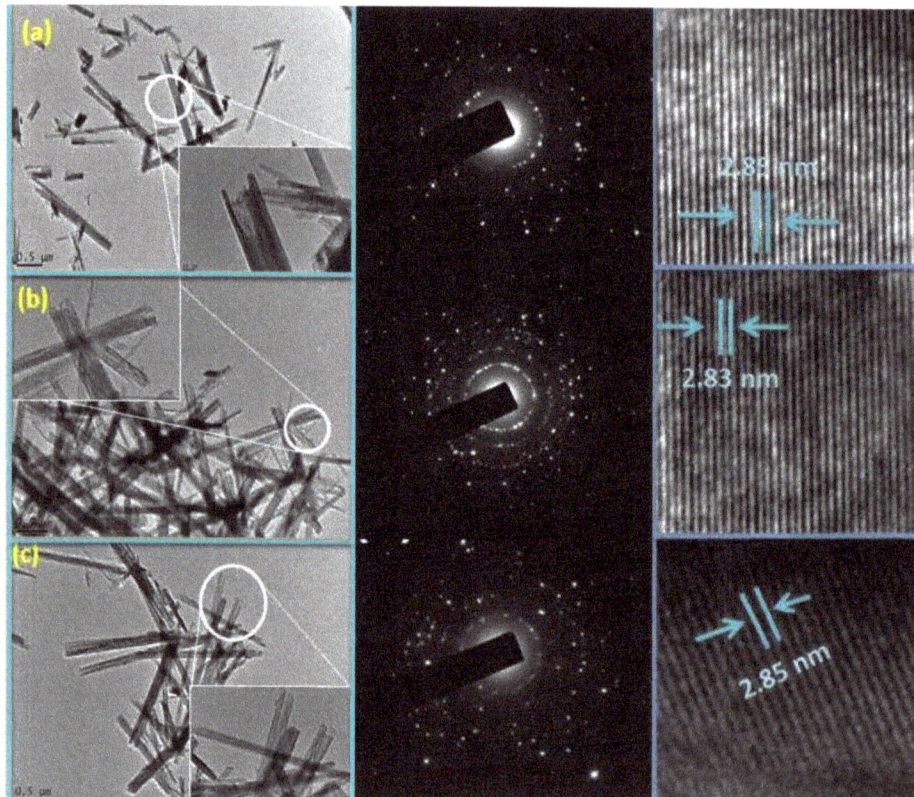

Figure 4. TEM images of (**a**) ZnO, (**b**) $Zn_{0.93}Co_{0.04}Gd_{0.03}O$, (**c**) $Zn_{0.92}Co_{0.04}Gd_{0.04}O$, their corresponding SEAD images (med panels) and HRTEM micrographs.

Because of the existence of positive and negative vacancies, a significant quantity of dipole moments (which are randomly oriented) will be produced, and only polarization may be caused by an external magnetic field in these dipole moments to cause RDP. Furthermore, space charge moving in the alternative direction of the electric fields will be entangled by interfacial defects in these nanocrystalline materials [35,36] As a result, SCP and dielectric constant are increased. As a result, grain boundaries become electrically active. The observed decrease in ε_r as (f) increases can be attributed to the fact that any species that helps to increase polarizability will always be behind the applied field. When particles get smaller, the interfacial area increases, which rises RDP and SCP, but Co co-doping improves the OVS, which raises RDP and SCP. As a result, both crystalline size and doping concentration have an effect on dielectric properties. As the concentration of Gd co-doping increases, the ε_r increases, as illustrated in Figure 5a,d, due to lattice distortion caused by small-sized Co^{2+} ions replacing Zn^{2+} ions, as originally observed [28].

3.2.2. Dielectric Loss

The variations in dielectric loss (ε'') with the (f) of doped and co-doped ZnO nanoparticles are shown in schematic 5(b). It is shown here that increasing the frequency causes ε'' to drop. All of the specimens behaved similarly, with dispersal at low(f) and independent at high (f). Ionic migration is responsible for the drop in ε'' noticed at higher frequencies. Ionic hopping and charge transfer conduction losses also contribute to the (ε'') at low and intermediate frequencies (f_s). Ionic polarization losses can play a part to these phenomena. Ion vibrations, however, might be the only cause at high (f_s).

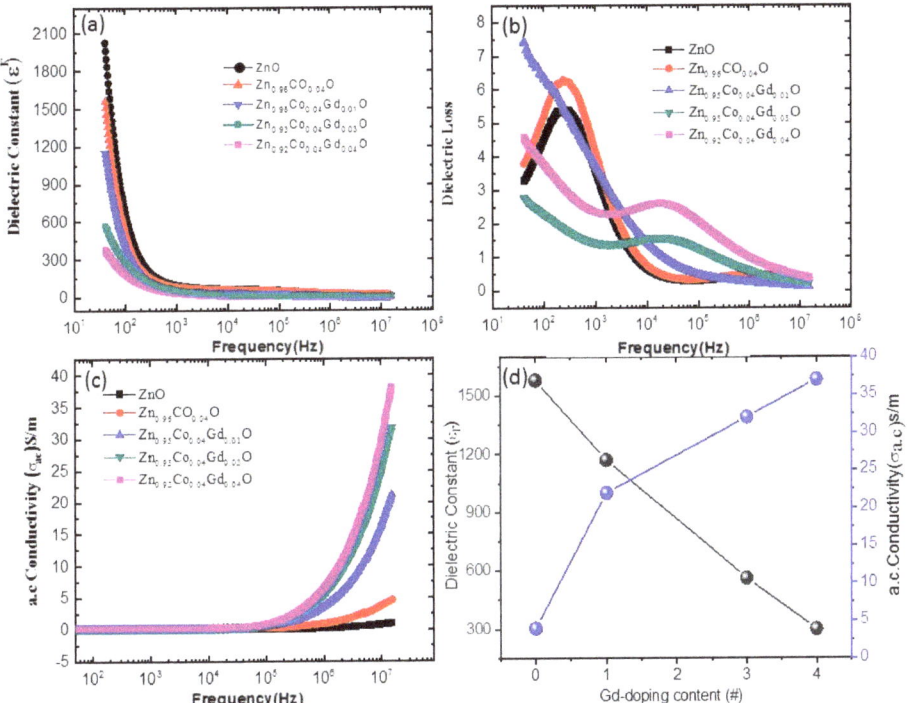

Figure 5. (**a**) Variation of ε_r versus (f) (**b**) the f dependence of ε'' curves, (**c**) the variation of $\alpha_{a.c.}$ with f curves and (**d**) Phase diagram of dielectric constant and conductivity of pure and Gd doped ZnO and (Co (fixed), Gd)-ZnO co-doped with 4% Gd and Co = 1, 3 and 4%.

3.2.3. Electrical Conductivity

Figure 5c illustrates the (f) AC conductivity ($\alpha_{a.c}$), AC, of (Co, Gd) co-doped ZnO samples. The σ_{ac} rises with (f) for each voltage, and these results are consistent from earlier observations [37]. Hence, at low frequencies, σ_{ac} grows gradually as f increases, whereas conductivity rises dramatically at higher frequencies. The hopping idea is responsible for this behavior; for f-dependent σ_{ac}, charge carriers must hop while being transported, yet at low frequency, the σ_{ac} remains constant despite traveling over infinite paths. σ_{ac} follows the relationship shown below for all specimens.

$$\alpha_{ac} = \varepsilon_0 \varepsilon'' \omega \tag{2}$$

where $\alpha_{a.c}$ is the alternating current electrical conductivity, ε'' is the imaginary, ε_r is the dielectric constant of free space, which is a component of the dielectric constant, and $\omega = 2\pi f$ is the frequency. f is the frequency. Equation (2) shows that $\alpha_{a.c}$ is only affected by the ε''. As a result, as f increases, the ε'' lowers but $\alpha_{a.c}$ increases. This conclusion is reliable with prior discoveries, It demonstrated that $\alpha_{a.c}$ increases as (f) increases because of the series resistance effect [38]. To explain this phenomenon, two theories have been proposed: (1) electric energy associated to the high (f) area encouraging charge carrier hopping, and (2) higher dielectric relaxation of ZnO Nanotubes polarization in the high (f) zone. Co-doped ZnO nanopowder is thus an excellent material for high-energy storage devices. Furthermore, the enhanced σ_{ac} discovered may be beneficial for industrial gas sensing applications because to higher electron transmission [39,40]. The phase diagram of the dielectric constant and conductivity vs doping was shown in Figure 5d. It demonstrated that doping promotes dielectric behavior and conductivity (as seen in Figure 4c,d) because the inclusion of impurity ions increases the quantity of free electrons, which aids in conduction.

3.3. Magnetic Properties

Figure 6a displays the magnetic hysteresis (M-H) loops of (Co, Gd) co-doped ZnO NTs observed at ambient temperature (a). At 300 K, the M-H loop in pure ZnO exhibits diamagnetism, whereas samples of $Zn_{0.96}Co_{0.04}O$ and $Zn_{0.96-x}Co_{0.04}Gd_xO$ (Co = 0, 0.01, 0.03, 0.04) show a ferromagnetic (FM) reaction that is clearly visible. Remanent magnetization (M_r) values are 0.0093 emu/g, 0.014 emu/g, 0.0165 emu/g, and 0.0163 emu/g for $Zn_{0.96-x}Co_{0.04}Gd_xO$ (Co = 0, 0.01, 0.03, and 0.04) samples, respectively, according to Figure 6a,b. Table 2 contains the saturation magnetization values, which are 1.02, 1.23, 2.47, and 1.64 (10^{-2} emu/g). The Table displays that the $Zn_{0.92}Co_{0.04}Gd_{0.04}O$ sample's M_r value is higher than that of the literature [24]. Figure 6a shows the change from the paramagnetic to the FM states. M_s readings for specimens of $Zn_{0.96-x}Co_{0.04}Gd_xO$ (Co = 0.03) are higher than those for pure ZnO. When O_2 annealing raises the quantity of Co and Gd ions doped into the host lattice, the number of defects increases. The room temperature ferromagnetism in the specimen with 1% Co co-doped might be attributed to both intrinsic and external magnetic sources. In contrast to extrinsic sources, intrinsic sources require the creation of groups of transition elements or secondary phases.

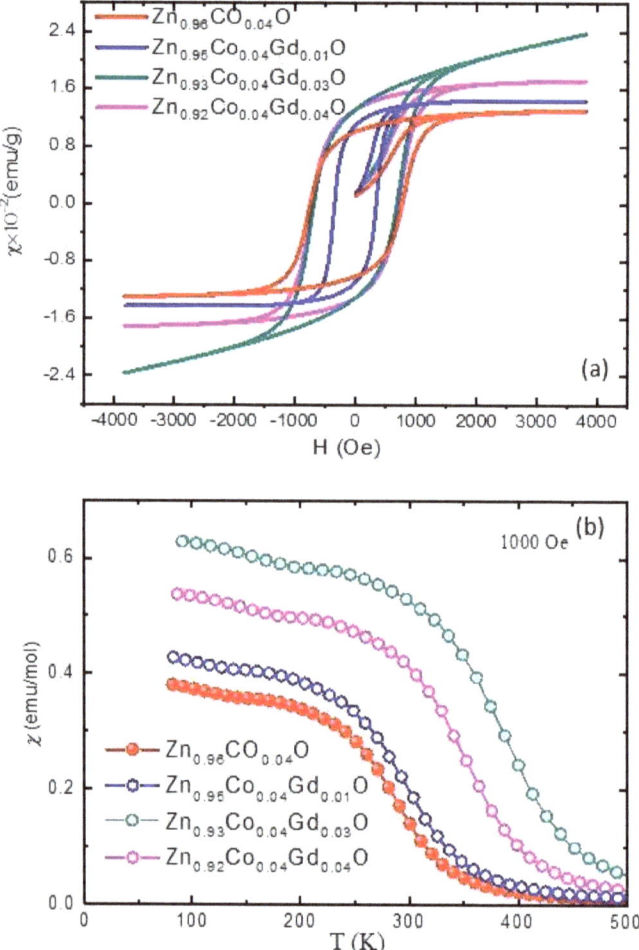

Figure 6. (**a**) The (Gd, Co) co-doped ZnO NTs' magnetic hysteresis (M-H) loops and (**b**) The related temperature-dependent magnetization.

Table 2. The (Gd, Co) co-doped ZnO magnetic NPs' magnetic parameters.

Sample	Remanent Magnetization (M_r) (emu/g)	Coercive Field (Oe)	Saturation Magnetization (M_s) (emu/g)
$Zn_{0.96}Co_{0.04}O$	0.93×10^{-2}	88	1.02×10^{-2}
$Zn_{0.95}Co_{0.04}Gd_{0.01}O$	1.41×10^{-2}	50	1.23×10^{-2}
$Zn_{0.93}Co_{0.04}Gd_{0.03}O$	1.65×10^{-2}	83	2.47×10^{-2}
$Zn_{0.92}Co_{0.04}Gd_{0.04}O$	1.63×10^{-2}	80	1.64×10^{-2}

The schematic 6(b) shows magnetization vs temperature charts in a 10^3 Oe magnetic field to further clarify the magnetic behavior. According to these findings, FM behavior is enhanced in samples of $Zn_{0.96-x}Co_{0.04}Gd_xO$ (Co = 0.03) compared to pure Gd or specimens with higher Co co-doping. It was discovered [39,40] that the earliest stages of FM in TM-ZnO were generated by the interaction of TM ions and bound polarons, which resulted in the creation of bound magnetic polarons. Experiments on defect-bound transporters for point defect hybridization are helpful in the process of developing RTFM in doped ZnO through the use of TM-doping. We demonstrate how O_2 annealing can make the process of replacing Gd and Co in the ZnO lattice more successful. The production of O_2 vacancies during annealing to maintain charge balance is what causes room temperature ferromagnetism in $Zn_{0.96-x}Co_{0.04}Gd_xO$ Nanotubes. This is accomplished by doping Gd^{3+} and Co^{2+} for Zn^{2+} and O_2. The method of preparing the specimens introduced these intriguing new phenomena.

The lower 350 K T_c is discernible in the $\chi(T)$ Co = 0.03 as shown in Figure 6c. It is possible that the structural features of the Co co-doped ZnO samples are attributable to the observed reduction in the magnetic moment for Co with higher dependency on the ZnO sample. The alterations in cell characteristics demonstrate that as cobalt content rises, lattice constants rise as well, leading to an increase in unit cell volume. Because of the increased volume of the unit cell, the cobalt ions that are closest to one another in the ZnO matrix are now further apart. This causes antiferromagnetic super-exchange interactions between the neighboring Co ions, which raises the magnetic moment. Between 2 and 300 K, temperature has an impact on the inverse magnetic susceptibility of nanoparticles. The inverse susceptibility is found to decrease linearly until 380 K, at which point it deviates from the Curie-Weiss line. As shown in Equation (3) below, the modified Curie-Weiss equation provides a close approximation of the susceptibility of the nanoparticles that were utilized in the course of our research.

$$\chi(T) = \chi_0 + (1/8)\mu_{eff}^2 \, x/T - \theta_c \qquad (3)$$

where χ_0 demonstrates the temperature-independent susceptibility, C the Curie constant and θ_C is the Curie–Weiss temperature, x is the concentration of Co ions and μ_{eff} is the effective moment. The calculated Curie–Weiss law parameters i.e., effective moment μ_{eff} = 2.45 μ_B and θ = −25 K. The effective moment values, which range from 2.46 to 2.49 μ_B., provide insight into the substitute integral j, an evaluation of the degree to which the magnetic ions interact with one another, but the results for the substituted $Zn_{0.96-x}Co_{0.04}Gd_xO$ samples (x = 0.0, 0.1, 0.3, and 0.4) drop as the concentration of Co co-doped increases in the sample. The fact that has negative values suggests that the magnetic dopants only have a weak antiferromagnetic interaction with one another [41,42]. The values for most other oxide compounds range from 24 to 27 K, which is nearly as low as the value for Co-doped ZnO. Magnetic characteristics discovered at low temperatures may be the outcome of manufacturing defects or the presence of impurity phases. Further research is needed to clarify this point.

4. Conclusions

This work effectively produced ZnO and ferromagnetic (Co, Gd) co-doped ZnO NTs. All of the (Co, Gd) co-doped samples had a tetragonal structure, according to XRD observations. The dielectric loss, dielectric constant and electrical conductivity rise when the concentration of either (f) or Gd dopant increases. Electrical conductivity was noticed to be increased in annealed Co and Gd-doped ZnO samples. This happened when Zn ions were substituted for Co and Gd ions, which boosted the number of charge carriers available. When ZnO was co-doped with (Co, Gd), the switch from diamagnetic to ferromagnetic caused a significant modification in the hysteresis loop. The increased O_2 vacancies and zinc interstitials are substantially related to the improved magnetic and dielectric responses of the samples. According to our research, ferromagnetism in ZnO Nanotubes can be generated via adding Co fixed and limiting Gd and annealing the material at 400 °C. The magnetic analysis has showed that the magnetic behavior of the as prepared ZnO:Co, Gd samples is dominated by a paramagnetic component over ferromagnetic component, which is an indicative of dominant uncoupled Co spins. As a result, it is thought that raising the ion concentrations to improve direct ion coupling may destabilise the polarons structure, weakening the irreversible magnetic behaviour.

Author Contributions: This paper was written collaboratively by M.A.M., R.K., S.A.O., K.A., S.S.A., N.R., M.S., S.I. and R.K., created the idea and submitted the paper. All authors have read and agreed to the published version of the manuscript.

Funding: The authors acknowledge the financial support from Taif University for sponsorship and support. We also acknowledge the support and guidance provided by Taif University.

Data Availability Statement: The datasets generated during and/or analyzed during the current study are available from the corresponding author on reasonable request. Also, the manuscript experimental work and wording is original. No part of the was manuscript found plagiarized. If the reviewer further needs the proof, then we will send the plagiarized proof.

Conflicts of Interest: The authors declare that they have no known competing financial interest or personal relationship that could have influenced the work reported in this paper.

References

1. Hao, Y.; Lou, S.; Zhou, S.; Wang, Y.; Chen, X.; Zhu, G.; Yuan, R.; Li, N. Novel magnetic behavior of Mn-doped ZnO hierarchical hollow spheres. *J. Nanoparticle Res.* **2012**, *14*, 659. [CrossRef]
2. Prinz, G.A. Magnetoelectronics. *Science* **1998**, *282*, 1660–1663. [CrossRef] [PubMed]
3. Khan, R.; Althubeiti, K.; Zulfiqar Afzal, A.M.; Rahman, N.; Fashu, S.; Zhang, W.; Khan, A.; Zheng, R. Structure and magnetic properties of (Co, Ce) co-doped ZnO-based diluted magnetic semiconductor nanoparticles. *J. Mater. Sci. Mater. Electron.* **2021**, *32*, 24394–24400. [CrossRef]
4. Li, P.; Wang, S.; Li, J.; Wei, Y. Structural and optical properties of Co-doped ZnO nanocrystallites prepared by a one-step solution route. *J. Lumin.* **2012**, *132*, 220–225. [CrossRef]
5. Khan, R.; Tirth, V.; Ali, A.; Irshad, K.; Rahman, N.; Algahtani, A.; Sohail, M.; Isalm, S. Effect of Sn-doping on the structural, optical, dielectric and magnetic properties of ZnO nanoparticles for spintronics applications. *J. Mater. Sci. Mater. Electron.* **2021**, *32*, 21631–21642. [CrossRef]
6. Wang, X.; Zhu, L.; Zhang, L.; Jiang, J.; Yang, Z.; Ye, Z.; He, B. Properties of Ni doped and Ni–Ga co-doped ZnO thin films prepared by pulsed laser deposition. *J. Alloy. Compd.* **2011**, *509*, 3282–3285. [CrossRef]
7. Hao, Y.-M.; Lou, S.-Y.; Zhou, S.-M.; Yuan, R.-J.; Zhu, G.-Y.; Li, N. Structural, optical, and magnetic studies of manganese-doped zinc oxide hierarchical microspheres by self-assembly of nanoparticles. *Nanoscale Res. Lett.* **2012**, *7*, 100. [CrossRef]
8. Stroppa, A.; Duan, X.; Peressi, M. Structural and magnetic properties of Mn-doped GaAs (1 1 0) surface. *Mater. Sci. Eng. B* **2006**, *126*, 217–221. [CrossRef]
9. Das, T.K.; Poater, A. Review on the use of heavy metal deposits from water treatment waste towards catalytic chemical syntheses. *Int. J. Mol. Sci.* **2021**, *22*, 13383. [CrossRef]
10. Das, T.K.; Das, N.C. Preparation of 1D, 2D, and 3D nanomaterials for water treatment. In *Nano-Enabled Technologies for Water Remediation*; Elsevier: Amsterdam, The Netherlands, 2022; pp. 1–22.
11. Fu, J.; Ren, X.; Yan, S.; Gong, Y.; Tan, Y.; Liang, K.; Du, R.; Xing, X.; Mo, G.; Chen, Z.; et al. Synthesis and structural characterization of ZnO doped with Co. *J. Alloys Compd.* **2013**, *558*, 212–221. [CrossRef]

12. Xu, X.; Cao, C. Structure and ferromagnetic properties of Co-doped ZnO powders. *J. Magn. Magn. Mater.* **2009**, *321*, 2216–2219. [CrossRef]
13. Szwacki, N.G.; Majewski, J.; Dietl, T. Aggregation and magnetism of Cr, Mn, and Fe cations in GaN. *Phys. Rev. B* **2011**, *83*, 184417.
14. Lu, Z.; Hsu, H.-S.; Tzeng, Y.; Huang, J.-C. Carrier-mediated ferromagnetism in single crystalline (Co, Ga)-codoped ZnO films. *Appl. Phys. Lett.* **2009**, *94*, 152507. [CrossRef]
15. Sun, L.; Yan, F.; Zhang, H.; Wang, J.; Wang, G.; Zeng, Y.; Li, J. Room-temperature ferromagnetism and in-plane magnetic anisotropy characteristics of nonpolar GaN: Mn films. *Appl. Surf. Sci.* **2009**, *255*, 7451–7454. [CrossRef]
16. Husnain, G.; Tao, F.; Yao, S.-D. Structural and magnetic properties of Co+ implanted n-GaN dilute magnetic semiconductors. *Phys. B Condens. Matter* **2010**, *405*, 2340–2343. [CrossRef]
17. Cui, Z.; Wu, H.; Bai, K.; Chen, X.; Li, E.; Shen, Y.; Wang, M. Fabrication of a g-C_3N_4/MoS_2 photocatalyst for enhanced RhB degradation. *Phys. E Low-Dimens. Syst. Nanostructures* **2022**, *144*, 115361. [CrossRef]
18. Cui, Z.; Yang, K.; Ren, K.; Zhang, S.; Wang, L. Adsorption of metal atoms on $MoSi_2N_4$ monolayer: A first principles study. *Mater. Sci. Semicond. Process.* **2022**, *152*, 107072. [CrossRef]
19. Cui, Z.; Zhang, S.; Wang, L.; Yang, K. Optoelectronic and magnetic properties of transition metals adsorbed Pd_2Se_3 monolayer. *Micro Nanostructures* **2022**, *167*, 207260. [CrossRef]
20. Zhang, L.; Cui, Z. Electronic, Magnetic, and Optical Performances of Non-Metals Doped Silicon Carbide. *Front. Chem.* **2022**, *10*, 898174. [CrossRef]
21. Fabbiyola, S.; Kennedy, L.J.; Aruldoss, U.; Bououdina, M.; Dakhel, A.; JudithVijaya, J. Synthesis of Co-doped ZnO nanoparticles via co-precipitation: Structural, optical and magnetic properties. *Powder Technol.* **2015**, *286*, 757–765. [CrossRef]
22. Hao, H.; Qin, M.; Li, P. Structural, optical, and magnetic properties of Co-doped ZnO nanorods fabricated by a facile solution route. *J. Alloys Compd.* **2012**, *515*, 143–148. [CrossRef]
23. Oshio, T.; Masuko, K.; Ashida, A.; Yoshimura, T.; Fujimura, N. Effect of Mn doping on the electric and dielectric properties of ZnO epitaxial films. *J. Appl. Phys.* **2008**, *103*, 093717. [CrossRef]
24. Khan, R.; Rahman, M.-U.; Fashu, S. Effect of annealing temperature on the dielectric and magnetic response of (Co, Zn) co-doped SnO_2 nanoparticles. *J. Mater. Sci. Mater. Electron.* **2017**, *28*, 2673–2679. [CrossRef]
25. Liu, C.; Meng, D.; Pang, H.; Wu, X.; Xie, J.; Yu, X.; Chen, L.; Liu, X. Influence of Fe-doping on the structural, optical and magnetic properties of ZnO nanoparticles. *J. Magn. Magn. Mater.* **2012**, *324*, 3356–3360. [CrossRef]
26. Saleh, R.; Prakoso, S.P.; Fishli, A. The influence of Fe doping on the structural, magnetic and optical properties of nanocrystalline ZnO particles. *J. Magn. Magn. Mater.* **2012**, *324*, 665–670. [CrossRef]
27. Khan, R.; Zaman, Y. Effect of annealing on structural, dielectric, transport and magnetic properties of (Zn, Co) co-doped SnO_2 nanoparticles. *J. Mater. Sci. Mater. Electron.* **2016**, *27*, 4003–4010. [CrossRef]
28. Khan, R.; Fashu, S.; Rahman, M.-U. Effects of Ni co-doping concentrations on dielectric and magnetic properties of (Co, Ni) co-doped SnO_2 nanoparticles. *J. Mater. Sci. Mater. Electron.* **2016**, *27*, 7725–7730. [CrossRef]
29. Khan, R.; Fashu, S. Effect of annealing on Ni-doped ZnO nanoparticles synthesized by the co-precipitation method. *J. Mater. Sci. Mater. Electron.* **2017**, *28*, 10122–10130. [CrossRef]
30. Cong, C.; Liao, L.; Liu, Q.Y.; Li, J.C.; Zhang, K.L. Effects of temperature on the ferromagnetism of Mn-doped ZnO nanoparticles and Mn-related Raman vibration. *Nanotechnology* **2006**, *17*, 1520. [CrossRef]
31. Khan, R.; Fashu, S.; Zaman, Y. Magnetic and dielectric properties of (Co, Zn) co-doped SnO_2 diluted magnetic semiconducting nanoparticles. *J. Mater. Sci. Mater. Electron.* **2016**, *27*, 5960–5966. [CrossRef]
32. Khan, R.; Zulfiqar; Fashu, S.; Rehman, Z.U.; Khan, A.; Rahman, M.U. Structure and magnetic properties of (Co, Mn) co-doped ZnO diluted magnetic semiconductor nanoparticles. *J. Mater. Sci. Mater. Electron.* **2018**, *29*, 32–37.
33. Lv, X.; Zhang, N.; Ma, Y.; Zhang, X.-X.; Wu, J. Coupling effects of the A-site ions on high-performance potassium sodium niobate ceramics. *J. Mater. Sci. Technol.* **2022**, *130*, 198–207. [CrossRef]
34. Khan, R.; Zulfiqar; Rahman, M.-U.; Rehman, Z.; Fashu, S. Effect of air annealing on the structure, dielectric and magnetic properties of (Co, Ni) co-doped SnO2 nanoparticles. *J. Mater. Sci. Mater. Electron.* **2016**, *27*, 10532–10540. [CrossRef]
35. Szu, S.-P.; Lin, C.-Y. AC impedance studies of copper doped silica glass. *Mater. Chem. Phys.* **2003**, *82*, 295–300. [CrossRef]
36. Gu, F.; Wang, S.F.; Lv, M.K.; Zhou, G.J.; Xu, D.; Yuan, D.R. Photoluminescence properties of SnO_2 nanoparticles synthesized by sol−gel method. *J. Phys. Chem. B* **2004**, *108*, 8119–8123. [CrossRef]
37. Pakma, O.; Serin, N.; Serin, T.; Altındal, Ş. Influence of frequency and bias voltage on dielectric properties and electrical conductivity of Al/TiO_2/p-Si/p+ (MOS) structures. *J. Phys. D Appl. Phys.* **2008**, *41*, 215103. [CrossRef]
38. Elilarassi, R.; Chandrasekaran, G. Synthesis and characterization of ball milled Fe-doped ZnO diluted magnetic semiconductor. *Optoelectron. Lett.* **2012**, *8*, 109–112. [CrossRef]
39. Lin, Y.; Jiang, D.; Lin, F.; Shi, W.; Ma, X. Fe-doped ZnO magnetic semiconductor by mechanical alloying. *J. Alloys Compd.* **2007**, *436*, 30–33. [CrossRef]
40. Yin, S.; Xu, M.X.; Yang, L.; Liu, J.F.; Rösner, H.; Hahn, H.; Gleiter, H.; Schild, D.; Doyle, S.; Liu, T.; et al. Absence of ferromagnetism in bulk polycrystalline $Zn_{0.9}Co_{0.1}O$. *Phys. Rev. B* **2006**, *73*, 224408. [CrossRef]

41. Fukumura, T.; Jin, Z.; Kawasaki, M.; Shono, T.; Hasegawa, T.; Koshihara, S.; Koinuma, H. Magnetic properties of Mn-doped ZnO. *Appl. Phys. Lett.* **2001**, *78*, 958–960. [CrossRef]
42. Yang, L.; Wu, X.L.; Huang, G.S.; Qiu, T.; Yang, Y.M. In situ synthesis of Mn-doped ZnO multileg nanostructures and Mn-related Raman vibration. *J. Appl. Phys.* **2005**, *97*, 014308. [CrossRef]

Disclaimer/Publisher's Note: The statements, opinions and data contained in all publications are solely those of the individual author(s) and contributor(s) and not of MDPI and/or the editor(s). MDPI and/or the editor(s) disclaim responsibility for any injury to people or property resulting from any ideas, methods, instructions or products referred to in the content.

Article

Processing and Investigation of $Cd_{0.5}Zn_{0.5}Fe_{2-x}Cr_xO_4$ ($0 \leq x \leq 2$) Spinel Nanoparticles

Reem Khalid Alharbi [1,*], Noura Kouki [1,*], Abdulrahman Mallah [1], Lotfi Beji [2,*], Haja Tar [1], Azizah Algreiby [1], Abrar S. Alnafisah [1] and Sobhi Hcini [3]

1. Department of Chemistry, College of Science, Qassim University, Buraydah 51452, Saudi Arabia; a.mallah@qu.edu.sa (A.M.); h.tar@qu.edu.sa (H.T.); grieby@qu.edu.sa (A.A.); alnafisaha@qu.edu.sa (A.S.A.)
2. Department of Physics, College of Sciences and Arts at ArRass, Qassim University, Buraidah 51452, Saudi Arabia
3. Faculty of Science and Technology of Sidi Bouzid, University Campus Agricultural City, University of Kairouan, Sidi Bouzid 9100, Tunisia; hcini_sobhi@yahoo.fr
* Correspondence: 411200348@qu.edu.sa (R.K.A.); n.kouki@qu.edu.sa (N.K.); l.beji@qu.edu.sa (L.B)

Abstract: This study presents the synthesis of $Cd_{0.5}Zn_{0.5}Fe_{2-x}Cr_xO_4$ nanoparticles via the sol–gel method, along with a comprehensive characterization of their morphological, structural, infrared, and magnetic properties. The X-ray diffraction pattern confirms the formation of the spinel structure, and the cation distribution is estimated using X-ray analysis and confirmed by magnetization measurements. The crystalline size, ranging from 152 to 189 nm, and lattice parameter, varying from 8.51134 Å to 8.42067 Å, decrease with increasing Cr content. The saturation magnetization decreases from 55 emu/g to 10.8 emu/g, while the remanent magnetization increases (3.5 emu/g $\leq M_r \leq$ 6.27 emu/g), and the coercivity increases (82 Oe $\leq H_C \leq$ 422.15 Oe) with the addition of Cr ions. Fourier transform infrared (FTIR) spectroscopy reveals two absorption bands at ν_1 and ν_2, located near 600 and 400 cm^{-1}, respectively, which correspond to the vibrations of the metal–oxygen bonds in the spinel structure.

Keywords: spinels; sol–gel method; XRD; FTIR; SEM; magnetic properties

1. Introduction

Ferrites with spinel structures of MFe_2O_4 (M = Cd, Zn, Ni, and Co) are among the most extensively investigated oxides in recent years. Physico-chemical investigations of these materials have drawn upon various disciplines, including magnetism, optics, electronics, and mechanics. These materials, which can exist as nanoparticles, aggregates, and nanostructured powders consisting of grains separated by grain boundaries, offer distinct advantages for manipulation and utilization in various applications, such as recording heads, antenna rods, loading coils, microwave devices, and core materials for power transformers in electronics and telecommunication applications [1–3].

Numerous synthesis techniques, such as electrochemical [4], hydrothermal [5], co-precipitation [6], sol–gel [7], plasma synthesis [8], citrate precursor [9], and reverse micelle [10] techniques, have been developed to produce ferrite materials. Among these methods, the sol–gel route is an efficient technique due to its simplicity and ability to regulate the properties of the final product, leading to a homogeneous material with a stoichiometric composition and nanoscale grain size [11–14]. The sol–gel method enables the adjustment of various parameters to enhance the physical and chemical characteristics of spinel ferrites, including pH, citric acid content, calcination temperature, and grain size [11,12].

Scientists can change ferrite materials by adding different ions or using different processes to make them better for specific uses. For example, the substitution of Cr^{3+} can improve magnetic properties like remanence magnetization and coercivity, which are essential for technology [15]. When natural chromite materials form in an environment with oxygen, they can mix Fe^{2+} and Fe^{3+} in different places, affecting their properties [16]. Lee and other researchers have studied how magnetic properties change when Cr^{3+} is substituted and found that magnetic moment and Curie temperature decrease with this substitution [17]. Other researchers have also looked at what happens when Fe^{3+} is replaced by Cr^{3+} [18,19]. The effect of Al^{3+} and Cr^{3+} substitution in cobalt ferrite has also been reported [20,21]. It was observed that substituting Al^{3+} and Cr^{3+} ions in the cobalt ferrite lattice leads to a decrease in the saturation magnetization values due to the lower magnetic moment of Al^{3+} and Cr^{3+} compared to Fe^{3+} ions.

Due to their potential applications, the Cd-Zn ferrites have gained significant attention in nanoscience and nanotechnology [22]. Various compositions of Cd-Zn ferrites have been extensively investigated and characterized in the literature [23–27], showing intriguing electrical, magnetic, and optical properties. Enhancing their properties is of great interest given the numerous applications of Cd-Zn ferrites. To this end, mixed Cd-Zn ($Cd_{0.5}Zn_{0.5}Fe_{2-x}Cr_xO_4$ with x ranging from 0 to 2) ferrites substituted with Cr were examined in this work. Equal concentrations of Cd and Zn were maintained. Cr^{3+} ions preferentially occupy the octahedral B-sites in $Cd_{0.5}Zn_{0.5}Fe_{2-x}Cr_xO_4$ ferrites, leading to their selection. The substitution of Fe^{3+} ions by Cr^{3+} ions with a different ionic radius alters the crystal geometry and modifies the materials' magnetic and dielectric characteristics. In this study, we report the synthesis of $Cd_{0.5}Zn_{0.5}Fe_{2-x}Cr_xO_4$ ($0 \leq x \leq 2$) samples using the sol–gel method and their morphological and structural characterization. In addition, the infrared and magnetic properties at room temperature were investigated. Our results showed that the prepared $Cd_{0.5}Zn_{0.5}Fe_{2-x}Cr_xO_4$ materials maintained a regular spinel cubic structure. These samples present several advantages, such as their good infrared and magnetic properties, low cost, and, above all, their easy synthesis. These features make the $Cd_{0.5}Zn_{0.5}Fe_{2-x}Cr_xO_4$ spinels a good candidate for magnetic devices and can be studied in perspective for other potential applications.

2. Experimental Section

2.1. Materials Synthesis

Cadmium, zinc, iron, and chromium nitrates were precursors to synthesize $Cd_{0.5}Zn_{0.5}Fe_{2-x}Cr_xO_4$ ($0 \leq x \leq 2$) nanoparticles. Stoichiometric amounts of the nitrates were weighed and dissolved in distilled water, which was heated to 90 °C. The metal cations were complexed with citric acid, which was added to each solution. Next, the pH was adjusted to around seven by adding ammonia to the solutions. Ethylene glycol, a polymerization agent, was added at this stage. After approximately 4 h, a viscous liquid (gel) began to form. To create a soft powder, the magnetic stirring temperature gradually increased to 250 °C. After grinding and annealing in the air for 12 h, the powders were subjected to an annealing temperature range of 700 °C to 1200 °C. All characterizations of $Cd_{0.5}Zn_{0.5}Fe_{2-x}Cr_xO_4$ spinels annealed at 1200 °C are presented in this study.

2.2. Materials Characterization Technics

The samples' X-ray diffraction (XRD) patterns were collected using the "Panalytical X'Pert Pro System" diffractometer, operating at a copper wavelength of 1.5406 Å. The measurements ranged from 10° to 80° with a step size of 0.02° and a counting period of 18 s per step. The morphology of the materials in the form of pellets was studied using Philips XL 30 scanning electron microscopy (SEM) equipped with an electron gun and a 15 kV accelerating voltage. The FTIR spectra in a wavenumber range of 400–1000 cm^{-1} were recorded using a Shimadzu Fourier Transform Infrared Spectrophotometer (FTIR-8400S).

3. Results and Discussions

3.1. SEM Micrographs

The samples were characterized using scanning electron microscopy (SEM). The resulting images and their corresponding grain size distributions are shown in Figure 1a–e.

The microscopic structure and morphology of $Cd_{0.5}Zn_{0.5}Fe_{2-x}Cr_xO_4$ with Cr substitution were also investigated. The SEM images revealed that the synthesized materials comprised an irregularly shaped group of tailed grains with a non-uniform grain size distribution. The particles exhibited a prismatic and pyramidal morphology.

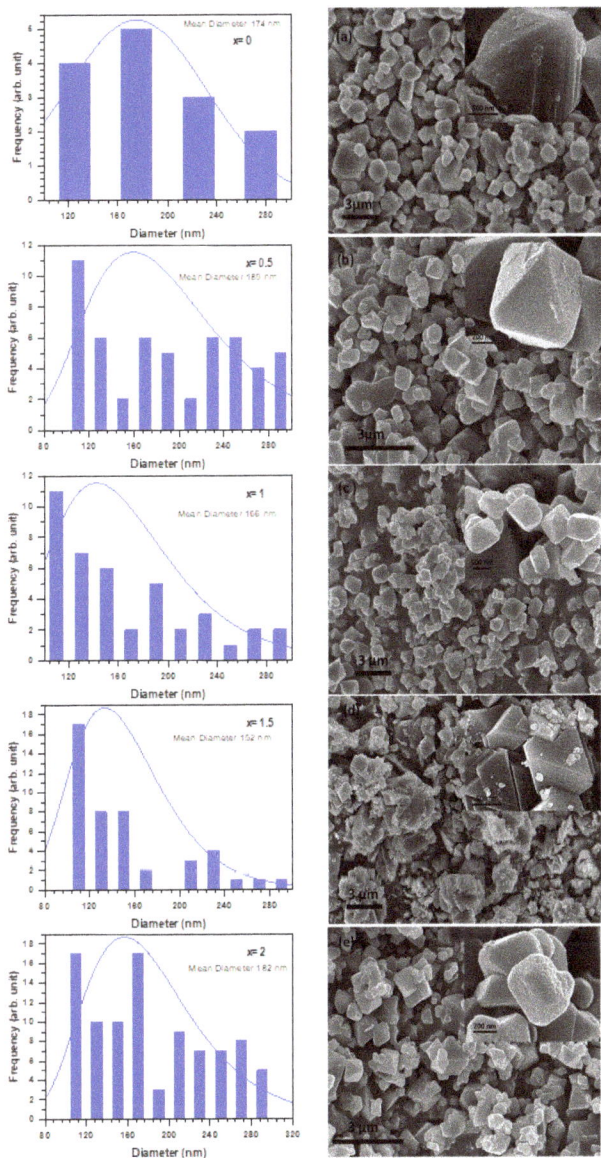

Figure 1. SEM micrographs and particle size distributions of $Cd_{0.5}Zn_{0.5}Fe_{2-x}Cr_xO_4$ spinels, elaborated by sol–gel method. Images labeled (**a–e**) correspond to x = 0, 0.5, 1, 1.5, and 2 Cr compositions, respectively. The inset images are the higher magnifications of micrographs.

Moreover, they were non-uniformly distributed, agglomerated, and inhomogeneous. Some massive particles were observed, along with smaller particles and increased agglomeration. All samples' average grain size values varied from 152 nm to 189 nm and were found to be random with a high Cr content [28,29].

3.2. Structural Properties and Cation Distributions

The XRD patterns of $Cd_{0.5}Zn_{0.5}Fe_{2-x}Cr_xO_4$ ferrites in Figure 2 reveal that a cubic spinel structure occurs in one phase. All diffraction lines are indexed in the cubic spinel structure, indicating the well-crystalline nature of the compounds. In addition, the absence of any reflection peak related to secondary phases confirms the purity of the samples. The cation distribution in the system was determined based on previous studies [30]. Mössbauer spectroscopic investigations have determined the cation distribution in ferrites with the general formula AB_2O_4. An investigation of Cd-Zn ferrites [24] and Cr-substituted ferrites [25] revealed that the tetrahedral A-sites were preferably occupied by Cd^{2+} and Zn^{2+} ions. In contrast, Cr^{3+} ions are distributed over the octahedral B-sites, and Fe^{3+} ions are distributed over both sites. This cation distribution has been confirmed in other studies [31,32]. Hence, the Rietveld refinement for $Cd_{0.5}Zn_{0.5}Fe_{2-x}Cr_xO_4$ samples was performed using the $\left(Cd_{0.5}^{2+}Zn_{0.5}^{2+}\right)_A [Fe_{2-x}^{3+}Cr_x^{3+}]_B O_4^{2-}$ cation distribution model. In this cation distribution, the A-sites are completely occupied by both Cd^{2+} and Zn^{2+} cations with equal concentrations (50 atom%). Hence, in the case of our samples, the Fe^{3+} and Cr^{3+} ions are distributed only over the octahedral B-sites. Furthermore, this cation distribution model confirms the absence of the inversion phenomenon and the non-occupation of the A-site by Fe^{3+} cations [33]. Figure 3 shows a typical example of the Rietveld refinement of $Cd_{0.5}Zn_{0.5}Fe_{2-x}Cr_xO_4$ spinel (x = 2). Table 1 outlines the various properties of the prepared compositions. The reliability factors (Bragg R_{Bragg}, profile R_p, experimental R_{exp}, and weighted profile R_{wp}) are all less than 10% in all cases. Rietveld fittings tend to be good, as shown by the $\chi^2 = R_{wp}/R_{exp}$ (goodness of fit) tendency towards unity. As a result, the refined occupancy factors for (Cd/Zn) and (Fe/Cr) at the A- and B-sites corresponded with the nominal values, supporting the suggested hypothesis. According to Table 1 and Figure 4, the decrease in lattice constant (a) and volume (V) appears to be caused by the replacement of a smaller radius of the Cr^{3+} ($r_{Cr}^{3+} = 0.63 Å$) ion for the Fe^{3+} ion radius ($r_{Fe}^{3+} = 0.67 Å$) [34]. Moreover, other Cr-doped ferrites have shown similar reductions in lattice parameters [35]. Furthermore, the atomic positions of oxygen exhibit the characteristic features of the spinel structure [13]. Alternatively, the cation–oxygen bond at the octahedral sites (d_{B-O}) is shorter with Cr substitution because of the decrease in the average ionic radius of the B-site $<r_B>$. Since the ionic radius of the A-site ($<r_A>$) remains the same, the length of the cation–oxygen bonds (d_{A-O}) remains almost constant. Table 1 also shows the bond angle values (φ_{A-O-B}) associated with A-O-B interactions in the produced samples. The bond angle for A-O-B is greater than that of B-O-B, according to Table 1. Thus, A-B exchange interactions are more potent than B-B exchange interactions [36,37]. Furthermore, the observed decrease in the bond angle (φ_{A-O-B}) indicates that A-B exchange interactions become less intense when Cr replacement is conducted. The XRD density was calculated using the following formula [14]:

$$d_x = \frac{8M}{Na^3} \quad (1)$$

where M is the molar mass, a is the cell parameter, and N is the Avogadro number (6.022×10^{23}). Table 1 (also Figure 4) shows that the XRD density increases with Cr substitution. This finding is consistent with previous reports in the literature [38]. The increase in XRD density may be due to the reduction in oxygen vacancies, which significantly impact densification kinetics [29]. It can also be attributed to the dominant effect of the reduction in the lattice parameter compared to the relatively small variation in molar mass resulting from the lower molar mass of Cr^{3+} ions (51.996 g/mol) compared to Fe^{3+} ions (55.847 g/mol).

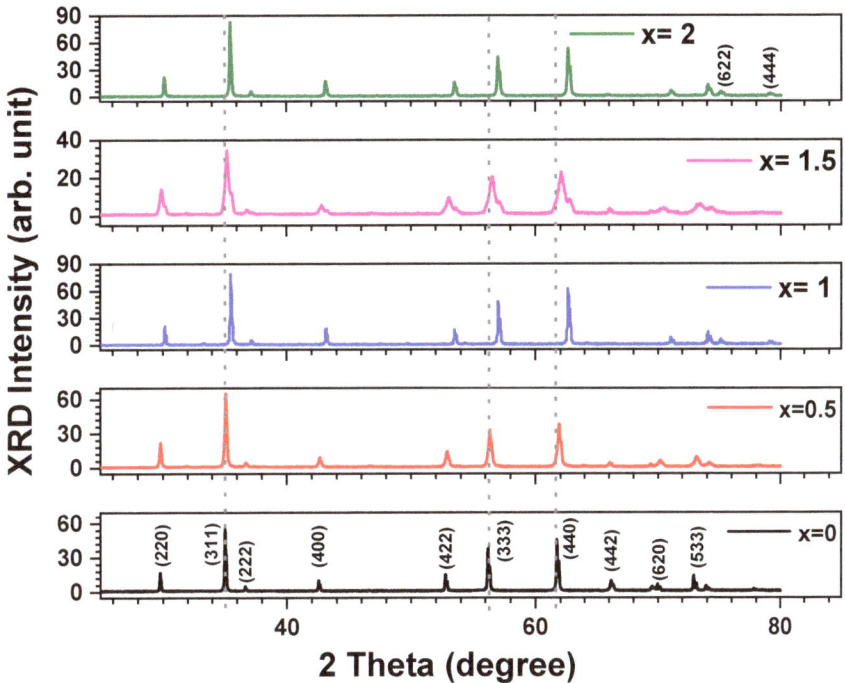

Figure 2. XRD patterns of $Cd_{0.5}Zn_{0.5}Fe_{2-x}Cr_xO_4$ spinels with ($0 \leq x \leq 2$).

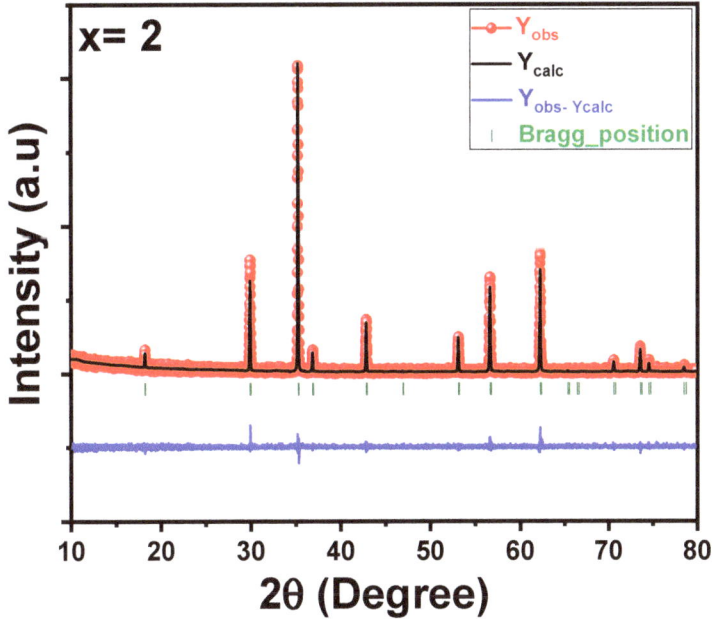

Figure 3. Typical example for the structural refinement of the XRD patterns using the Rietveld method for $Cd_{0.5}Zn_{0.5}Fe_{2-x}Cr_xO_4$ spinels with ($x = 2$).

Table 1. Structural parameters for $Cd_{0.5}Zn_{0.5}Fe_{2-x}Cr_xO_4$ spinels with ($0 \leq x \leq 2$) obtained following the structural refinement by the Rietveld method. a: cell parameter; V: cell volume; B_{iso}: isotropic thermal agitation parameter. Definitions of structural parameters are given in the text.

Cr Content				0	0.5	1	1.5	2
Space group				Fd$\bar{3}$m				
Cell parameters		a (Å)		8.5113 (4)	8.4745 (4)	8.4586 (4)	8.4395 (4)	8.4207 (4)
		V (Å^3)		616.59 (4)	608.61 (5)	605.20 (4)	601.10 (4)	597.09 (4)
Atoms	Tetrahedral A-site (Cd/Zn)	Wyckoff positions		4c	4c	4c	4c	4c
		Site symmetry		−43m	−43m	−43m	−43m	−43m
		Atomic positions	x = y = z	1/8	1/8	1/8	1/8	1/8
		Occupancy factors		0.50 (1)/ 0.49 (1)	0.51 (1)/ 0.50 (1)	0.50 (1)/ 0.50 (1)	0.49 (1)/ 0.50 (1)	0.50 (1)/ 0.50 (1)
		B_{iso} (Å^2)		1.19	1.22	1.35	1.18	1.27
	Octahedral B-site [Fe/Cr]	Wyckoff positions		16d	16d	16d	16d	16d
		Site symmetry		−3m	−3m	−3m	−3m	−3m
		Atomic positions	x = y = z	1/2	1/2	1/2	1/2	1/2
		Occupancy factors		2.01 (2)/ 0	1.51 (2)/0.49 (2)	1.01 (2)/1.02 (2)	0.50 (2)/1.48 (2)	0/2.02(2)
		B_{iso} (Å^2)		1.46	1.14	1.22	0.94	1.34
	O	Wyckoff positions		32e	32e	32e	32e	32e
		Site symmetry		3m	3m	32e	32e	32e
		Atomic positions	x = y = z	0.2553 (1)	0.2551 (8)	0.2548 (8)	0.2545 (8)	0.2541 (8)
		Occupancy factors		4	4	4	4	4
		B_{iso} (Å^2)		1.42	1.54	1.42	1.65	1.58
Structural parameters		d_{A-O} (Å)		1.905 (8)	1.903 (7)	1.901 (9)	1.898 (7)	1.896 (8)
		d_{B-O} (Å)		2.058 (9)	2.053 (7)	2.045 (8)	2.041 (7)	2.036 (7)
		φ_{A-O-B} (°)		124.8 (5)	124.5 (3)	123.7 (4)	123.4 (3)	123.1 (3)
		φ_{B-O-B} (°)		92.4 (5)	91.2 (3)	91.0 (4)	90.8 (3)	90.3 (3)
		d_x (g·cm^{-3})		5.7004	5.7331	5.7443	5.7622	5.7795
Agreement factors		R_p (%)		6.41	5.47	5.44	5.63	5.48
		R_{wp} (%)		8.25	7.52	7.35	7.42	7.25
		R_{exp} (%)		7.14	7.33	7.47	7.12	7.04
		R_{Bragg} (%)		3.83	3.34	3.83	2.94	2.72
		χ^2 (%)		1.13	1.19	1.23	1.32	1.18

The values of the crystallite size (D_{XRD}) and the lattice strain (ε) were determined by the Williamson–Hall method as a function of Cr content. This method, developed by G.K. Williamson and his student W.H. Hall [39], utilizes the full width at half maximum (FWHM) of Bragg peaks ($\Delta\theta$, in radians) and the angle of peak position (θ), as well as the X-ray wavelength (λ = 1.5406 Å), to calculate the average crystallite size (D) and lattice strain (ε). The relationship is given by $\Delta\theta \cos\theta = \frac{k\lambda}{D} + 4\varepsilon\sin\theta$, where k is a constant value (0.94) obtained by assuming the spherical nature of the powders. By plotting $\Delta\theta \times \cos\theta$ versus $4\sin\theta$, the strain component (ε) can be determined from the slope, and the size component can be determined from the intercept ($\frac{k\lambda}{D}$). This plot is known as a Williamson–Hall plot. Figure 5a–e depicts the variations in (($\Delta\theta_{hkl}$)$\cos\theta_{hkl}$) as a function of ($4\sin\theta_{hkl}$) for $Cd_{0.5}Zn_{0.5}Fe_{2-x}Cr_xO_4$ spinels (x = 0; x = 0.5; x = 1; x = 1.5; x = 2). The estimated values of D_{XRD} and ε are (150 nm, 1.93×10^{-4}), (120 nm, 1.12×10^{-3}), (103 nm, 1.32×10^{-4}), (108 nm, 3.10×10^{-3}), and (95 nm, 3.4×10^{-3}) for $Cd_{0.5}Zn_{0.5}Fe_{2-x}Cr_xO_4$ spinels (x = 0; x = 0.5; x = 1; x = 1.5; x = 2), respectively. These results suggest a small variation in crystallite size due to Cr substitution, consistent with the values obtained from SEM analysis. The lattice strain increases while the crystallite size decreases approximately with an increasing Cr content.

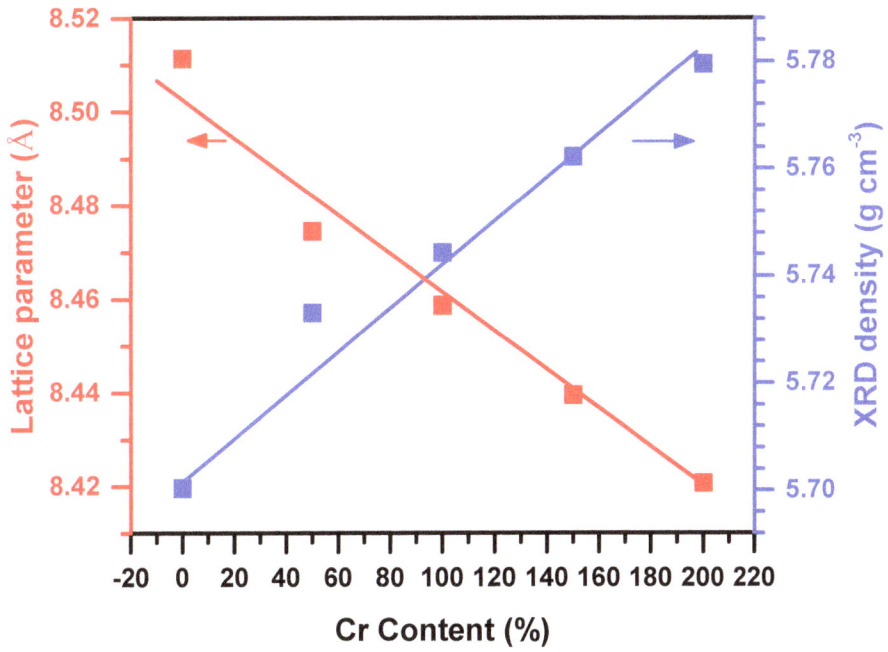

Figure 4. Lattice parameter and XRD density of $Cd_{0.5}Zn_{0.5}Fe_{2-x}Cr_xO_4$ ($0 \leq x \leq 2$) spinels as a function of Cr content.

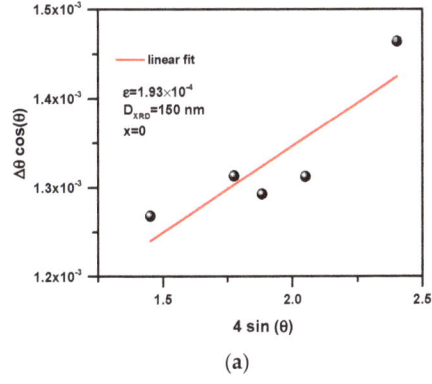

(a)

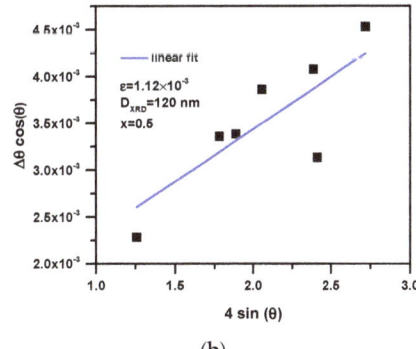

(b)

Figure 5. *Cont.*

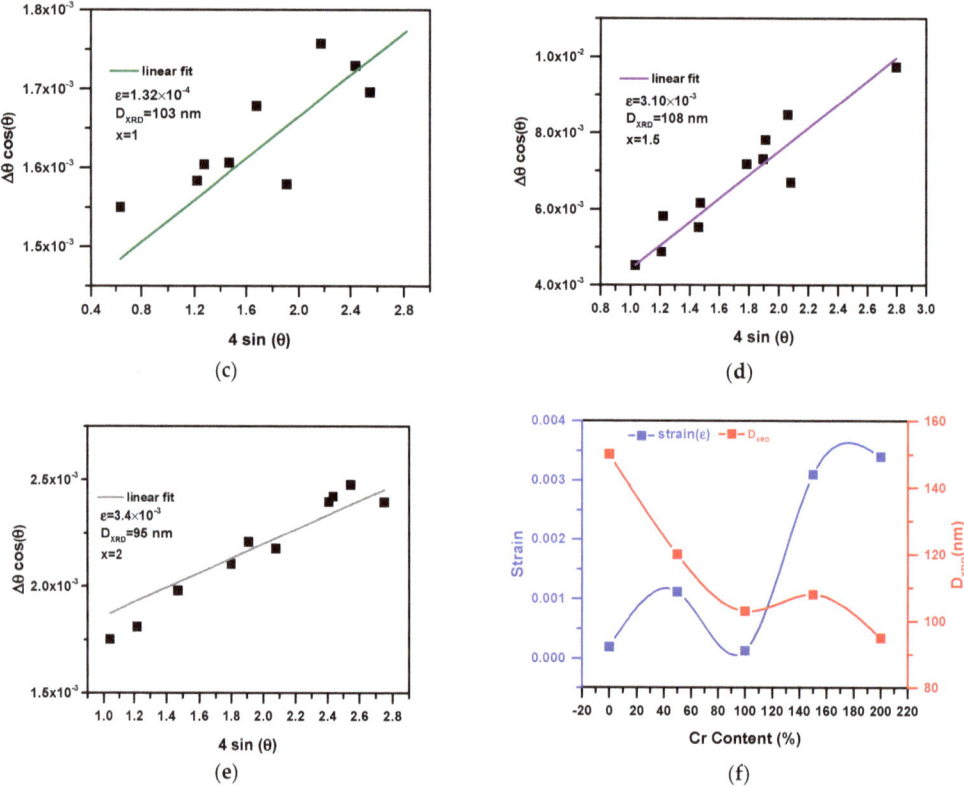

Figure 5. (**a**–**e**) Williamson–Hall plots of (Δθcosθ) vs. (4sinθ) of $Cd_{0.5}Zn_{0.5}Fe_{2-x}Cr_xO_4$ spinels (x = 0; x = 0.5; x = 1; x = 1.5; x = 2). (**f**) Values of the crystallite's size (DXRD) and the lattice strain (ε) calculated using the Williamson–Hall methods.

3.3. FTIR Spectra

The infrared (IR) spectra provide valuable information about the crystal lattice's valence state and vibrational modes. Table 2 presents the band positions obtained from the IR spectra of the $Cd_{0.5}Zn_{0.5}Cr_xFe_{2-x}O_4$ series. Figure 6 shows the IR spectra of this series, with the high-frequency band υ_1 observed in the 524–586 cm^{-1} range and a small band in the low-frequency band υ_2 in the 420–424 cm^{-1} range. These absorption bands indicate the formation of a single-phase spinel structure. The two major absorption bands at υ_1 and υ_2 are due to vibrations of the oxygen bonds with positive ions at A- and B-sites [40]. The small band at low-frequency band υ_2 is constant for all samples except for x = 2, where it disappears. The vibrational bands υ_1 and υ_2 are assigned to intrinsic vibrations of the tetrahedral and octahedral sites, respectively [41].

Table 2. Band positions (υ_1 and υ_2) and force constants (K_O and K_T) of $Cd_{0.5}Zn_{0.5}Fe_{2-x}Cr_xO_4$.

x	υ_1	υ_2	$K_T \times 10^5$ (dyne cm^{-1})	$K_O \times 10^5$ (dyne cm^{-1})
0	524	423	1.86	1.06
0.5	569	424	2.19	1.05
1	584	421	2.31	1.02
1.5	586	420	2.33	1.01
2	586	420	2.33	1.00

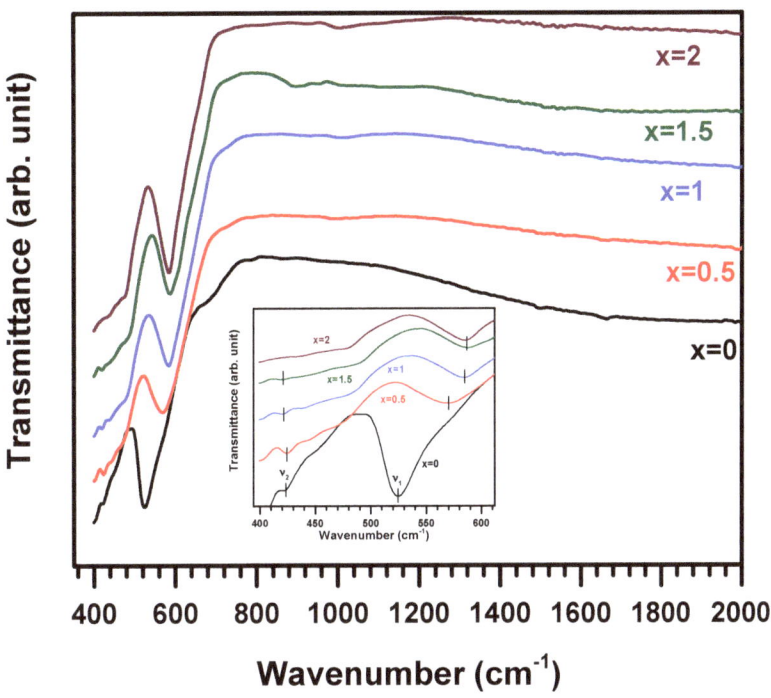

Figure 6. FTIR spectra at room temperature of $Cd_{0.5}Zn_{0.5}Fe_{2-x}Cr_xO_4$ spinel ferrites. Spectra is vertically translated for more clarity details. The inset figure is the higher magnification of the region between 400 and 600 cm^{-1} represented in semi-log scale.

The infrared spectra of the Cr-substituted ferrite system prepared through the ceramic route exhibit similar features, as reported in the literature [42]. The intensity of the absorption band corresponding to the tetrahedral complex (υ_1) increases and shifts towards a higher frequency with an increased Cr content, while the octahedral complex (υ_2) exhibits weaker absorption bands. This behavior can be attributed to the first selection rule, which states that transitions between d orbitals in a complex with a center of symmetry are forbidden. As the tetrahedral complex possesses a center of symmetry, its absorption bands are more intense than those of the octahedral complex, which lacks a center of symmetry and thus allows more transitions to occur between d orbitals [43].

The observed shift in the band position in the IR spectra is attributed to the change in the Fe^{3+}-O_2^{2-} distance for the tetrahedral and octahedral complexes. The slight frequency change in band υ_2 and the significant shift of band υ_1 towards a higher frequency are due to the substitution of Cr^{3+} ions, which replace Fe^{3+} ions only at the octahedral B-site, leading to no significant change in the size of the octahedral site. As the Fe_B^{3+}-O_2^{2-} complex numbers decrease, metal–oxygen vibrational energies increase, prompting a decrease in the Fe_B^{3+}-O_2^{2-} intermolecular distance. This phenomenon is observed due to the increased number of Cr^{3+}-O_2^{2-} complexes [38] and the creation of $Me^{3+}O_2^{2-}$ complexes at A-sites. As Cd^{2+}-O^{2-} and Zn^{2+}-O^{2-} bonds are stretched at the A-sites and Fe^{3+}-O^{2-} and Cr^{3+}-O^{2-} bonds are stretched at the B-sites, these bands are produced. The two bands may exhibit different positions for various reasons, including differences in ionic radius, the average distance between metal and oxygen, and electronegativity. It has been found that similar results have been obtained for other ferrite systems [44–46]. Assuming that the other independent parameters are constant, the force constant would be the second derivative of the potential energy based on the site radius. Based on Waldron's method [41], we calculated force constants for tetrahedral and octahedral sites. For each site, Waldron gives the force constants K_T and K_O as follows:

$$K_T = 7.62 M_1 v_1^2 10^{-3} \left(\frac{dyne}{cm}\right) \quad (2)$$

$$K_O = 10.62 \left(\frac{M_2}{2}\right) v_2^2 10^{-3} \left(\frac{dyne}{cm}\right) \quad (3)$$

Assuming that M_1 and M_2 refer to the molecular weight of the cations at sites A and B, respectively. Based on the cation distribution for the prepared samples, tetrahedral M_1 and octahedral M_2 molecular weights have been calculated. Table 2 contains the force constants K_T and K_O. With an increasing Cr content, force constants K_T and K_O increase. According to IR studies, bond length and the force constant inversely relate [47].

3.4. Magnetic Properties

To obtain the magnetic hysteresis curves, a magnetic field (± 50 kOe) is applied to the prepared samples at room temperature (see Figure 7). Samples at low magnetic fields exhibit nonlinear magnetization and become saturated at high magnetic fields, revealing ferromagnetism. Table 3 summarizes saturation magnetization (M_s), remanent magnetization (M_r), and coercivity (H_c) results. The synthesized samples have low H_c values. Therefore, the samples could be classified as soft magnetic spinels. As a result, the $Cd_{0.5}Zn_{0.5}Cr_xFe_{2-x}O_4$ spinels have the potential to be applied in some magnetic applications such as recording heads, spintronic devices, microwave devices, transformers, induction cores, telecommunication systems, electromagnetic devices, and magnetic recording field sensors [48–50]. As the Cr content increases, the H_c also increases, indicating an increase in the resistive nature against spin inversion. The anisotropy constant increases with an increasing Cr content but decreases when the Cr content is more significant than 0.5. The anisotropy constant K depends on the substituted ion concentration [51], which can be evaluated using the corresponding relation.

$$H_c = 0.98 \frac{K}{M_S} \quad (4)$$

Furthermore, saturation magnetization is related to H_c through Brown's relation [52], and $H_c = \frac{2K}{\mu_0 M_S}$, states that H_c is inversely proportional to M_s. This is consistent with our experimental results.

Table 3. Values of the spontaneous magnetization (M_s), remanent magnetization (M_r), coercivity (H_c), H_s magnetic field, and anisotropy constant K.

x.	M_r (emu/g)	M_s (emu/g)	H_c (Oe)	H_s	K (erg/cm^3)
0	3.5	55	82	4950	4602
0.5	8.2	37.8	237	4500	9141
1	9	27.45	311.35	4478	8536
1.5	10.8	17	402	4423	6973
2	6.27	10.8	422.15	4387	4652

Table 3 illustrates the decrease in the M_s value with Cr replacement, consistent with other spinel systems [53,54]. There is a correlation between the increase in M_s values and Neel's theory [55] and the cations distribution between A- and B-sites. According to Neel's model, ferrimagnet materials interact in three ways: A-A, B-B, and A-B sublattices. A-A and B-B interactions within the sublattice are dominated by the super-exchange interaction between A- and B-sites. Consequently, the net magnetic moment consists of the vector sum of magnetic moments on sublattices A and B [56]:

$$n_B^{cal} = |M_B - M_A| \quad (5)$$

M_B and M_A represent B and A sublattice magnetic moments in Bohr magneton (μ_B), respectively. When Cr^{3+} replaces Fe^{3+} at the octahedral site, saturation magnetization decreases since the Cr^{3+} ion (3 μ_B) has a smaller magnetic moment than Fe^{3+} (5 μ_B) [57]. Accordingly, the magnetic properties of the prepared samples are closely related to their predicted cation distribution.

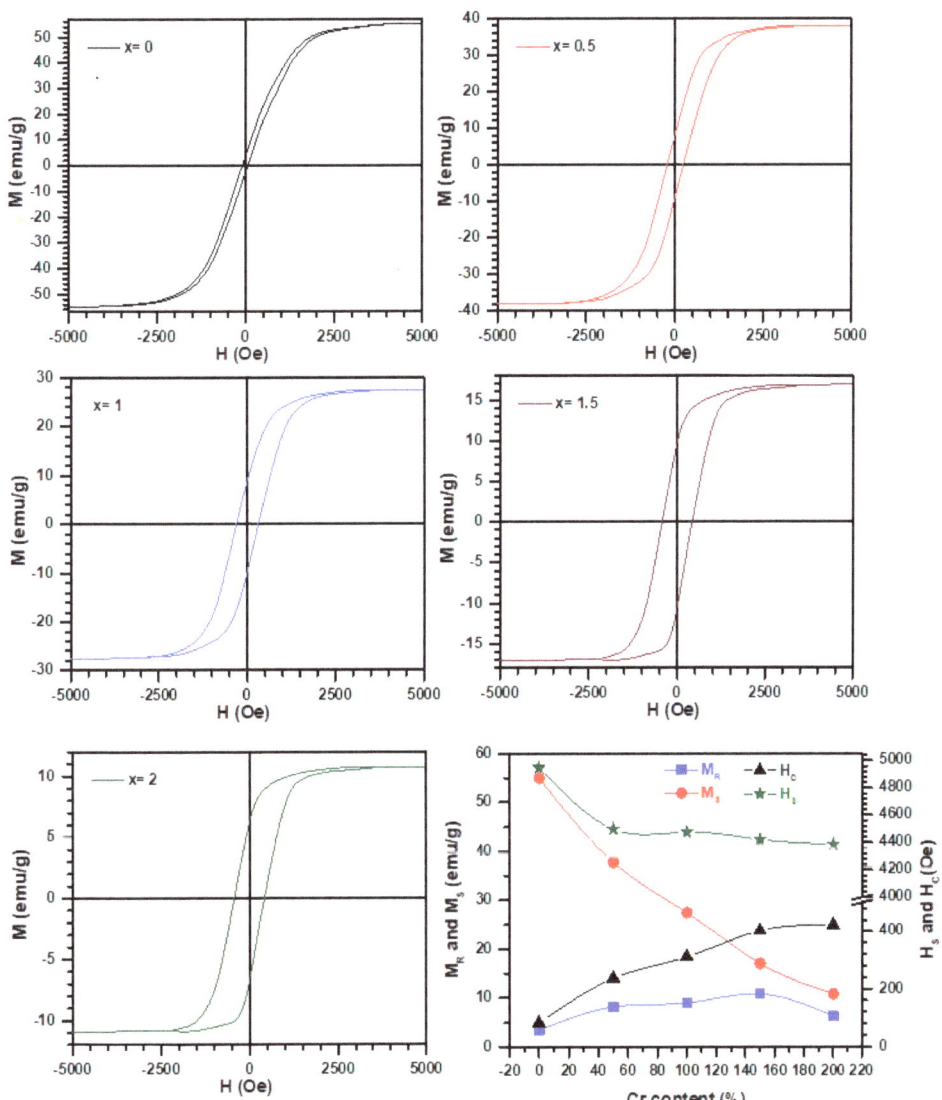

Figure 7. Magnetization loop M-H of $Cd_{0.5}Zn_{0.5}Fe_{2-x}Cr_xO_4$ spinels recorded at 300 K The values M_r, M_s, H_c, and H_s were extracted from M-H curves and plotted as a function of Cr content.

The inset of Figure 8 shows variations in the anisotropy constant "K" and static susceptibility "χ_S" with Cr. The ratio M_s/H_s, named static susceptibility χ_S, increases linearly with the Cr content. In contrast, the anisotropic coefficient increases with the Cr content. It reaches a maximum when the Cr composition is equal to that of Fe and regains the same value as the beginning ferrite when the material becomes chromite.

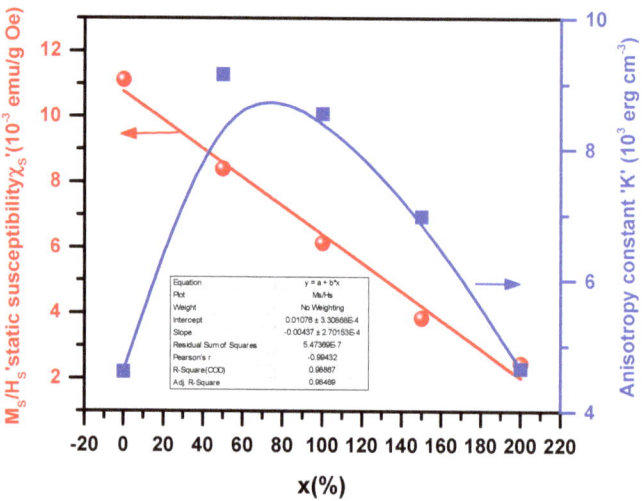

Figure 8. Variation in anisotropy constant "K" and static susceptibility "χ_S" with Cr content.

4. Conclusions

$Cd_{0.5}Zn_{0.5}Fe_{2-x}Cr_xO_4$ ($0 \leq x \leq 2$) ferrites synthesized via the sol–gel method exhibit a cubic Fd3m spinel structure. Substituting Cr for Fe reduces cell parameters, average grain size, spontaneous magnetization, and conductivity compared to the parent compound $Cd_{0.5}Zn_{0.5}Fe_2O_4$. The FTIR spectra reveal two principal absorption bands that increase with Cr substitution. These materials show potential for use in various magnetic and electronic applications. The significant findings of this work show that the examined materials have regular spinel cubic structures and low coercive fields, allowing them to be used in magnetic devices.

Author Contributions: Conceptualization, S.H., A.M. and L.B.; Methodology, R.K.A., N.K. and A.M.; Software, A.S.A. and H.T.; Formal analysis, L.B., S.H. and A.S.A.; Investigation, L.B. and S.H.; Writing—original draft, R.K.A., L.B., A.A., A.S.A. and S.H.; Writing—review & editing, N.K., A.A. and H.T.; Supervision, N.K. and A.M. All authors have read and agreed to the published version of the manuscript.

Funding: This research received no external funding.

Data Availability Statement: Not applicable.

Conflicts of Interest: The authors declare no conflict of interest.

References

1. Adam, J.D.; Davis, L.E.; Dionne, G.F.; Schloemann, E.F.; Stitzer, S.N. Ferrite Devices and Materials. *IEEE Trans. Microw. Theory Tech.* **2002**, *50*, 721–737. [CrossRef]
2. Kulikowski, J. Soft Magnetic Ferrites—Development or Stagnation? *J. Magn. Magn. Mater.* **1984**, *41*, 56–62. [CrossRef]
3. Harris, V.G.; Geiler, A.; Chen, Y.; Yoon, S.D.; Wu, M.; Yang, A.; Chen, Z.; He, P.; Parimi, P.V.; Zuo, X. Recent Advances in Processing and Applications of Microwave Ferrites. *J. Magn. Magn. Mater.* **2009**, *321*, 2035–2047. [CrossRef]
4. Mazarío, E.; Herrasti, P.; Morales, M.P.; Menéndez, N. Synthesis and Characterization of CoFe2O4 Ferrite Nanoparticles Obtained by an Electrochemical Method. *Nanotechnology* **2012**, *23*, 355708. [CrossRef] [PubMed]
5. Chen, K.; Jia, L.; Yu, X.; Zhang, H. A Low Loss NiZnCo Ferrite, Prepared Using a Hydrothermal Method, for Antenna Applications. *J. Appl. Phys.* **2014**, *115*, 17A520. [CrossRef]
6. Rahman, S.; Nadeem, K.; Anis-ur-Rehman, M.; Mumtaz, M.; Naeem, S.; Letofsky-Papst, I. Structural and Magnetic Properties of ZnMg-Ferrite Nanoparticles Prepared Using the Co-Precipitation Method. *Ceram. Int.* **2013**, *39*, 5235–5239. [CrossRef]
7. Sun, L.; Zhang, R.; Ni, Q.; Cao, E.; Hao, W.; Zhang, Y.; Ju, L. Magnetic and Dielectric Properties of $Mg_xCo_{1-x}Fe_2O_4$ Ferrites Prepared by the Sol-Gel Method. *Phys. B Condens. Matter* **2018**, *545*, 4–11. [CrossRef]

8. Safari, A.; Gheisari, K.; Farbod, M. Structural, Microstructural, Magnetic and Dielectric Properties of Ni-Zn Ferrite Powders Synthesized by Plasma Arc Discharge Process Followed by Post-Annealing. *J. Magn. Magn. Mater.* **2019**, *488*, 165369. [CrossRef]
9. Soibam, I.; Phanjoubam, S.; Prakash, C. Mössbauer and Magnetic Studies of Cobalt Substituted Lithium Zinc Ferrites Prepared by Citrate Precursor Method. *J. Alloys Compd.* **2009**, *475*, 328–331. [CrossRef]
10. Singh, C.; Jauhar, S.; Kumar, V.; Singh, J.; Singhal, S. Synthesis of Zinc Substituted Cobalt Ferrites via Reverse Micelle Technique Involving in Situ Template Formation: A Study on Their Structural, Magnetic, Optical and Catalytic Properties. *Mater. Chem. Phys.* **2015**, *156*, 188–197. [CrossRef]
11. Hankare, P.P.; Jadhav, S.D.; Sankpal, U.B.; Chavan, S.S.; Waghmare, K.J.; Chougule, B.K. Synthesis, Characterization and Effect of Sintering Temperature on Magnetic Properties of MgNi Ferrite Prepared by Co-Precipitation Method. *J. Alloys Compd.* **2009**, *475*, 926–929. [CrossRef]
12. Ebrahimi, S.A.S.; Masoudpanah, S.M. Effects of PH and Citric Acid Content on the Structure and Magnetic Properties of MnZn Ferrite Nanoparticles Synthesized by a Sol–Gel Autocombustion Method. *J. Magn. Magn. Mater.* **2014**, *357*, 77–81. [CrossRef]
13. Kouki, N.; Hcini, S.; Boudard, M.; Aldawas, R.; Dhahri, A. Microstructural Analysis, Magnetic Properties, Magnetocaloric Effect, and Critical Behaviors of $Ni_{0.6}Cd_{0.2}Cu_{0.2}Fe_2O_4$ Ferrites Prepared Using the Sol–Gel Method under Different Sintering Temperatures. *RSC Adv.* **2019**, *9*, 1990–2001. [CrossRef] [PubMed]
14. Hcini, S.; Kouki, N.; Omri, A.; Dhahri, A.; Bouazizi, M.L. Effect of Sintering Temperature on Structural, Magnetic, Magnetocaloric and Critical Behaviors of Ni-Cd-Zn Ferrites Prepared Using Sol-Gel Method. *J. Magn. Magn. Mater.* **2018**, *464*, 91–102. [CrossRef]
15. Mane, D.R.; Patil, S.; Birajdar, D.D.; Kadam, A.B.; Shirsath, S.E.; Kadam, R.H. Sol–Gel Synthesis of Cr^{3+} Substituted $Li_{0.5}Fe_{2.5}O_4$: Cation Distribution, Structural and Magnetic Properties. *Mater. Chem. Phys.* **2011**, *126*, 755–760. [CrossRef]
16. Mitra, S.; Bidyananda, M.; Samanta, A.K. Cation Distribution in Cr-Spinels from the Sittampundi Layered Complex and Their Intracrystalline Thermodynamics. *Curr. Sci.* **2006**, *90*, 435–439.
17. Lee, S.H.; Yoon, S.J.; Lee, G.J.; Kim, H.S.; Yo, C.H.; Ahn, K.; Lee, D.H.; Kim, K.H. Electrical and Magnetic Properties of $NiCr_xFe_{2-x}O_4$ Spinel ($0 \leq X \leq 0.6$). *Mater. Chem. Phys.* **1999**, *61*, 147–152. [CrossRef]
18. Ikehara, T.; Yamaguchi, H.; Hosokawa, K.; Miyamoto, H.; Aizawa, K. Effects of ELF Magnetic Field on Membrane Protein Structure of Living HeLa Cells Studied by Fourier Transform Infrared Spectroscopy. *Bioelectromagn. J. Bioelectromagn. Soc. Soc. Phys. Regul. Biol. Med. Eur. Bioelectromagn. Assoc.* **2003**, *24*, 457–464. [CrossRef]
19. Gismelseed, A.M.; Yousif, A.A. Mössbauer Study of Chromium-Substituted Nickel Ferrites. *Phys. B Condens. Matter* **2005**, *370*, 215–222. [CrossRef]
20. Singhal, S.; Sharma, R.; Namgyal, T.; Jauhar, S.; Bhukal, S.; Kaur, J. Structural, Electrical and Magnetic Properties of $Co_{0.5}Zn_{0.5}Al_xFe_{2-x}O_4$ (X = 0, 0.2, 0.4, 0.6, 0.8 and 1.0) Prepared via Sol–Gel Route. *Ceram. Int.* **2012**, *38*, 2773–2778. [CrossRef]
21. Singhal, S.; Jauhar, S.; Singh, J.; Chandra, K.; Bansal, S. Investigation of Structural, Magnetic, Electrical and Optical Properties of Chromium Substituted Cobalt Ferrites ($CoCr_xFe_{2-x}O_4$, $0 \leq X \leq 1$) Synthesized Using Sol Gel Auto Combustion Method. *J. Mol. Struct.* **2012**, *1012*, 182–188. [CrossRef]
22. Valan, M.F.; Manikandan, A.; Antony, S.A. Microwave Combustion Synthesis and Characterization Studies of Magnetic $Zn_{1-x}Cd_xFe_2O_4$ ($0 \leq X \leq 0.5$) Nanoparticles. *J. Nanosci. Nanotechnol.* **2015**, *15*, 4543–4551. [CrossRef] [PubMed]
23. Harish, K.N.; Naik, H.S.B.; Viswanath, R. Synthesis, Enhanced Optical and Photocatalytic Study of Cd–Zn Ferrites under Sunlight. *Catal. Sci. Technol.* **2012**, *2*, 1033–1039. [CrossRef]
24. Gupta, M.; Gupta, M.; Mudsainiyan, R.K.; Randhawa, B.S. Physico-Chemical Analysis of Pure and Zn Doped Cd Ferrites ($Cd_{1-x}Zn_xFe_2O_4$) Nanofabricated by Pechini Sol–Gel Method. *J. Anal. Appl. Pyrolysis* **2015**, *116*, 75–85. [CrossRef]
25. Chakrabarti, M.; Sanyal, D.; Chakrabarti, A. Preparation of $Zn_{(1-x)}Cd_xFe_2O_4$ (X = 0.0, 0.1, 0.3, 0.5, 0.7 and 1.0) Ferrite Samples and Their Characterization by Mössbauer and Positron Annihilation Techniques. *J. Phys. Condens. Matter* **2007**, *19*, 236210. [CrossRef]
26. Siddique, M.; Anwar-ul-Islam, M.; Butt, N.M.; Abbas, T. Composition Dependence of Quadrupole Splitting in Cd—Zn Ferrites. *Phys. Status Solidi* **1999**, *216*, 1069–1072. [CrossRef]
27. Arean, C.O.; Diaz, E.G.; Gonzalez, J.M.R.; Garcia, M.A.V. Crystal Chemistry of Cadmium-Zinc Ferrites. *J. Solid State Chem.* **1988**, *77*, 275–280. [CrossRef]
28. Weil, L.; Bertaut, F.; Bochirol, L. Propriétés Magnétiques et Structure de La Phase Quadratique Du Ferrite de Cuivre. *J. Phys. le Radium* **1950**, *11*, 208–212. [CrossRef]
29. Mane, D.R.; Birajdar, D.D.; Shirsath, S.E.; Telugu, R.A.; Kadam, R.H. Structural and Magnetic Characterizations of Mn—Ni—Zn Ferrite Nanoparticles. *Phys. Status Solidi* **2010**, *207*, 2355–2363. [CrossRef]
30. Cvejic, Z.; Rakic, S.; Kremenovic, A.; Antic, B.; Jovalekic, C.; Colomban, P. Nanosize Ferrites Obtained by Ball Milling: Crystal Structure, Cation Distribution, Size-Strain Analysis and Raman Investigations. *Solid State Sci.* **2006**, *8*, 908–915. [CrossRef]
31. Hakim, M.A.; Nath, S.K.; Sikder, S.S.; Maria, K.H. Cation Distribution and Electromagnetic Properties of Spinel Type Ni–Cd Ferrites. *J. Phys. Chem. Solids* **2013**, *74*, 1316–1321. [CrossRef]
32. Khalaf, K.A.M.; Al Rawas, A.D.; Gismelssed, A.M.; Al Jamel, A.; Al Ani, S.K.J.; Shongwe, M.S.; Al Riyami, K.O.; Al Alawi, S.R. Influence of Cr Substitution on Debye-Waller Factor and Related Structural Parameters of $ZnFe_{2-x}Cr_xO_4$ Spinels. *J. Alloys Compd.* **2017**, *701*, 474–486. [CrossRef]
33. Hossain, A.K.M.A.; Mahmud, S.T.; Seki, M.; Kawai, T.; Tabata, H. Structural, Electrical Transport, and Magnetic Properties of $Ni_{1-x}Zn_xFe_2O_4$. *J. Magn. Magn. Mater.* **2007**, *312*, 210–219. [CrossRef]

34. Shannon, R.D. Revised Effective Ionic Radii and Systematic Studies of Interatomic Distances in Halides and Chalcogenides. *Acta Crystallogr. Sect. A Cryst. Phys. Diffr. Theor. Gen. Crystallogr.* **1976**, *32*, 751–767. [CrossRef]
35. Patange, S.M.; Shirsath, S.E.; Lohar, K.S.; Algude, S.G.; Kamble, S.R.; Kulkarni, N.; Mane, D.R.; Jadhav, K.M. Infrared Spectral and Elastic Moduli Study of $NiFe_{2-X}Cr_XO_4$ Nanocrystalline Ferrites. *J. Magn. Magn. Mater.* **2013**, *325*, 107–111. [CrossRef]
36. Kumar, G.; Kotnala, R.K.; Shah, J.; Kumar, V.; Kumar, A.; Dhiman, P.; Singh, M. Cation Distribution: A Key to Ascertain the Magnetic Interactions in a Cobalt Substituted Mg–Mn Nanoferrite Matrix. *Phys. Chem. Chem. Phys.* **2017**, *19*, 16669–16680. [CrossRef]
37. Sharma, R.; Thakur, P.; Kumar, M.; Thakur, N.; Negi, N.S.; Sharma, P.; Sharma, V. Improvement in Magnetic Behaviour of Cobalt Doped Magnesium Zinc Nano-Ferrites via Co-Precipitation Route. *J. Alloys Compd.* **2016**, *684*, 569–581. [CrossRef]
38. Hemeda, O.M.; Amer, M.A.; Aboul-Enein, S.; Ahmed, M.A. Effect of Sintering on X-Ray and IR Spectral Behaviour of the $MnAl_xFe_{2-x}O_4$ Ferrite System. *Phys. Status Solidi* **1996**, *156*, 29–38. [CrossRef]
39. Williamson, G.K.; Hall, W.H. X-Ray Line Broadening from Filed Aluminium and Wolfram. *Acta Metall.* **1953**, *1*, 22–31. [CrossRef]
40. AlArfaj, E.; Hcini, S.; Mallah, A.; Dhaou, M.H.; Bouazizi, M.L. Effects of Co Substitution on the Microstructural, Infrared, and Electrical Properties of $Mg_{0.6-x}Co_xZn_{0.4}Fe_2O_4$ Ferrites. *J. Supercond. Nov. Magn.* **2018**, *31*, 4107–4116. [CrossRef]
41. Waldron, R.D. Infrared Spectra of Ferrites. *Phys. Rev.* **1955**, *99*, 1727. [CrossRef]
42. Amer, M.A.; Ahmed, M.A.; El-Nimr, M.K.; Mostafa, M.A. Mössbauer and Infrared Studies of the Cu-Cr Ferrites. *Hyperfine Interact.* **1995**, *96*, 91–98. [CrossRef]
43. Shaeel, A.; Al-Thabaiti. Communications de la Facult 'e des Sciences de l'Universit 'e d'Ankara B; La Faculté: Leiden, The Netherlands, 2003; Volume 49, pp. 5–14.
44. Mohammed, K.A.; Al-Rawas, A.D.; Gismelseed, A.M.; Sellai, A.; Widatallah, H.M.; Yousif, A.; Elzain, M.E.; Shongwe, M. Infrared and Structural Studies of $Mg_{1-X}Zn_XFe_2O_4$ Ferrites. *Phys. B Condens. Matter* **2012**, *407*, 795–804. [CrossRef]
45. Khalaf, K.A.M.; Al-Rawas, A.D.; Widatallah, H.M.; Al-Rashdi, K.S.; Sellai, A.; Gismelseed, A.M.; Hashim, M.; Jameel, S.K.; Al-Ruqeishi, M.S.; Al-Riyami, K.O. Influence of Zn^{2+} Ions on the Structural and Electrical Properties of $Mg_{1-X}Zn_XFeCrO_4$ Spinels. *J. Alloys Compd.* **2016**, *657*, 733–747. [CrossRef]
46. Yoon, S.J.; Lee, S.H.; Kim, K.H.; Ahn, K.S. Electrical and Magnetic Properties of Spinel $ZnCr_{2-X}Fe_XO_4$ ($0 \leq X \leq 1.0$). *Mater. Chem. Phys.* **2002**, *73*, 330–334. [CrossRef]
47. Pradeep, A.; Priyadharsini, P.; Chandrasekaran, G. Sol–Gel Route of Synthesis of Nanoparticles of $MgFe_2O_4$ and XRD, FTIR and VSM Study. *J. Magn. Magn. Mater.* **2008**, *320*, 2774–2779. [CrossRef]
48. Shokrollahi, H.; Janghorban, K. Soft Magnetic Composite Materials (SMCs). *J. Mater. Process. Technol.* **2007**, *189*, 1–12. [CrossRef]
49. Chakrabarti, P.K.; Nath, B.K.; Brahma, S.; Das, S.; Das, D.; Ammar, M.; Mazaleyrat, F. Magnetic and Hyperfine Properties of Chemically Synthesized Nanocomposites of $(Al_2O_3)_x(Ni_{0.2}Zn_{0.6}Cu_{0.2}Fe_2O_4)_{(1-x)}$ (X = 0.15, 0.30, 0.45). *Solid State Commun.* **2007**, *144*, 305–309. [CrossRef]
50. Modak, S.; Ammar, M.; Mazaleyrat, F.; Das, S.; Chakrabarti, P.K. XRD, HRTEM and Magnetic Properties of Mixed Spinel Nanocrystalline Ni–Zn–Cu-Ferrite. *J. Alloys Compd.* **2009**, *473*, 15–19. [CrossRef]
51. Kambale, R.C.; Shaikh, P.A.; Bhosale, C.H.; Rajpure, K.Y.; Kolekar, Y.D. The Effect of Mn Substitution on the Magnetic and Dielectric Properties of Cobalt Ferrite Synthesized by an Autocombustion Route. *Smart Mater. Struct.* **2009**, *18*, 115028. [CrossRef]
52. Coey, J.M.D. *Rare Earth Permanent Magnetism*; John Wiley and Sons: Hoboken, NJ, USA, 1996. [CrossRef]
53. Patange, S.M.; Shirsath, S.E.; Jadhav, S.S.; Jadhav, K.M. Cation Distribution Study of Nanocrystalline $NiFe_{2-X}Cr_XO_4$ Ferrite by XRD, Magnetization and Mössbauer Spectroscopy. *Phys. Status Solidi* **2012**, *209*, 347–352. [CrossRef]
54. Patange, S.M.; Shirsath, S.E.; Toksha, B.G.; Jadhav, S.S.; Jadhav, K.M. Electrical and Magnetic Properties of Cr^{3+} Substituted Nanocrystalline Nickel Ferrite. *J. Appl. Phys.* **2009**, *106*, 023914. [CrossRef]
55. Néel, L. Magnetism and Local Molecular Field. *Science* **1971**, *174*, 985–992. [CrossRef] [PubMed]
56. Torkian, S.; Ghasemi, A.; Razavi, R.S. Cation Distribution and Magnetic Analysis of Wideband Microwave Absorptive $Co_xNi_{1-X}Fe_2O_4$ Ferrites. *Ceram. Int.* **2017**, *43*, 6987–6995. [CrossRef]
57. Satyanarayana, G.; Nageswara Rao, G.; Babu, K.V.; Santosh Kumar, G.V.; Dinesh Reddy, G. Effect of Cr_{3+} Substitution on the Structural, Electrical and Magnetic Properties of $Ni_{0.7}Zn_{0.2}Cu_{0.1}Fe_{2-x}Cr_xO_4$ Ferrites. *J. Korean Phys. Soc.* **2019**, *74*, 684–694. [CrossRef]

Disclaimer/Publisher's Note: The statements, opinions and data contained in all publications are solely those of the individual author(s) and contributor(s) and not of MDPI and/or the editor(s). MDPI and/or the editor(s) disclaim responsibility for any injury to people or property resulting from any ideas, methods, instructions or products referred to in the content.

Article

Preparation and Characterization of Nano-Sized Co(II), Cu(II), Mn(II) and Ni(II) Coordination PAA/Alginate Biopolymers and Study of Their Biological and Anticancer Performance

Maged S. Al-Fakeh [1,2,*], Munirah S. Alazmi [1] and Yassine EL-Ghoul [1,3]

[1] Department of Chemistry, College of Science, Qassim University, Buraidah 51452, Saudi Arabia; m-alazmi@qu.edu.sa (M.S.A.); y-elghoul@qu.edu.sa (Y.E.-G.)
[2] Taiz University, Taiz 3086, Yemen
[3] Textile Engineering Laboratory, University of Monastir, Monastir 5019, Tunisia
* Correspondence: m.alfakeh@qu.edu.sa

Citation: Al-Fakeh, M.S.; Alazmi, M.S.; EL-Ghoul, Y. Preparation and Characterization of Nano-Sized Co(II), Cu(II), Mn(II) and Ni(II) Coordination PAA/Alginate Biopolymers and Study of Their Biological and Anticancer Performance. *Crystals* **2023**, *13*, 1148. https://doi.org/10.3390/cryst13071148

Academic Editor: Waldemar Maniukiewicz

Received: 19 June 2023
Revised: 14 July 2023
Accepted: 21 July 2023
Published: 23 July 2023

Copyright: © 2023 by the authors. Licensee MDPI, Basel, Switzerland. This article is an open access article distributed under the terms and conditions of the Creative Commons Attribution (CC BY) license (https:// creativecommons.org/licenses/by/ 4.0/).

Abstract: Four of the crosslinked sodium alginate and polyacrylic acid biopolymers based nanoscale metal natural polysaccharides, [M(AG-PAA)Cl(H$_2$O)$_3$], where M = Co(II), Cu(II), Mn(II) and Ni(II), AG = sodium alginate and PAA = polyacrylic acid, have been synthesized and structurally characterized. Because of their numerous biological and pharmacological activities of polysaccharides, including antimicrobial, immunomodulatory, antitumor, antidiabetic, antiviral, antioxidant, hypoglycemic and anticoagulant activities, polysaccharides are one of the near-promising candidates in the biomedical and pharmaceutical fields. The complexity of the polymeric compounds has been verified by carbon and nitrogen analysis, magnetic and conductance measurements, FT-IR spectra, electronic spectral analysis and thermal analysis (DTA, TG). All the synthesized complexes were non-electrolytes with magnetic moments ranging from 1.74 to 5.94 BM. The polymeric complexes were found to be of octahedral geometry. The developed coordination polymeric was found to be crystalline using X-ray powder diffraction examinations, which is confirmed by the SEM analysis. As a result, the crystallite size of all polymeric nanocrystals was in the range of 14 - 69 nm. The test of four compounds exhibits a broad spectrum of antimicrobial activity against both Gram-positive and Gram-negative bacteria and fungal *Candida albicans*. Using DPPH as a substrate, studies on radical scavenging tests are carried out. The findings demonstrated the antioxidant activities of each complex. In addition, results showed that the two chosen polymeric complexes had a good ability to kill cancer cells in a dose-dependent way. The copper(II) polymeric complex showed to its superior functionality as evidenced by microbial activity. After 72 h of interaction with the normal human breast epithelial cells (MCF10A), the synthesized polymeric compounds of Cu(II) and Co(II) showed exceptional cytocompatibility with the different applied doses. Compared to poly-AG/PAA/Co(II), poly-AG/PAA/Cu(II) exhibits a greater anticancer potential at various polymeric dosages.

Keywords: crosslinked sodium alginate and polyacrylic acid biopolymers; XRD; scanning electron microscopy; antimicrobial activity; antioxidants and anticancer

1. Introduction

As biopolymers (BPs) are composed of living organisms such as microbes and plants, they are a renewable materials resource, unlike most polymers which are petroleum-based coordination polymers. Generally, biopolymeric complexes are degradable. They find use in different industries from food industries to packaging, manufacturing and biomedical engineering [1–3]. BPs are heartening materials owing to their properties such as biocompatibility, unique properties and abundance such as non-toxicity, etc. With some nano-sized reinforcements to promote their characteristics and experimental applications, BPs are being researched for their use in more and more methods possible. Biopolymer compounds are divided into four main categories: those made from microorganisms, biomass products (agro-biopolymers), biotechnological products and petrochemical products [4]. Numerous different substances, including polysaccharides (starches, celluloses, alginates, pectin's, gums and chitosan), are found in biopolymers made from biomass products [5–8]. Other applications of biopolymers are used in the production of nanomaterials. Nanotechnology, also known as 1 nm to 100 nm scale manipulation, is the process of changing the size and shape of structures, systems and electronics [9–11]. Due to their small size, they have higher reactivity than the corresponding bulk forms, more significant surface areas and a variety of properties that can be tuned [12–14]. These unique characteristics have accelerated the development of nanoscience and the use of nanoparticles in a diversity of industries, including cosmetics, electronics, food analysis, the environment, biomedicine and other applications [15–19]. Important characteristics of biopolymers include their biodegradability, biocompatibility, stability, sustainability, bioresorbability, flexibility, renewability and antibacterial and antifungal activity [20–24]. They are also less toxic, non-immunogenic, easier to extract, carbon-neutral, non-carcinogenic and non-thrombogenic [25–27]. Because of the numerous biological and pharmacological activities of polysaccharides, including antitumor, antioxidant, anticoagulant, immunomodulatory, antimicrobial, hypoglycemia activities, antidiabetic and antiviral activities, polysaccharides are one of the maximally promising candidates in pharmaceutical and biomedical applications. Polysaccharides (PSD) are fundamental macromolecules that nearly exist in all living forms and have significant biological functions. There are numerous sources of polysaccharides, including plants, microorganisms, algae and animals [28,29]. Alginates are a type of polysaccharide that is primarily derived from brown algae and bacteria found in seaweed. This bio-material is a naturally occurring polysaccharide that also occurs as capsular (PSD) in some bacteria, for example, Azotobacter and Pseudomonas, and as structural elements in the cell walls of marine brown algae such as Phaeophyceae. Alginates (AGs) constitute a strain of linear binary unbranched co-polymers consisting of α-L-guluronic acid (monomer G) and 1,4-linked β-D-mannuronic acid (monomer M) residues. Sodium alginate (Na-AG) is the sodium salt composed of alginic acid and gum mainly extracted by the cell walls of brown algae, with chelating activity (Figure 1) [30,31]. AGs, natural multifunctional polymers, have gained prominence as desirable compounds in pharmaceutical industries and biomedicine over the past few decades as a result of their distinctive physicochemical characteristics and diverse biological activities [32,33]. They are non-antigenic, biocompatible, non-toxic and biodegradable [34,35]. Additionally, alginates are used in food manufacturing as a gelling, thickening, stabilizing or emulsifying agent, as a by-product of microbial and viral protection and to coat fruits and vegetables, whereas alginate substantially contributes to the sustained release of drug-delivery products [36].

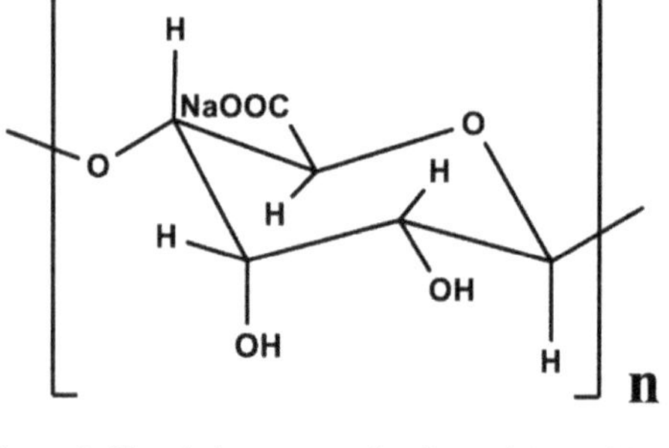

Figure 1. Chemical structure of sodium alginate (Na-AG).

Polyacrylic acid (PAA) is the most basic acrylate polymer [37]. The structure of (PAA), which has a carbon backbone and an ionizable COOH group as a side chain in each reiterates unit, is depicted in (Figure 2). In water with a pH of 7, (PAA) is an anionic polymer, meaning that many of its COOH groups will lose their H^+ and develop a negative charge.

As a polyelectrolyte, PAA can therefore absorb and hold onto water, expanding to many times its initial volume [38]. In recent years, interest in PAA, a non-toxic, biodegradable and biocompatible polymer, has increased significantly. By chemically altering carboxyl groups, it is possible to create PAA nano-derivatives, which have better chemical properties than unaltered PAA [39]. An acrylic acid (AA), polymer with a (-COOH) on each of monomer unit end is known as PAA, also referred to as a carbomer. PAA, a thermoplastic polymer, has a high bioavailability due to its numerous carboxyl groups, thus serving as a surface variation for biological nano-materials [40]. Cross-linked PAA has also been used in the processing of household products, inclusively floor cleaners. The neutralized (PAA) gels are appropriate to obtain biocompatible matrices used for medical applications, for example, gels for skin disease treatment products or skin care [41].

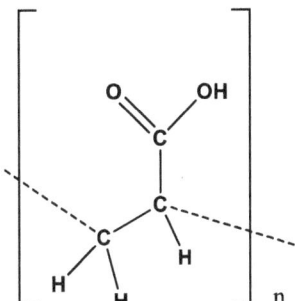

Figure 2. Chemical structure of polyacrylic acid (PAA).

Previously, Ni(II) and Co(II) alginate biopolymer complexes and polyacrylic acid nanoplatforms were reported to show antimicrobial, anticancer and biosensing [42,43]. Therefore, we meditate that polymeric compounds of nano-sized Co(II), Cu(II), Mn(II) and Ni(II)-PAA/alginate biopolymers will show more antioxidant, antibacterial and anticancer activity. In light of these results, we created a work on metal-PAA/alginate biopolymers compounds, which included a review of their NPs and testing for antifungal, antibacterial, antioxidative and anticancer properties. The primary goal of this paper is to synthesize and analyze the physicochemical properties of nanoscale polymeric complexes of cobalt(II), copper(II), manganese(II) and nickel(II) that have been synthesized in both the solid and solution states. CHN, UV–vis, FT-IR, spectra, molar conductance, magnetic, TG, DTA, XRD and scanning electron microscopy are used to depict the framework of the examined compounds. Further, the metal polymeric coordination compounds' antibacterial, antifungal and antioxidant characteristics as well as their anticancer research are presented.

2. Materials and Methods

The chemicals listed below were analytical-grade substances that were used directly after delivery without further purification. Sodium Alginate, poly-(acrylic acid), methanol, dimethyl sulfoxide (DMSO), Co(II) chloride hexahydrate, Cu(II) chloride dehydrate, Mn(II) chloride tetrahydrate and Ni(II) chloride hexahydrate were all purchased from Sigma-Aldrich company.

2.1. Preparation of the Crosslinked Polymeric Ligand (Poly-PAA/AG)

First, we vigorously stirred 100 mL of distilled water with 3 g of sodium alginate (AG) for an hour at 80 °C in a round bottom flask. Then, we added dropwise a polyacrylic acid (PAA) solution that had been previously made by combining 3 g of PAA with 100 mL of H_2O, stirring continuously and heating at 70 °C. Under N_2 pressure, the mixture was stirred at 120 °C for two hours. The reaction solution was then preserved at 90 °C overnight while being stirred. The resultant suspension was then brought to 25 °C before being concentrated at a lower pressure. The final step was to dry the obtained product at 65 °C while maintaining a high vacuum, producing a white color stable powder (90%). Anal. Calc. for $C_9H_{10}O_7$: C, 46.97; H, 4.38. Found: C, 46.93; H, 4.98. IR data: $\nu_{(OH)}$ 3222, $\nu_{(C-H)}$ 2922, $\nu_{(C=O)}$ 1725, $\nu_{(COO)}$ 1413, $\nu_{(CO)}$ 1263, $\nu_{(C-O-C)}$ 1025.

2.2. Synthesis of Metal Polymeric Complexes Nanoparticles

2.2.1. [Co(AG-PAA)Cl(H$_2$O)$_3$]

This compound was made by dissolving 1 g (0.2 mmol) of the (AG/PAA) ligand in 20 mL of methanol, boiling it for 5 min and then stirring while adding 20 mL of pure water until the ligand dispersed. We allowed the mixture to cool somewhat before adding 0.82 g (0.2 mmol) of the metal (CoCl$_2$.6H$_2$O), which had been dissolved in 10 mL of distilled water. Stirring continuously until the purple precipitate forms, it is then dried over CaCl$_2$ in a desiccator. Anal. Calc. for $C_9H_{15}CoClO_{10}$: C, 28.62; H, 4.00. Found: C, 29.06; H, 4.28. IR data: $\nu_{(OH)}$ 3225, $\nu_{(C-H)}$ 2930, $\nu_{(COO)sym}$ 1715, $\nu_{(C=O)}$ 1715, $\nu_{(COO)asym}$ 1415, $\nu_{(CO)}$ 1157, $\nu_{(C-O-C)}$ 1022, $\nu_{(M-O)}$ 519 cm^{-1}, m.p. 230 °C and molar conductance 18.63 S cm^2 mol^{-1}.

2.2.2. [Cu(AG-PAA)Cl(H$_2$O)$_3$]

This complex was prepared by dissolving 0.5 g (0.1 mmol) of the (AG/PAA) ligand in 20 mL of methanol and heating it for 5 min before adding 20 mL of distilled water and heating and stirring until the ligand was dissolved. We allowed it to cool somewhat before adding 0.37 g (0.1 mmol) of the metal salt (CuCl$_2$.2H$_2$O) that was dissolved in 10 mL of distilled water. The resultant mixture was stirred, and then the mixture solution containing copper(II) complex was sonicated for 30 min; then, the green polycrystalline powder was obtained, filtered and then cooled to room temperature. The green precipitate was washed with distilled water and ethanol and then dried over CaCl$_2$ in a desiccator. Anal. Calc. for $C_9H_{15}CuClO_{10}$: C, 28.28; H, 3.95. Found: C, 28.61; H, 3.98. IR data: $\nu_{(OH)}$ 3199, $\nu_{(C-H)}$ 2932, $\nu_{(COO)sym}$ 1703, $\nu_{(C=O)}$ 1703, $\nu_{(COO)asym}$ 1421, $\nu_{(CO)}$ 1273, $\nu_{(C-O-C)}$ 1026, $\nu_{(M-O)}$ 582 cm^{-1}, m.p. 190 °C and molar conductance 17.85 S cm^2 mol^{-1}.

2.2.3. [Mn(AG-PAA)Cl(H$_2$O)$_3$]

This compound was synthesized in 2 steps:

Step 1: In a beaker, 0.8 g (0.2 mmol) of AG-PAA was dissolved in 20 mL of methanol, and the mixture was heated for five minutes. Next, 20 mL of distilled water was added, and the mixture was stirred and heated for another 20 min.

Step 2: We added the metal solution (0.68 g, 0.2 mmol) of tetra-hydrated manganese chloride, which was dissolved in 10 mL of distilled water. After allowing it to cool down a bit, we stirred it until the light brown precipitate formed and then dried it in calcium chloride anhydrous. Anal. Calc. for $C_9H_{15}MnClO_{10}$: C, 28.93; H, 4.04. Found: C, 29.10; H, 4.47. IR data: $\nu_{(OH)}$ 3444, $\nu_{(C-H)}$ 2931, $\nu_{(COO)sym}$ 1696, $\nu_{(C=O)}$ 1696, $\nu_{(COO)asym}$ 1419, $\nu_{(CO)}$ 1103, $\nu_{(C-O-C)}$ 1027, $\nu_{(M-O)}$ 523 cm^{-1}, m.p. 210 °C and molar conductance 14.18 S cm^2 mol^{-1}.

2.2.4. [Ni(AG-PAA)Cl(H$_2$O)$_3$]

The 0.8 g of the (AG/PAA) ligand was dissolved in 20 mL of methanol, heated for 5 min, and stirred until the ligand was dissolved. Next, 20 mL of distilled H$_2$O was added, and the mixture was heated and stirred for an additional minute. We allowed it to cool slightly before adding 0.65 g of the metal (NiCl$_2$·6H$_2$O), which was dissolved in 10 mL of distilled H$_2$O. We continued stirring until the light green precipitate forms, after which it is dried in the calcium chloride anhydrous. Anal. Calc. for $C_9H_{15}NiClO_{10}$: C, 28.64; H, 4.00. Found: C, 28.92; H, 4.33. IR data: $\nu_{(OH)}$ 3216, $\nu_{(C-H)}$ 2925, $\nu_{(COO)sym}$ 1700, $\nu_{(C=O)}$ 1700, $\nu_{(COO)asym}$ 1414, $\nu_{(CO)}$ 1164, $\nu_{(C-O-C)}$ 1018, $\nu_{(M-O)}$ 515 cm^{-1}, m.p. 196 °C and molar conductance 19.15 S cm^2 mol^{-1}.

2.3. Physical Measurements

A Gmbh Vario El analyzer was used to determine the elemental analyzers. A Thermo Nicolet (6700), Fourier-transform infrared (FT-IR) spectrophotometer with a wavenumber range of (400–4000 cm^{-1}) was used to collect structural data from FT-IR spectra. Using a Shimadzu (UV-2101) PC spectrophotometer, the UV–Vis spectra were collected. The sonication experiments were performed on sonicator type Q 700, 20 KHz, output 700 W. On a magnetic susceptibility balance of the kind (MSB-Auto), measurements of magnetic susceptibility were made. Using a conductivity meter made by JENWAY, model 4310, the complexes' conductance was measured. On a Shimadzu (DTG 60-H) thermal analyzer heated to a rate of 10 °C per minute, thermal analysis of the polymeric complexes was performed in dynamic air. On an XRD diffractometer Model (PW 1720 Philips, Eindhoven, The Netherlands), measurements of the X-ray diffraction (XRD) were collected at room temperature with Cu-Kα radiation (λ_{Cu} = 0.154059 Å). In the structural refinement procedure of our samples, we have followed the standard steps of the Rietveld method which consists of following the sequence:

(i) Refinement of overall Scale factor + background coefficients (all other parameters are kept fixed);
(ii) The same + refinement of detector zero offsets (or sample displacement in Bragg-Brentano geometry) + refinement of lattice parameters;
(iii) The same + refinement of shape parameters + refinement of asymmetry parameters;
(iv) The same + refinement of atomic positions + refinement of global DebyeWaller parameter or thermal agitation factors;
(v) The same + refinement of site occupancy rate. In our refinements, we have taken care to respect this sequence of steps to release the different parameters. This ensures the stability of the refinement with all the parameters released.

By computing indicators such as the goodness of fit "χ^2" and the R factors (R_{wp} = weighted profile R-factor, R_B = Bragg factor and R_{exp} = expected R factor), the fitting quality of the experimental data is evaluated.

Using a scanning electron microscope, the morphology and structure of the prepared materials were examined.

2.4. Microbial Species and Culture Media

In this article, various metal polymeric complexes, such as Co(II), Cu(II), Mn(II) and Ni(II), were tested against Gram-positive (+) and Gram-negative (−) bacterial strains to gain insight into their broad-spectrum effect. Two Gram-positive strains, *Micrococcus luteus* NCIMB 8166 (S4) and *Staphylococcus aureus* ATCC 25923 (S1), and two Gram-negative strains, *Salmonella thyphimurium* ATCC 14028(S10) and *Escherichia coli* ATCC 35218 (S5), were the pathogenic strains that were used. A pathogenic reference strain of the yeast *Candida albicans* ATCC 90028 (9C) was used to test the antifungal activity. The strains were cultured on nutrient agar (Oxoid) for 24 h at 37 °C as well as in nutrient broth (Oxoid) for 24 h at 37 °C. For 24 h, the yeast strain was grown at 25 °C in Sabouraud Chloramphenicol broth (Oxoid) and grown for 24 h at 37 °C on Sabouraud Chloramphenicol agar (Oxoid). The different species are listed in Table 1.

Table 1. The used microbial strains.

Microbe Type	Strain	Reference
Gram-positive bacteria	S1	*Staphylococcus aureus* ATCC 25923
	S4	*Micrococcus luteus* NCIMB 8166
Gram-negative bacteria	S5	*Escherichia coli* ATCC 35218
	S10	*Salmonella thyphimurium* ATCC 14028
Yeast	9C	*Candida albicans* ATCC 90028

2.5. Antimicrobial Activity

All complexes were tested for antimicrobial activity using the agar disk diffusion method. A total of 50 mg of each extract was dissolved in 1 mL of a 5% solution of dimethyl sulfoxide, or "DMSO," prior to testing. The strains were cultured in Mueller–Hinton (MH) broth (Oxoid) at 37 °C for 24 h, and suspensions were calibrated to 0.5 McFarland standard turbidity. Afterward, 100 µL of each precultured suspension was spread onto plates containing MH agar. Sterile filter paper discs (6 mm in diameter) were impregnated with 20 µL of the different extracts and placed on agar. The treated plates were put for 1 h at 4 °C and then incubated for 24 h at 37 °C. After incubation, the diameter of the inhibition zone (clear halo) about the discs was measured. Each sample was tested in duplicate.

2.6. Antioxidant Assays

DPPH Radical Scavenging Assay

According to the method proposed by Mahdhi et al. [44], the free radical scavenging influence of the extracts was assessed using the next criteria: 1 mL of sample (5 mg/mL) was combined with 3 mL of DPPH (2,2-diphenyl-1-picrylhydrazyl) (300 µM) methanolic solution. The reaction mixture was vortexed and left to sit at 25 °C for 30 min. The solution's absorbance was determined at 517 nm. The standard was ascorbic acid. The following Equation (1) was used to determine the inhibitory % of DPPH:

$$\text{DPPH Scavenging effect (\%)} = [1 - (\text{Abs sample}/\text{Abs control})] \times 100 \quad (1)$$

2.7. Cell Viability and Anticancer Assays

For the cell culture, human breast cancer cells (MCF-7) and human normal breast epithelial cells (MCF-10A) were obtained from American Type Culture Collection (Manassas, WV, USA). The cells were grown in Dulbecco Modified Eagle's Medium with the addition of a synthesized solution of fetal bovine serum (10%) and a mixture of penicillin (100 IU/mL) and streptomycin (100 g/mL) as antibiotics. The medium was then incubated in a 5% CO_2 atmosphere, 100% relative humidity at 37 °C. The anticancer activity and the cell viability assays were performed via the MTT tetrazolium standardized test with some modifications. This test is based on the ability of live cells to proliferate and thus metabolize and reduce the yellow MTT tetrazolium salt into purple formazan structured by a typical absorbance at 570 nm. The cultivated cells were seeded into 96-well plates at a density of 2104 cells/well. After 24 h of incubation, samples with different concentrations were added to each cell culture medium. After incubation for 72 h, we proceed with the MTT assay. Cell viability was assessed as the average % of relative formazan crystals formed taking into account the control culture. Tests were carried out in triplicate for each test.

3. Results and Discussion

The biopolymers were synthesized by the reaction of AG-PAA with Co(II), Cu(II), Mn(II) and Ni(II) chlorides (dissolved in MeOH and distilled water). The prepared compounds were found to react in the molar ratio of 1: 1: metal: AG-PAA. The polymeric complexes are air-stable and partially soluble in dimethyl sulphoxide. The electrical conductivity of the polymeric compounds adequately confirmed their non-electrolytic nature.

3.1. Elemental Analyses

The results of elemental analyses (carbon and hydrogen) of the ligand and its polymeric complexes along with the proposed molecular formula and physical characteristics were shown in the experimental section.

3.2. Molar Conductance

The molar conductivity of the polymeric complexes was measured in dimethylsulphoxide at 25 °C using 10^{-3} M solutions of polymeric compounds. All complexes had low conductivity readings, proving that they were nonionic, according to the results in Table 2.

Table 2. FTIR spectral bands and their assignments of the ligand and its metal complexes.

Compound	$\nu_{(O-H)}$	$\nu_{(C-H)}$	$\nu_{(COO)sym}$	$\nu_{(COO)asym}$	$\upsilon\Delta$	$\nu_{(C=C)}$	$\nu_{(CO)}$	$\nu_{(C-O-C)}$	$\nu_{(M-O)}$	$\nu_{(M-Cl)}$
AG-PAA	3222	2922	1725	1413	312	1600	1263	1025	-	-
Co(II) complex	3225	2930	1715	1415	300	1599	1157	1022	519	417
Cu(II) complex	3199	2932	1703	1421	282	1592	1273	1026	582	471
Mn(II) complex	3444	2931	1696	1419	277	1595	1103	1027	523	409
Ni(II) complex	3216	2925	1700	1414	286	1604	1164	1018	515	422

$\nu_{(C=O)}$ represents the frequencies of the ester carbonyl obtained after condensation of the AG and the PAA.

3.3. Fourier Transform Infrared Spectra

In order to pinpoint the functional groups of the components and comprehend the binding process, FT-IR spectrophotometric analysis was carried out. In this case, IR analysis was conducted to evaluate the preparation of the polymeric ligand (poly-PAA/AG) via the crosslinking process. However, this method would be crucial in establishing the existence of the various compounds between the created coordination polymer and the several metals examined in our study. Figure 3 displays the various spectra that were obtained. In relation to the synthesis of the polymeric ligand, we observe the appearance of a new band at about 1413 cm^{-1} indicating the fashioning of the ester group when comparing the spectra of the ligand and the alginate or polyacrylic acid biopolymer. This ester fashioning was between the COOH groups of the PAA polycarboxylic acid and the OH groups of the glycosidic moiety of the (AG) biopolymer. Therefore, we conclude that the polymeric ligand (poly-PAA/AG) was obtained throw a poly-esterification reaction between the alginate(AG) and the PAA crosslinking agent. The FT-IR spectrum of the (poly-PAA/AG) ligand showed several characteristic bands also existent in the spectra of the initial coordination polymers. Here, we notice the presence of a large band thereabout at 3222 cm^{-1} which is referred to as the hydroxyl group stretching (O-H). The (C-O-C) stretching vibration of the glycosidic structure was visible in a strong band near 1025 cm^{-1} [45]. The pyranose's C-O-C glycosidic bonds are attributed to absorption bands with a wavelength between 1000 and 1100 cm^{-1}. We observe some apparent shifts in the spectra of the compounds when compared to the polymeric ligand (poly-PAA/AG) alone, particularly at 3222 cm^{-1} (O-H), which suggests that the hydroxyl groups of the AG and the various metals have a coordinating binding (Table 2). Furthermore, for $\nu_{(COO)sym}$ at about 1413 cm^{-1} and $\nu_{(COO)asym}$ at around 1600 cm^{-1}, we notice clear varied shifts with the used metal revealing the existence of interaction bounds between the different carboxylates of the AG and the various transition elements. In addition, at these wavenumber frequencies, the IR spectra of the complexes showed separation values $\Delta\upsilon$ (representing the difference between $\nu_{(COO)asym}$ and $\nu_{(COO)sym}$) which are around 277–312 cm^{-1}, pointing to a monodentate mode of coordination for the carboxylate group in a network of hydrogen bonds [46]. The IR spectra of the polymeric complexes show a band at 409–471 cm^{-1} allocated to (M-Cl) [47]. The weak bands at 409–471 cm^{-1} in the far IR regions were observed for complexes and are attributed to chloride anions coordinated in a trans-octahedral geometry. The existence of coordinated M-O is assured by the band at 515–582 cm^{-1} that appears with the four complexes [48]. We can deduce that the various functional groups of the prepared crosslinked coordination polymer, the poly-PAA/AG, which is mainly hydroxyl (OH), carboxylic (COOH) and ester groups, are able to coordinate with the different transition element ions and especially in a bidentate type.

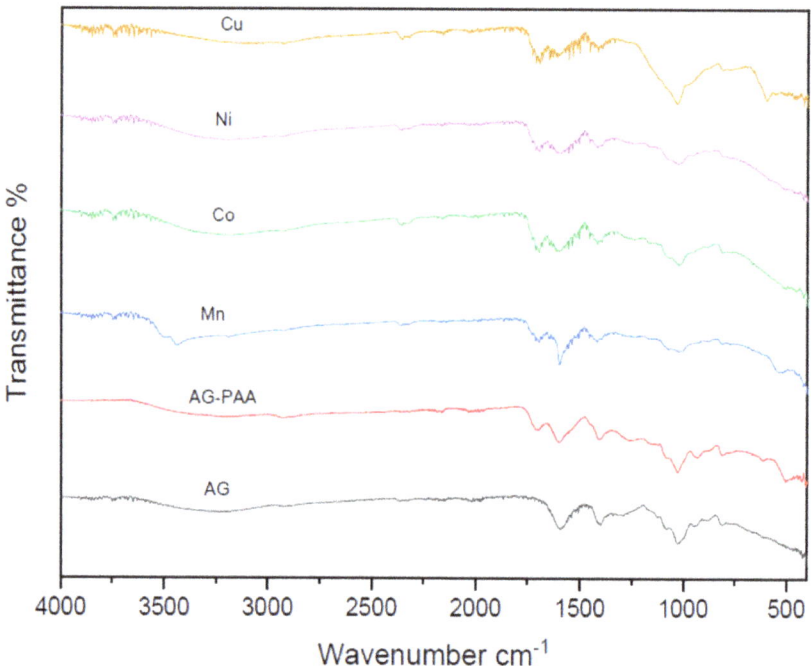

Figure 3. FT-IR spectra of the alginate (AG), polymeric ligand (poly-PAA/AG) and the different synthesized Co(II), Cu(II), Mn(II) and Ni(II) polymeric compounds.

3.4. Electronic Spectra

The electronic spectra of transition metal polymeric and their free ligands (poly-PAA/AG) are measured in the presence of DMSO (10^{-3} M). The outcomes show up in Table 3. The UV–Vis spectrum of (AG) peaks at 254 nm are appropriated to ($\pi \rightarrow \pi^*$) [49]. PAA displayed bands of absorbance at 270 nm appropriated to $n \rightarrow \pi^*$ [50]. As for the electronic spectra of polymeric compounds for cobalt(II), copper(II), manganese(II) and nickel(II), it has special bands, in the visible zone of spectra assigned to d-d transition scopes at 492–580 nm. As for the bands identical to $\pi \rightarrow \pi^*$ and $n \rightarrow \pi^*$, transitions were shown in the scopes of 258–277 nm and 273–298 nm, respectively, which indicates a bonding between the ligand (poly-PAA/AG) and the transition metals. The electronic spectrum of the Mn(II) compound in DMSO solution exhibited a peak around 510 nm assignable to the $^6A_{1g} \rightarrow {}^4E_g, {}^4A_{1g}$ transition. These data along with magnetic data are compatible with an octahedral structure around the transition metal ion [51]. The cobalt(II) polymeric complex displayed a weaker broad absorption band at 560 nm appointed to $^4T_{1g}$ (F) $\rightarrow {}^4T_{1g}$ (P) transition of octahedral geometry [52]. The nickel(II) polymeric complex showed two bands located at 512 and 580 nm, which may be referred to as $^3A_{2g}(F) \rightarrow {}^3T_{1g}(P)(\upsilon 3)$ and $^3A_{2g}(F) \rightarrow {}^3T_{1g}(F)(\upsilon 2)$, which are characteristic of the nickel(II) ion, in an octahedral geometry [53]. The electronic spectrum of copper(II) showed a low-intensity broad band at 690 nm corresponding to the $^2E_g \rightarrow {}^2T_{2g}$ transition, suggesting a distorted octahedral structure around Cu(II) [54].

Table 3. Electronic spectral data of free ligand and its polymeric compounds and magnetic moments.

Ligands and the Complexes	λmax. (nm)	ῡ (cm^{-1})	Assignment	μ$_{eff}$ (B.M.)	Geometry
PAA	270	37,037	n→π*	-	-
AG	254	39,370	π→π*	-	-
Co(II) complex	560	17,857	$^4T_{1g}(F) \rightarrow {}^4T_{1g}(P)$	4.82	Octahedral
	273	36,630	n→π*		
	258	38,759	π→π*		
Cu(II) complex	690	14,490	$^2E_g \rightarrow {}^2T_{2g}$	1.74	Octahedral
	277	36,101	n→π*		
	263	38,022	π→π*		
Mn(II) complex	510	19,607	$^6A_{1g} \rightarrow {}^4E_g, {}^4A_{1g}$	5.94	Octahedral
	278	35,971	n→π*		
	263	38,022	π→π*		
Ni(II) complex	512	19,531	$^3A_{2g}(F) \rightarrow {}^3T_{1g}(P)(\upsilon 3)$	2.88	Octahedral
	580	17,241	$^3A_{2g}(F) \rightarrow {}^3T_{1g}(F)(\upsilon 2)$		
	286	34,965	n→π*		
	251	39,840	π→π*		

3.5. Magnetic Moment Analysis

The determination of magnetic susceptibilities (μ$_{eff}$) of the polymeric compounds was carried out at 25 °C. The data of the effective magnetic moment (μ$_{eff}$) were specified by the data of diamagnetic corrections and Pascal's constants according to Equation (2):

$$\mu_{eff} = 2828(\chi_M T)^{1/2} \text{ (in B.M.)} \quad (2)$$

where μ$_{eff}$ is the effective magnetic moment (in Bohr Magneton, B.M.), T is the temperature (K) and χ_M is the molar magnetic susceptibility after correction. The value of μeff for the Mn(II) complex was 5.94 μB, which suggested an octahedral geometry, whereas, in the case of the Co(II) complex, the magnetic measurements illustrated a 4.82 μB value, which is sufficient for octahedral geometry [55,56]. As for the Ni(II) compound, magnetic measurements revealed that the compound has a μeff value of 2.88 μB, which is compatible with the range of the expected octahedral geometry of the nickel(II) compound [57]. Moreover, it has been found that the Cu(II) compound stabilizes in an octahedral geometry with a value of 1.74 μB [58].

The suggested structures for nano-sized Co(II), Cu(II), Mn(II) and Ni(II) coordination PAA/alginate biopolymers are shown in Figure 4.

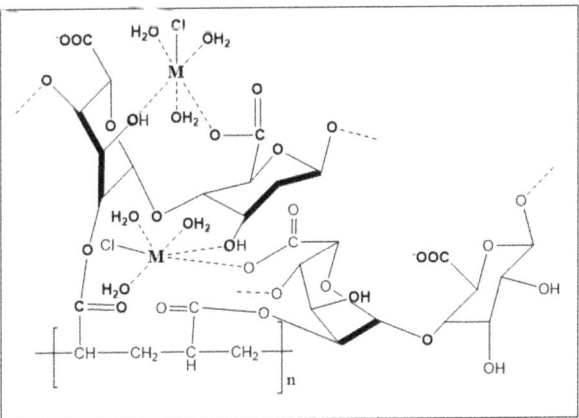

Figure 4. Suggested structure of metallic coordination polymers.

Where M = Cobalt(II), copper(II), manganese(II) or nickel(II).

3.6. Thermal Analysis

Thermal analyses of the manufactured compounds were performed up to 500 °C. Using this method, complex composition, temperature stability and the presence or absence of water molecules (if any) inside or outside the compounds inner coordination sphere can all be determined. TGA thermograms of the synthesized ligand and transition metal polymeric complexes showed a gradual weight reduction, indicating disintegration by fragmentation as the temperature increased. The findings revealed a good agreement between the calculated data, and the suggested formulae for weight loss (Table 4).

Table 4. Thermal analysis of polymeric complexes.

Compounds	Stage	Temp. Range (°C)	TGA (Wt. Loss) (%) Found	Calcd.	Assignment
Co(II) complex	1st	65–150	13.94	14.32	Loss of three water molecules
	2nd	152–220	9.05	9.38	Loss of chloride atom
	3rd	222–350	59.83	60.94	Decomposition rest of the organic ligand AG-PAA with
	4th	352–500			the formation of cobalt oxide
Cu(II) complex	1st	60–118	13.98	14.15	Loss of three water molecules
	2nd	120–205	9.18	9.27	Loss of chloride atom
	3rd	207–326	59.32	60.20	Decomposition rest of the organic ligand AG-PAA with
	4th	328–500			the formation of copper oxide
Mn(II) complex	1st	70–142	13.90	14.46	Loss of three water molecules
	2nd	144–202	9.12	9.48	Loss of chloride atom
	3rd	204–295	59.88	61.59	Decomposition rest of the organic ligand AG-PAA with
	4th	297–500			the formation of manganese oxide
Ni(II) complex	1st	64–138	13.65	14.31	Loss of three water molecules
	2nd	140–230	9.08	9.39	Loss of chloride atom
	3rd	232–368	59.72	60.98	Decomposition rest of the organic ligand AG-PAA with
	4th	370–500			the formation of nickel oxide

3.6.1. [Co(AG-PAA)Cl(H$_2$O)$_3$] Complex

Four decomposition stages were observed for the thermolysis curve of the Co(II) polymeric complex. These occur in the temperature ranges 65–150, 152–220, 222–350 and 352–500 °C. The first mass loss correlates well with the release of three H$_2$O molecules. This may be referred to as an ion-dipole interaction between cobalt and water. The corresponding mass loss was found to 13.94% (calc. 14.32%). For this step, a DTG midpoint appears at 92 °C with an endothermic peak in the DTA curve, at 94 °C. In the subsequent steps, the decomposition of the complex proceeds. The ultimate product was characterized as cobalt oxide (calc. 19.84%, found 17.18%). Scheme 1 depicts these decompositions.

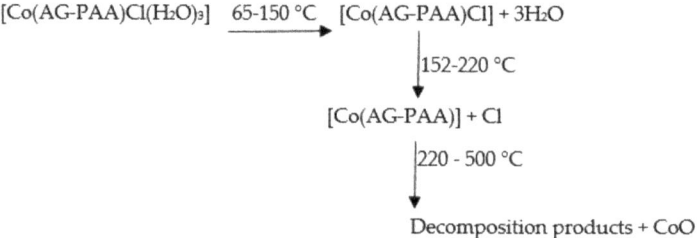

Scheme 1. Steps for decomposition of Co(II) polymeric complex.

3.6.2. [Cu(AG-PAA)Cl(H$_2$O)$_3$] Complex

The TG thermogram of [Cu(AG-PAA)Cl(H$_2$O)$_3$] exhibits four inflection points corresponding to four decomposition steps. Three water molecules are released in the first step, (calc. 14.15%, found 13.98%). For this step (thermal gravimetric analysis) (peak at 96 °C), an endothermic effect is observed at 98 °C in the differential thermal analysis trace. The observed mass loss in the 2nd step (120–205 °C) agrees well with the expected loss of chloride atoms (calc. 9.27%, found 9.18%) (DTG peak at 180 °C). This step is marked on the differential thermal analysis (DTA) curve as an exothermic peak at 182 °C. Then, the decomposition products are produced in the rest of the steps. The 3rd and 4th steps are consistent with the decomposition of rest products (calc. 60.20%, found 59.32%) (thermal gravimetric analysis) (peaks at 290 and 386 °C), for which exothermic peaks at 292 and 388 °C are recorded in the DTA trace. The residue was suggested to be CuO on the basis of mass loss consideration (calc. 20.80%, found 17.52%).

3.6.3. [Mn(AG-PAA)Cl(H$_2$O)$_3$] Complex

There were four distinct stages of decomposition visible in the Mn(II) complex thermograms. The first stage of decomposition occurs between 70 °C and 142 °C, where the loss of three H$_2$O molecules results in a rated mass loss of 13.90% (calculated to be 14.46%) (thermal gravimetric analysis TG) (peak at 98 °C), for which a broad, endothermic peak manifests in the DTA curve at 100 °C. In the second stage, chloride atoms are lost with a mass loss of 9.12% percent (calculated as 9.48 percent) between 144 and 202 °C. At this stage, a DTG peak manifests at 182 °C, and a broad endothermic effect is listed in the differential thermal analysis trace at 184 °C. The third and fourth stages of decomposition took place at 204–295 °C and 297–500 °C, respectively, as a result of the remain of the organic ligand AG-PAA decomposing and forming the final product (MnO), which experienced mass losses of 59.88% (calculated at 61.59%) and 17.10% (calculated at 18.98%). These stages are manifested in the DTG curve as peaks at 244 and 384 °C, and the DTA trace furnishes a broad exothermic effect at 246 and 386 °C. (Figure 5).

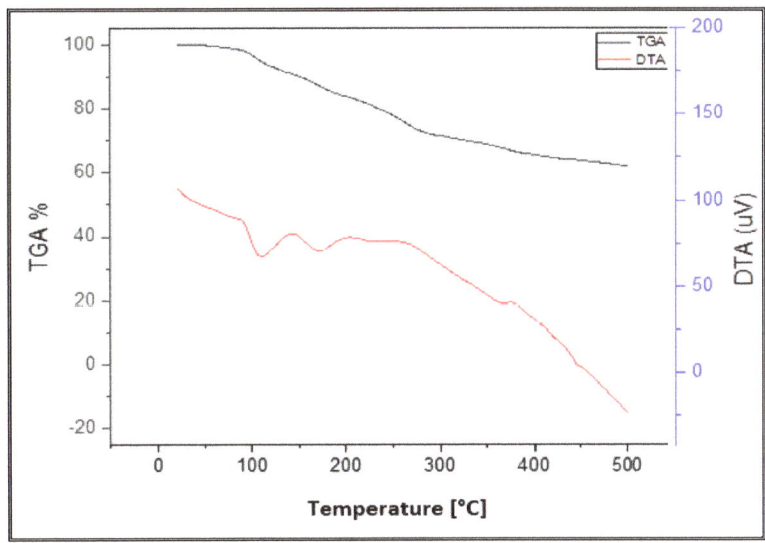

Figure 5. Thermal gravimetric analysis and differential thermal analysis thermograms of Mn(II) compound.

3.6.4. [Ni(AG-PAA)Cl(H$_2$O)$_3$] Complex

The decomposition of [Ni(AG-PAA)Cl(H$_2$O)$_3$] proceeds in four distinct steps in the temperature ranges 64–138, 140–230, 232–368 and 370–500 °C. The 1st mass loss correlates well with the relief of three H$_2$O molecules (calc. 14.31%, found 13.65%) (thermal gravimetric analysis) (peak at 98 °C) with an endothermic broad peak in the differential thermal analysis trace at 100 °C. The second mass loss is compatible with the expulsion of the chloride atom (calc. 9.39%, found 9.08%), A thermal gravimetric analysis peak at 202 °C and an exothermic effect at 204 °C in the differential thermal analysis trace are noticed. The third and fourth mass loss represents the release of the rest of organic (AG-PAA) ligands (calc. 60.98%, found 59.72%). The final product is consistent with NiO (calc. 19.79%, found 17.55%).

3.7. X-ray Powder Diffraction

This XRD study is another piece of proof for the emergence of metal–ligand complexes. The compounds' XRD patterns were documented. Table 5 lists the crystal structures of various compounds. The comparison of the diffraction patterns of the ligands with the obtained complexes.

Table 5. X-ray powder diffraction data of the different polymeric compounds (Cu-Kα radiation (λ_{Cu} = 0.154059 Å)).

Parameters	Crosslinked Polymeric Ligand (Poly-PAA/AG)	Co(II) Complex	Cu(II) Complex	Mn(II) Complex	Ni(II) Complex
Empirical formula	C$_9$H$_{10}$O$_7$	C$_9$H$_{15}$CoClO$_{10}$	C$_9$H$_{15}$CuClO$_{10}$	C$_9$H$_{15}$MnClO$_{10}$	C$_9$H$_{15}$NiClO$_{10}$
Formula Weight	230.12	377.60	382.21	373.60	377.36
a (Å)	20.22	7.967	7.8447	7.4014	5.673
b (Å)	11.58	7.967	7.8447	8.7901	5.655
c (Å)	20.74	7.967	26.358	3.6889	8.004
Alfa (o)	90.00	90.00	90.00	90.00	90.35
Beta (o)	110.64	90.00	90.00	98.165	90.72
gamma (o)	90.00	90.00	120.00	90.00	89.99
Crystal system	Monoclinic	Cubic	Hexagonal	Monoclinic	Triclinic
Space group	C12/m1	Fm-3m	R-3m	C2/m	I-1
Volume of unit cell (Å^3)	4498	505.8	1404.7	237.57	256.7
Particle size (nm)	171	14	38	33	68

The XRD diffractogram of the ligand film indicates the presence of an amorphous structure, while the prepared Co(II), Cu(II), Mn(II) and Ni(II) compounds were crystalline. All polymeric complexes have sharp peaks in Figure 6. On the other hand, the Mn(II) compound has a monoclinic crystal system while the Co(II) compound has a cubic crystal system. The Ni(II) and Cu(II) compounds have a triclinic, and a hexagonal crystal system, respectively. The significant broadening of the peaks illustrates that the particles are of nm dimensions. Scherrer's Equation (3) was used to estimate the particle size of the polymeric complexes.

$$D = K\lambda / \beta \cos\theta \quad (3)$$

where K is the shape factor, λ is the X-ray wavelength typically 1.54 Å, β is the line broadening at half the maximum intensity in radians and θ is Bragg angle. D is the mean size of the ordered (crystalline) domains, which may be smaller or equal to the grain size. Scherrer's equation is limited to nanoscale particles. The average size of the particles lies in the range of 14–68 nm which is in agreement with that noticed by scanning electron microscopy.

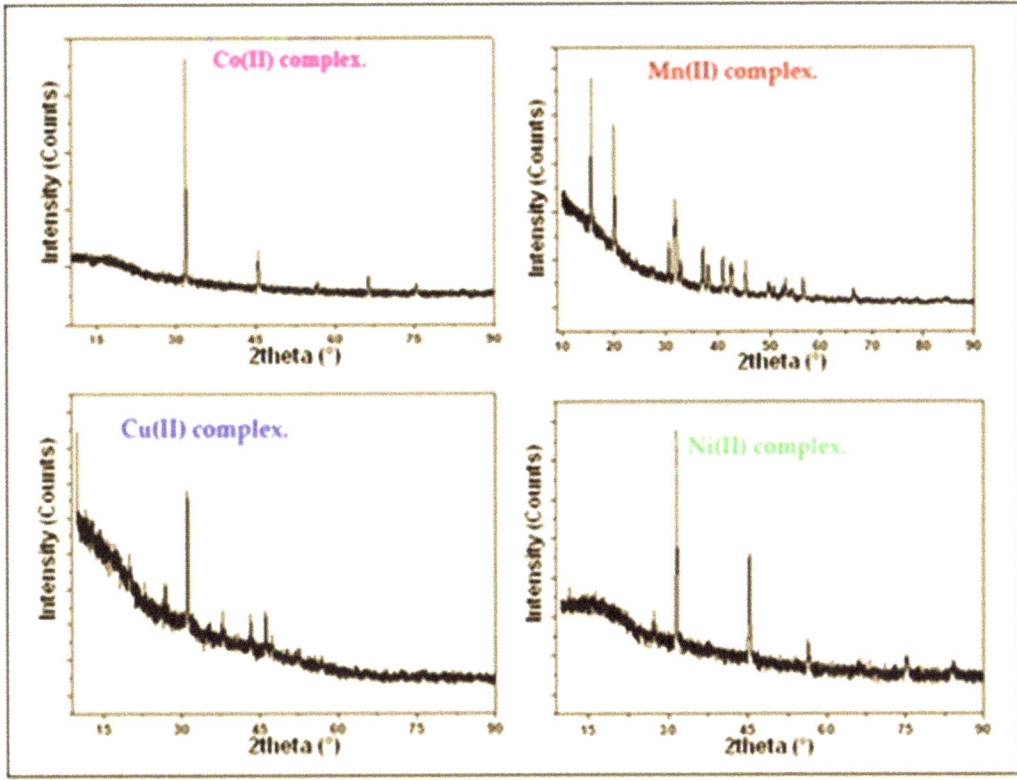

Figure 6. X-ray diffraction patterns of the different synthesized polymeric complexes (Cu-Kα radiation (λ_{Cu} = 0.154059 Å)).

3.8. SEM Morphological Analysis

Electron microscopy (EM) has played a vital function in polymeric complexes properties and analysis. In addition to extremely more magnification levels, it can provide us with significant information containing composition and novel materials' morphology. It can also show distinguishing contaminant materials that are only existent in polymers in trace amounts. Figure 7 displays as a general note an evident difference in surface morphology which is diverse depending on the type of transition element used in prepared polymeric complexes. The crosslinked polymeric ligand (poly-PAA/AG) revealed a porous and irregular surface reflecting its considerable hydrophilic property. This surface property could be efficient in the fashioning of ligand (poly-PAA/AG-metal coordination) polymers, and given that polymeric complexes are readily available in solution, continuous crosslinking into nanostructures is expected. While such particles have been used in the biomedical industry [59], their presence does support the aim to synthesize homogenous Co(II), Cu(II), Mn(II) and Ni(II) coordination PAA/alginate biopolymer NPs. A microscopic image of Mn(II) and Co(II) polymeric compounds illustrates small particles with a diameter between 20 and 29 nm. The surface morphology of the Ni(II) compound appears as uniform spherical particles with a diameter between 60 and 65 nm. In the Cu(II) complex, the SEM shows different shapes with diameters between 34 and 40 nm.

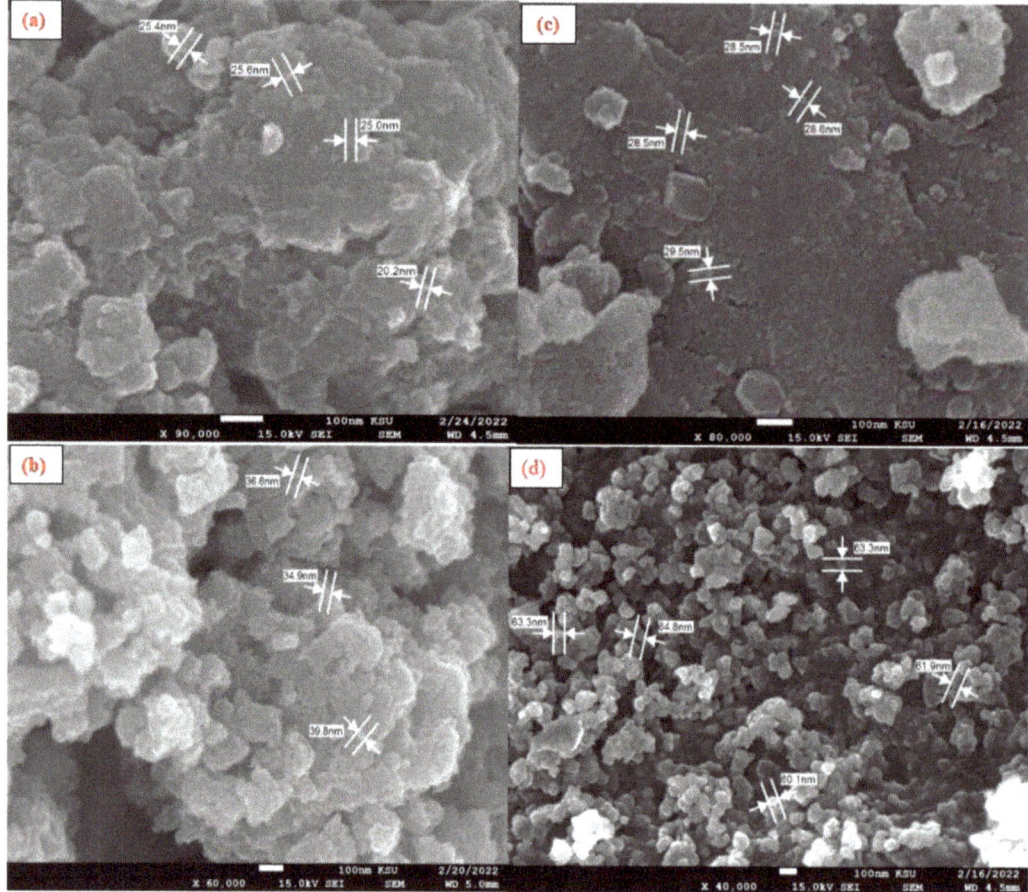

Figure 7. SEM of the different prepared polymeric complexes, (**a**) Co(II) complex, (**b**) Cu(II) complex, (**c**) Mn(II) complex and (**d**) Ni(II) complex.

3.9. Antimicrobial and Antioxidant Assays

The test polymeric complex NPs exhibit a broad spectrum of antimicrobial activity because they are active against both Gram-positive and Gram-negative bacteria, but especially against Gram-positive strains, according to the antimicrobial results summarized in Table 6 and Figure 8a–d. The inhibitory zone diameter of the other extracts reached 5 cm, demonstrating their strong antimicrobial activity. The strongest are copper(II) and cobalt(II) compounds, which have a strain-dependent antimicrobial effect. The antioxidant effect of the different polymeric complexes is displayed in Figure 9.

Table 6. Antimicrobial and antioxidant effect (Inhibitory zone expressed in cm ± SD).

Compound	Antimicrobial					Antioxidant
	S1	S4	S5	S10	9C	
Co(II) complex	2.1	1.3	2.2	1.4	3.1	71 ± 1.4
Cu(II) complex	5.4	1.4	1.7	2.9	4.1	82 ± 1.4
Mn(II) complex	1	1.65	1.4	1.4	1.3	61 ± 0.6
Ni(II) complex	1.9	1.3	1.9	1.2	1.35	70.5 ± 0.7

SD: Standard Deviation.

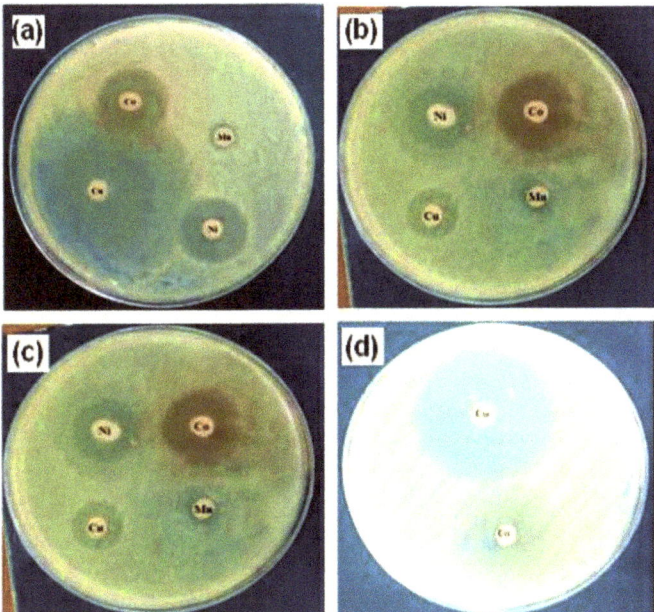

Figure 8. Microbiological activity against (**a**) *Staphylococcus aureus*, (**b**) *Escherichia coli*, (**c**) *Salmonella typhimurium* and (**d**) *Candida albicans*.

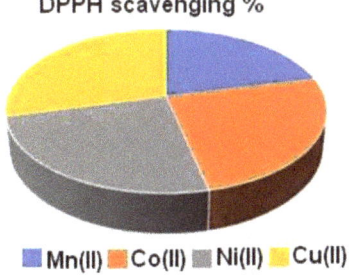

Figure 9. Antioxidant effect of the different polymeric complexes.

3.10. In Vitro Anticancer Assay

In vitro, cell assays proceeded to investigate both the biocompatibility and anticancer activity of the prepared polymeric compounds. Human breast cancer cells (MCF-7) were used to assess the anticancer potential of the crosslinked polysaccharide ligand (poly-AG/PAA) and its two metal coordination polymers, the poly-AG/PAA/copper(II) and the poly-AG/PAA/cobalt(II). The assessment of the anticancer activity was carried out via the treatment of the cancer cells using the MTT assay by varying the concentration of the different polymeric vectors. The outcomes in Figure 10 displayed no cytotoxicity of the polymeric ligand in the approach with normal epithelial cells. This was reported previously by various research investigations that showed the excellent biocompatibility of the naturally extracted polysaccharides [60,61]. Therefore, the crosslinked polymeric ligand based on alginate biopolymer can be considered as a safe drug delivery carrier and other pharmaceutical applications. In addition, the synthesized polymeric compounds of Cu(II) and Co(II) revealed excellent cytocompatibility after 72 h in contact with the normal human breast epithelial cells(MCF10A), with the various applied concentrations. This was an important stage; meanwhile, the potential of anticancer drugs significantly concentrated on their biocompatibility with the normal counterparts of the analyzed cancer cells. The anticancer valuation in Figure 11 exhibited that the polymeric ligand without metals has the ability to kill cancer cells, and this is from a dose of 50 g/mL. The anticancer performance increases by sample concentration, to reach 51% at a concentration of 200 g/mL, which is in concordance with some previous studies [62].

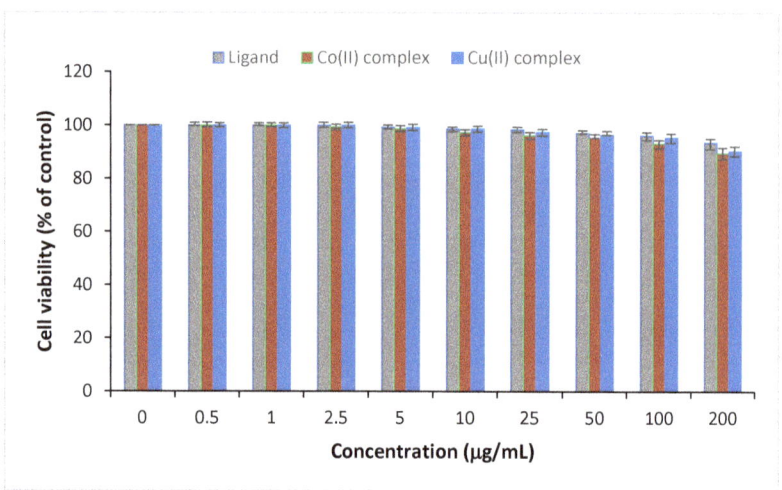

Figure 10. Cell viability assessments in human normal beast cells of the ligand and the polymeric compounds via cell growth inhibition rates by varying the concentration after 72 h incubation on MCF-10A and MCF-7 cell lines.

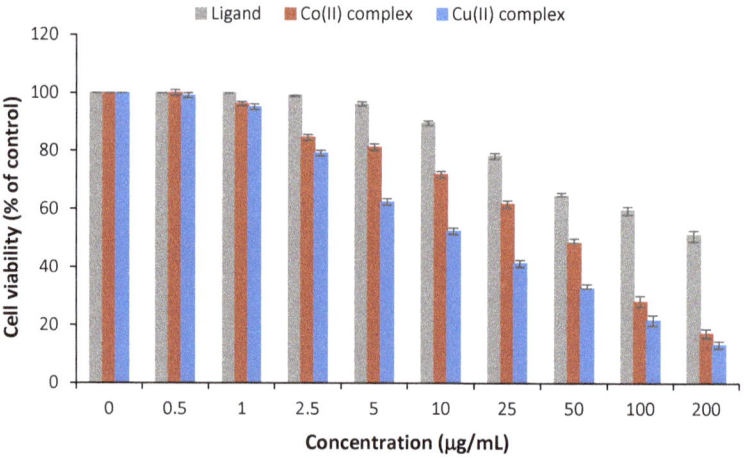

Figure 11. Anticancer activity in human breast cancer cells of the ligand and the polymeric compounds via cell growth inhibition rates by varying the concentrate ion after 72 h incubation on MCF-10A and MCF-7 cell lines.

Concerning the two selected polymeric complexes, results showed a good ability to kill the cancer cells in a dose-dependent method. The copper(II) polymeric compound confirms its excellent performance shown in microbial activity and demonstrates a higher anticancer potential compared to the Co(II) complex with the different polymeric doses. After a period of incubation fixed at 72 h and at a dose of 200 g/mL, the transition metal complexation significantly lowered the viability of the MCF-7 cancer cells up to $17.1 \pm 1.21\%$

and 13.2 ± 1.52% for the copper(II) and the cobalt(II) polymeric compounds, respectively. In summary, both the polymeric ligand (poly-AG/PAA) and the Cu(II) and Co(II) compounds exhibited good cytocompatibility with normal human epithelial cells. Furthermore, the in vitro anticancer evaluation showed the high anticancer performance of the Cu(II) and Co(II) polymeric compounds with a potential rising with the used polymeric concentration.

4. Conclusions

Polymeric complex NPs were prepared from polyacrylic acid (PAA) and sodium alginate (AG) with cobalt(II), copper(II), manganese(II) and nickel(II) chlorides. The various physico-chemical analyses included elemental analysis (C and H), FT-IR, UV–Vis spectra, TG, DTA, XRD and SEM. The magnetic studies suggested the octahedral geometrical structure for all produced polymeric complexes. Alginate and polyacrylic acid were used because they are biocompatible materials. As it is known, (PAA) gels are suitable biocompatible matrices for medical applications, such as gels for skin care products. PAA films can be deposited on orthopedic implants to protect them from corrosion. Crosslinked hydrogels of PAA and gelatin have also been used as in medical glue. In addition, Alginate (AG), which is rich in sources, is non-toxic and has excellent biocompatibility, biodegradability and safety, has been widely used in medical wound dressings, drug delivery carriers and delivery of bioactive substances in tissue engineering and skeleton materials. Therefore, nano-sized metallic polymers were prepared from these biopolymers, which really gave excellent results with the microbes that were tested, whether Gram-positive or Gram-negative bacteria or fungi. In summary, both the polymeric ligand (poly-AG/PAA) and the two metal complexes exhibited good cytocompatibility with normal human epithelial cells. Furthermore, the in vitro anticancer evaluation showed the high anticancer performance of the poly-AG/PAA/Cu(II) and poly-AG/PAA/Co(II) compounds with a potential rise with the used polymeric concentration.

Author Contributions: Data curation, M.S.A.-F., M.S.A. and Y.E.-G.; formal analysis, M.S.A.-F., M.S.A. and Y.E.-G.; investigation, M.S.A.-F., M.S.A. and Y.E.-G.; methodology, M.S.A.-F. and Y.E.-G.; project administration, M.S.A.-F.; software, M.S.A.-F. and M.S.A.; supervision, M.S.A.-F. and Y.E.-G.; validation, M.S.A.-F. and Y.E.-G.; writing—original draft, M.S.A.-F. and M.S.A.; writing—review and editing, M.S.A.-F. All authors have read and agreed to the published version of the manuscript.

Funding: This research received no external funding.

Data Availability Statement: The data are contained within the article.

Conflicts of Interest: The authors declare no conflict of interest.

References

1. Ezeoha, S.L. Production of Biodegradable Plastic Packaging Film from Cassava Starch. *IOSR J. Eng.* **2013**, *3*, 14–20. [CrossRef]
2. Rebelo, R.; Fernandes, M.; Fangueiro, R. Biopolymers in medical implants: A brief review. *Procedia Eng.* **2017**, *200*, 236–243. [CrossRef]
3. Bala, I.A.; Abdullahi, M.R.; Bashir, S.S. A review on formulation of enzymatic solution for biopolymer hydrolysis. *J. Chem.* **2017**, *6*, 9–13.
4. Pattanashetti, N.A.; Heggannavar, G.B.; Kariduraganavar, M.Y. Smart biopolymers and their biomedical applications. *Procedia Manuf.* **2017**, *12*, 263–279. [CrossRef]
5. Qasim, U.; Osman, A.I.; Al-Muhtaseb, A.H.; Farrell, C.; Al-Abri, M.; Ali, M.; Vo, D.V.N.; Jamil, F.; Rooney, D.W. Renewable cellulosic nanocomposites for food packaging to avoid fossil fuel plastic pollution: A review. *Environ. Chem. Lett.* **2021**, *19*, 613–641. [CrossRef]
6. Nessi, V.; Falourd, X.; Maigret, J.-E.; Cahier, K.; D'Orlando, A.; Descamps, N.; Gaucher, V.; Chevigny, C.; Lourdin, D. Cellulose nanocrystals-starch nanocomposites produced by extrusion: Structure and behavior in physiological conditions. *Carbohydr. Polym.* **2019**, *225*, 115123. [CrossRef]
7. Barclay, T.G.; Day, C.M.; Petrovsky, N.; Garg, S. Review of polysaccharide particle-based functional drug delivery. *Carbohydr. Polym.* **2019**, *221*, 94–112. [CrossRef]
8. Reis, A.V.; Guilherme, M.R.; Cavalcanti, O.A.; Rubira, A.F.; Muniz, E.C. Synthesis and characterization of pH-responsive hydrogels based on chemically modified Arabic gum polysaccharide. *Polymer* **2006**, *47*, 2023–2029. [CrossRef]

9. Farooqi, Z.U.R.; Qadeer, A.; Hussain, M.M.; Zeeshan, N.; Ilic, P. Characterization and physicochemical properties of nanomaterials. In *Nanomaterials: Synthesis, Characterization, Hazards and Safety*; Elsevier: Amsterdam, The Netherlands, 2021; pp. 97–121. [CrossRef]
10. Salem, S.S.; Fouda, A. Green Synthesis of Metallic Nanoparticles and Their Prospective Biotechnological Applications: An Overview. *Biol. Trace Elem. Res.* **2021**, *199*, 344–370. [CrossRef]
11. Nikalje, A.P. Nanotechnology and its applications in medicine. *Med. Chem.* **2015**, *5*, 081–089. [CrossRef]
12. Barik, T.K.; Maity, G.C.; Gupta, P.; Mohan, L.; Santra, T.S. Nanomaterials: An Introduction. *Nanomater. Their Biomed. Appl.* **2021**, *16*, 1–11.
13. Singh, B.K.; Lee, S.; Na, K. An overview on metal-related catalysts: Metal oxides, nanoporous metals and supported metal nanoparticles on metal organic frameworks and zeolites. *Rare Metals* **2020**, *39*, 751–766. [CrossRef]
14. Salem, S.S.; Fouda, M.M.G.; Fouda, A.; Awad, M.A.; Al-Olayan, E.M.; Allam, A.A.; Shaheen, T.I. Antibacterial, Cytotoxicity and Larvicidal Activity of Green Synthesized Selenium Nanoparticles Using Penicillium corylophilum. *J. Clust. Sci.* **2021**, *32*, 351–361. [CrossRef]
15. Khan, S.; Mansoor, S.; Rafi, Z.; Kumari, B.; Shoaib, A.; Saeed, M.; Alshehri, S.; Ghoneim, M.M.; Rahamathulla, M.; Hani, U.; et al. A review on nanotechnology: Properties, applications, and mechanistic insights of cellular uptake mechanisms. *J. Mol. Liquids* **2021**, *347*, 118008. [CrossRef]
16. Pérez-Hernández, H.; Pérez-Moreno, A.; Sarabia-Castillo, C.; García-Mayagoitia, S.; Medina-Pérez, G.; López-Valdez, F.; Campos-Montiel, R.; Jayanta-Kumar, P.; Fernández-Luqueño, F. Ecological Drawbacks of Nanomaterials Produced on an Industrial Scale: Collateral Effect on Human and Environmental Health. *Water Air Soil Pollut.* **2021**, *232*, 435. [CrossRef] [PubMed]
17. Singh, R.; Singh, S. Nanomanipulation of consumer goods: Effects on human health and environment. In *Nanotechnology in Modern Animal Biotechnology*; Springer: Berlin/Heidelberg, Germany, 2019; pp. 221–254. [CrossRef]
18. Pathakoti, K.; Goodla, L.; Manubolu, M.; Hwang, H.-M. Nanoparticles and Their Potential Applications in Agriculture, Biological Therapies, Food, Biomedical, and Pharmaceutical Industry: A Review. In *Nanotechnology and Nanomaterial Applications in Food, Health, and Biomedical Sciences*; Apple Academic Press: Palm Bay, FL, USA, 2019; pp. 121–162. ISBN 9780429425660.
19. Elkodous, M.A.; El-Husseiny, H.M.; El-Sayyad, G.S.; Hashem, A.H.; Doghish, A.S.; Elfadil, D.; Radwan, Y.; El-Zeiny, H.M.; Bedair, H.; Ikhdair, O.A.; et al. Recent advances in waste-recycled nanomaterials for biomedical applications: Waste-to-wealth. *Nanotechnol. Rev.* **2021**, *10*, 1662–1739. [CrossRef]
20. Parker, G. Measuring the environmental performance of food packaging: Life cycle assessment. In *Environmentally Compatible Food Packaging*; Elsevier: Amsterdam, The Netherlands, 2008; pp. 211–237.
21. Shankar, S.; Rhim, J.-W. Bionanocomposite films for food packaging applications. In *Reference Module in Food Science*; Elsevier: Amsterdam, The Netherlands, 2018; pp. 1–10.
22. Hassan, M.E.; Bai, J.; Dou, D.Q. Biopolymers; Definition, classification and applications. *Egypt. J. Chem.* **2019**, *62*, 1725–1737. [CrossRef]
23. Soldo, A.; Mileti'c, M.; Auad, M.L. Biopolymers as a sustainable solution for the enhancement of soil mechanical properties. *Sci. Rep.* **2020**, *10*, 267. [CrossRef]
24. Song, J.; Winkeljann, B.; Lieleg, O. Biopolymer-based coatings: Promising strategies to improve the biocompatibility and functionality of materials used in biomedical engineering. *Adv. Mater. Interfaces* **2020**, *7*, 2000850. [CrossRef]
25. Jummaat, F.; Yahya, E.B.; Khalil H.P.S., A.; Adnan, A.S.; Alqadhi, A.M.; Abdullah, C.K.; A.K., A.S.; Olaiya, N.G.; Abdat, M. The Role of Biopolymer-Based Materials in Obstetrics and Gynecology Applications: A Review. *Polymers* **2021**, *13*, 633. [CrossRef] [PubMed]
26. Reddy, M.S.B.; Ponnamma, D.; Choudhary, R.; Sadasivuni, K.K. A Comparative Review of Natural and Synthetic Biopolymer Composite Scaffolds. *Polymers* **2021**, *13*, 1105. [CrossRef]
27. Olivia, M.; Jingga, H.; Toni, N.; Wibisono, G. Biopolymers to improve physical properties and leaching characteristics of mortar and concrete: A review. In *IOP Conference Series: Materials Science and Engineering*; Institute of Physics Publishing: Bristol, UK, 2018; Volume 345, p. 012028.
28. Darge, H.F.; Andrgie, A.T.; Tsai, H.C.; Lai, J.Y. Polysaccharide and polypeptide based injectable thermo-sensitive hydrogels for local biomedical applications. *Int. J. Biol. Macromol.* **2019**, *133*, 545–563. [CrossRef]
29. Mohammed, A.S.A.; Naveed, M.; Jost, N. Polysaccharides; classification, chemical properties, and future perspective applications in fields of pharmacology and biological medicine (a review of current applications and upcoming potentialities). *J. Polym. Environ.* **2021**, *29*, 2359–2371. [CrossRef]
30. Grasdalen, H.; Larsen, B.; Smidsrød, O. 13C-NMR studies of alginate. *Carbohydr. Res.* **1977**, *56*, C11–C15. [CrossRef]
31. Grasdalen, H.; Larsen, B.; Smidsrød, O. A pmr study of the composition and sequence of uronate residues in alginates. *Carbohydr. Res.* **1979**, *68*, 23–31. [CrossRef]
32. Kothale, D.; Verma, U.; Dewangan, N.; Jana, P.; Jain, A.; Jain, D. Alginate as Promising Natural Polymer for Pharmaceutical, Food, and Biomedical Applications. *Curr. Drug Deliv.* **2020**, *17*, 755–775. [CrossRef]
33. EL-Ghoul, Y.; Al-Fakeh, M.S.; Al-Subai, N. Synthesis and Characterization of a New Alginate/Carrageenan crosslinked biopolymer and Study of the Antibacterial, Antioxidant, and Anticancer Performance of its Mn(II), Fe(III), Ni(II), and Cu(II) polymeric complexes. *Polymers* **2023**, *15*, 2511. [CrossRef]

34. da Silva, T.L.; Vidart, J.M.M.; da Silva, M.G.C.; Gimenes, M.L.; Vieira, M.G.A. Alginate and sericin: Environmental and pharmaceutical applications. *Biol. Act. Appl. Mar. Polysacch.* **2017**, 57–86. [CrossRef]
35. Draget, K.I.; Taylor, C. Chemical, physical and biological properties of alginates and their biomedical implications. *Food Hydrocolloids* **2011**, *25*, 251–256. [CrossRef]
36. Gheorghita Puscaselu, R.; Lobiuc, A.; Dimian, M.; Covasa, M. Alginate: From food industry to biomedical applications and management of metabolic disorders. *Polymers* **2020**, *12*, 2417. [CrossRef]
37. Orwoll, R.A.; Chong, Y.S. Poly(acrylic acid). In *Polymer Data Handbook*; James, M., Ed.; Oxford University Press: Oxford, UK, 1999; pp. 252–253.
38. Zhuo, W. *Synthesis, Characterization, and Self-Assembly of Amphiphilic Copolymer Based on Poly (Acrylic Acid)*; National University of Singapore: Singapore, 2015.
39. Arkaban, H.; Barani, M.; Akbarizadeh, M.R.; Pal Singh Chauhan, N.; Jadoun, S.; Dehghani Soltani, M.; Zarrintaj, P. Polyacrylic Acid Nanoplatforms: Antimicrobial, Tissue Engineering, and Cancer Theranostic Applications. *Polymers* **2022**, *14*, 1259. [CrossRef] [PubMed]
40. Xu, M.; Zhu, J.; Wang, F.; Xiong, Y.; Wu, Y.; Wang, Q.; Weng, J.; Zhang, Z.; Chen, W.; Liu, S. Improved in vitro and in vivo biocompatibility of graphene oxide through surface modification: Poly (acrylic acid)-functionalization is superior to PEGylation. *ACS Nano* **2016**, *10*, 3267–3281. [CrossRef] [PubMed]
41. Mahon, R.; Balogun, Y.; Oluyemi, G.; Njuguna, J. Swelling performance of sodium polyacrylate and poly(acrylamide-co-acrylic acid) potassium salt. *SN Appl. Sci.* **2020**, *2*, 117–132. [CrossRef]
42. Dumitrașcu, A.-M.; Caraș, I.; Tucureanu, C.; Ermeneanu, A.-L.; Tofan, V.-C. Nickel (II) and Cobalt (II) Alginate Biopolymers as a "Carry and Release" Platform for Polyhistidine-Tagged Proteins. *Gels* **2022**, *8*, 66. [CrossRef] [PubMed]
43. Sharafshadeh, M.S.; Tafvizi, F.; Khodarahmi, P.; Ehtesham, S. Preparation and physicochemical properties of cisplatin and doxorubicin encapsulated by niosome alginate nanocarrier for cancer therapy. *Int. J. Biol. Macromol.* **2023**, *235*, 123686. [CrossRef]
44. Mahdhi, A.; Leban, N.; Chakroun, I.; Chaouch, M.A.; Hafsa, J.; Fdhila, K.; Mahdouani, K.; Majdoub, H. Extracellular polysaccharide derived from potential probiotic strain with antioxidant and antibacterial activities as a prebiotic agent to control pathogenic bacterial biofilm formation. *Microb. Pathogenesis* **2017**, *109*, 214–220. [CrossRef]
45. Ammar, C.; Alminderej, F.M.; EL-Ghoul, Y.; Jabli, M.; Shafiquzzaman, M. Preparation and Characterization of a New Polymeric Multi-Layered Material Based K-Carrageenan and Alginate for Efficient Bio-Sorption of Methylene Blue Dye. *Polymers* **2021**, *13*, 411. [CrossRef]
46. Deacon, G.B.; Phillips, R.J. Relationships between the carbon-oxygen stretching frequencies of carboxylato complexes and the type of carboxylate coordination. *Coord. Chem. Rev.* **1980**, *33*, 227–250. [CrossRef]
47. Al-Fakeh, M.S.; Alsikhan, M.A.; Alnawmasi, J.S. Physico-Chemical Study of Mn(II), Co(II), Cu(II), Cr(III), and Pd(II) Complexes with Schiff-Base and Aminopyrimidyl Derivatives and Anti-Cancer, Antioxidant, Antimicrobial Applications. *Molecules* **2023**, *28*, 2555. [CrossRef]
48. Al-Fakeh, M.S.; Messaoudi, S.; Alresheedi, F.I.; Albadri, A.E.; El-Sayed, W.A.; Saleh, E.E. Preparation, Characterization, DFT Calculations, Antibacterial and Molecular Docking Study of Co(II), Cu(II), and Zn(II) Mixed Ligand Complexes. *Crystals* **2023**, *13*, 118. [CrossRef]
49. Yue, W.; Zhang, H.H.; Yang, Z.N.; Xie, Y. Preparation of low-molecular-weight sodium alginate by ozonation. *Carbohydr. Polym.* **2021**, *251*, 117104. [CrossRef] [PubMed]
50. Vetriselvi, V.; Santhi Raj, R.J.J. Synthesis and characterization of poly acrylic acid modified with dihydroxy benzene-redox polymer. *Res. J. Chem. Sci.* **2014**, *4*, 78–86.
51. Parihari, R.K.; Patel, R.K.; Patel, R.N. Synthesis and characterization of metal complexes of manganese-, cobalt-and zinc (II) with Schiff base and some neutral ligand. *J. Indian Chem. Soc.* **2000**, *77*, 339.
52. Mohanan, K.; Murukan, B. Complexes of manganese (II), iron (II), cobalt (II), nickel (II), copper (II), and zinc (II) with a bishydrazone. *Synth. React. Inorg. Met. Org. Nano Met. Chemistry* **2005**, *35*, 837–844. [CrossRef]
53. Gupta, L.K.; Bansal, U.; Chandra, S. Spectroscopic and physicochemical studies on nickel(II) complexes of isatin 3,2′ quinolylhydrazones and their adducts. *Spectrochim. Acta Part A Mol. Biomol. Spectro.* **2007**, *66*, 972–975. [CrossRef]
54. Ajaykumar, D. Kulkarni, et, Synthesis, spectral, electrochemical and biological studies of Co(II), Ni(II) and Cu(II) complexes with Schiff bases of 8-formyl-7-hydroxy-4-methyl coumarin. *J. Coord. Chem.* **2009**, *62*, 481–492.
55. Al-Fakeh, M.S.; Allazzam, G.A.; Yarkandi, N.H. Ni (II), Cu (II), Mn (II), and Fe (II) Metal Complexes Containing 1, 3-Bis (diphenylphosphino) propane and Pyridine Derivative: Synthesis, Characterization, and Antimicrobial Activity. *Int. J. Biomater.* **2021**, *2021*, 4981367. [CrossRef]
56. Al-Fakeh, M.S.; Alsaedi, R.O. Synthesis, characterization, and antimicrobial activity of CoO nanoparticles from a Co (II) complex derived from polyvinyl alcohol and aminobenzoic acid derivative. *Sci. World J.* **2021**, *2021*, 6625216. [CrossRef]
57. Al-Fakeh, M.S. Synthesis, characterization and anticancer activity of NiO nanoparticles from a Ni (II) complex derived from chitosan and pyridine derivative. *Bulg. Chem. Commun.* **2021**, *53*, 321–326.
58. Al-Fakeh, M.S. Synthesis, thermal stability and kinetic studies of copper (II) and cobalt (II) complexes derived from 4-aminobenzohydrazide and 2-mercaptobenzothiazole. *Eur. Chem. Bulletin* **2020**, *9*, 403–409. [CrossRef]
59. Zhang, H.; Cheng, J.; Ao, Q. Preparation of Alginate-Based Biomaterials and Their Applications in Biomedicine. *Mar. Drugs* **2021**, *19*, 264. [CrossRef] [PubMed]

60. Geetha Bai, R.; Tuvikene, R. Potential Antiviral Properties of Industrially Important Marine Algal Polysaccharides and Their Significance in Fighting a Future Viral Pandemic. *Viruses* **2021**, *13*, 1817. [CrossRef] [PubMed]
61. Liu, C.; Jiang, F.; Xing, Z.; Fan, L.; Li, Y.; Wang, S.; Ling, J.; Ouyang, X.-K. Efficient Delivery of Curcumin by Alginate Oligosaccharide Coated Aminated Mesoporous Silica Nanoparticles and In Vitro Anticancer Activity against Colon Cancer Cells. *Pharmaceutics* **2022**, *14*, 1166. [CrossRef] [PubMed]
62. Gutiérrez-Rodríguez, A.G.; Juárez-Portilla, C.; Olivares-Bañuelos, T.; Zepeda, R.C. Anticancer activity of seaweeds. *Drug Discov. Today* **2018**, *23*, 434–447. [CrossRef]

Disclaimer/Publisher's Note: The statements, opinions and data contained in all publications are solely those of the individual author(s) and contributor(s) and not of MDPI and/or the editor(s). MDPI and/or the editor(s) disclaim responsibility for any injury to people or property resulting from any ideas, methods, instructions or products referred to in the content.

Article

Non-Collinear Phase in Rare-Earth Iron Garnet Films near the Compensation Temperature

Dmitry A. Suslov [1,*], Petr M. Vetoshko [1,2], Alexei V. Mashirov [1], Sergei V. Taskaev [3], Sergei N. Polulyakh [2], Vladimir N. Berzhansky [2] and Vladimir G. Shavrov [1]

1. Kotelnikov Institute of Radioengineering and Electronic, Russian Academy of Sciences, 125009 Moscow, Russia; pvetoshko@mail.ru (P.M.V.); a.v.mashirov@mail.ru (A.V.M.); shavrov@cplire.ru (V.G.S.)
2. V. I. Vernadsky Crimean Federal University, 295007 Simferopol, Russia; s.polulyakh@gmail.com (S.N.P.); v.n.berzhansky@gmail.com (V.N.B.)
3. Chelyabinsk State University, 454001 Chelyabinsk, Russia; s.v.taskaev@gmail.com
* Correspondence: sda_53@mail.ru

Abstract: The experimental discovery of the suppression effect of the non-collinear phase in strong magnetic fields near the compensation point in ferrimagnetic structures was made. The observations were carried out using the magneto-optical method by creating a lateral temperature gradient in the plane of the epitaxial films of iron garnets. The non-collinear phase is absent in weak magnetic fields. If an external magnetic field exceeds the first critical value, the non-collinear phase arises near the compensation point. The temperature range of the non-collinear phase expands due to the field increase up to the second critical value. Further field increases conversely reduce the temperature range of the non-collinear phase so that the field above the second critical value causes the disappearance of the non-collinear phase. The effect of the occurrence and suppression of the non-collinear phase is demonstrated on samples of two types of iron garnet films with two and three magnetic sublattices. Phase diagrams of the magnetic states in the vicinity of the critical point are constructed, and it is shown that the region of existence of the non-collinear phase in a two-sublattice magnet is smaller than in a three-sublattice one.

Keywords: magneto-optical method; compensation point; magnetic phases; ferrimagnetic materials

Citation: Suslov, D.A.; Vetoshko, P.M.; Mashirov, A.V.; Taskaev, S.V.; Polulyakh, S.N.; Berzhansky, V.N.; Shavrov, V.G. Non-Collinear Phase in Rare-Earth Iron Garnet Films near the Compensation Temperature. *Crystals* **2023**, *13*, 1297. https://doi.org/10.3390/cryst13091297

Academic Editors: John A. Mydosh and Raphaël P. Hermann

Received: 27 June 2023
Revised: 18 August 2023
Accepted: 20 August 2023
Published: 23 August 2023

Copyright: © 2023 by the authors. Licensee MDPI, Basel, Switzerland. This article is an open access article distributed under the terms and conditions of the Creative Commons Attribution (CC BY) license (https://creativecommons.org/licenses/by/4.0/).

1. Introduction

Currently, ferrimagnetic materials with a compensation temperature are well known [1]. The magnetic compensation temperature T_m is a special point on the temperature dependence of the magnetization of a ferrimagnet at which the magnetizations of oppositely directed magnetic sublattices fully compensate one other and the total spontaneous magnetization of the material becomes zero. The theoretical analysis of the behavior of a multi-sublattice ferrimagnet in an external magnetic field allows the prediction of the existence of three magnetic phases that exist in certain temperature ranges [2]. At low temperatures, the magnetizations of non-equivalent sublattices are antiparallel to each other and collinear with the magnetic field; the resulting magnetization decreases with increasing temperature. At temperatures above T_m, a collinear magnetic structure is realized, with magnetization directions of the sublattices inverted with respect to the low-temperature interval.

In the vicinity of the compensation point, the occurrence of a non-collinear (canted) phase is predicted. In this case, the sublattice magnetizations are non-collinear to each other as well as to the external magnetic field.

Thin films of compensated ferrimagnets based on the combination of rare-earth and transition metals having alloy or multilayer forms are currently considered key materials for spintronics and other modern technology [3]. Thin films of Mn_4N also demonstrate non-collinear ferrimagnetism and are interesting materials for spintronics applications [4]. Modern investigations of great interest for fundamental physics and applications relate to ultrafast magnetism and optical switching. The prospective materials to explore these phenomena in the non-collinear phase near the magnetization compensation point are thin ferrimagnetic films based on a combination of rare-earth and transition metals [5], such as on cation-substituted iron garnets [6–8].

According to molecular field theory, the temperature range of the existence of the non-collinear magnetic structure expands with increasing magnetic field up to the disappearance of collinear magnetic structures [2,9,10]. However, as far as the authors are aware, this finding has not been experimentally confirmed.

The aim of this work was to experimentally detect and investigate various magnetic phases near the spin-reorientation transition caused by temperature. We studied thin magnetic films of cation-substituted iron garnet using by two types of samples. The samples of the first type had three magnetic sublattices related to iron ions in octahedral garnet positions, iron in tetrahedral positions, and rare-earth gadolinium ions in dodecahedral positions. The second type of samples had two magnetic sublattices represented by octahedral and tetrahedral iron, while all other cations, including rare-earth metals in dodecahedral positions, were non-magnetic ones. To visualize the magnetic phases, we applied the magneto-optical Faraday effect, following the approach proposed in [11]. In order to enhance the Faraday rotation, we selected samples of bismuth-containing garnets. To produce different magnetic phases, we applied the lateral temperature gradient in the film plane. The controlled temperature gradient provides the possibility of observing different magnetic phases in the same sample at the same external magnetic field values.

2. Samples

Two series of 5 μm thick and 10×10 mm bismuth iron garnet films were prepared for experiments using liquid-phase epitaxy [12]. The first type of film, $(BiGd)_3(FeGa)_5O_{12}$, contained magnetic Gd^{3+} (spin I = 5/2) ions in the dodecahedral garnets' positions and magnetic ions Fe^{3+} (spin I = 3/2) in both the tetrahedral and octahedral positions. As a result, there were three magnetic sublattices. The second type film, $(BiYLu)_3(FeGa)_5O_{12}$, had only two magnetic sublattices caused by iron ions.

In both compositions, a small amount of the diamagnetic Ga^{3+} ions substituted the magnetic iron ions. Gallium ions have a preference to occupy tetrahedral positions and dilute the largest iron sublattice. Such diamagnetic dilutions play a primary role in precise tuning the compensation temperature in the second type of samples where the dodecahedral sublattice remained completely non-magnetic. The content of magnetic Gd^{3+} ions in dodecahedral positions was the main contributor to the compensation temperature for the first type of sample. Bismuth, in both compositions, enhances the Faraday rotation, which has achieved record values in ferrites based on it [12–14].

A monocrystalline plate of $(GdCa)_3(GaMgZr)_5O_{12}$ garnet was used as the substrate for the epitaxial synthesis of the bismuth gadolinium iron garnet (first type sample). For the second-type film, a plate of $Gd_3Ga_5O_{12}$ garnet crystal was used as the substrate. Both substrates were plates oriented in the (111) crystal plane.

The magnetic state of the epitaxial films of iron garnet significantly depends on the mismatch between film and substrate lattice parameters. The parameters of the crystal structure of the film, a_F, and of the substrate, a_S, give the mismatch parameter $\Delta a = a_F - a_S$. Large values of the mismatch parameter Δa lead to the appearance of large magnetoelastic stresses and a change in the magnetic anisotropy. We minimized this parameter by choosing the type of substrate and the concentration of ions with the required values of ionic radii. Thus, to synthesize the films of the first type of garnet having elements of a large ionic radius, we used a $(GdCa)_3(GaMgZr)_5O_{12}$ substrate with a large cell parameter a = 1.2495 nm. Gadolinium gallium garnet $Gd_3Ga_5O_{12}$ with a smaller cell parameter, a = 1.2373 nm, was the substrate for the second type of film having the minimum number of big ions.

To measure the mismatch parameter Δa for the studied samples, we applied the X-ray diffraction method, which resulted in $\Delta a_1 = 0.0006$ nm for the first type of sample and $\Delta a_2 = 0.0003$ nm for the second type of sample. To estimate crystalline quality, we utilized X-ray rocking curves, which broaden the diffraction peaks. The half-width of the rocking curves did not exceed 15.0″–15.5″ for the used films. Some synthesis condition details and the XRD crystal quality assessment are provided in the Supplementary Materials (Tables S1 and S2, Figure S1).

The ferromagnetic resonance measurements show that both magnetic films demonstrated magnetic anisotropy of the "easy–axis" type at room temperature so that the direction of the easy axis was perpendicular to the film plane. This result is in good agreement with the shape of the domain structure observed by means of the Faraday effect at room temperature too.

The chemical composition of the magnetic films (Tables 1 and 2) was determined via electron probe microanalysis using a Jeol JSM-6480LV (Tokyo, Japan) scanning electron microscope with an INCA X-Maxn energy-dispersive spectrometer. The analysis was carried out at an accelerating voltage of 10 kV and electric current of 1.4 nA. Standardization of the emission lines was carried out using the standards of $GdPO_4$, $ScPO_4$, Fe, YPO_4, Pt, and $BiTe_2$ for the elements O and Gd, Sc, Fe, Y, Pt, and Bi, respectively. Standard deviations for the elements with concentrations larger than 10% in mass did not exceed 2%. Averaging over six points on each sample was applied.

Table 1. Chemical composition of the sample $(BiGd)_3(FeGa)_5O_{12}$. Metal atoms content recalculated per 12 oxygen atoms (according formula unit).

	Al	Cr	Fe	Cu	Ga	Gd	Pt	Bi	O
1	0.03	0.02	4.22	0.02	0.56	2.40	0.11	0.62	12.00
2	0.04	0.02	4.20	0.02	0.57	2.40	0.11	0.60	12.00
3	0.04	0.02	4.21	0.02	0.58	2.41	0.11	0.59	12.00
4	0.04	0.02	4.23	0.02	0.59	2.39	0.10	0.58	12.00
5	0.03	0.02	4.22	0.02	0.58	2.40	0.11	0.60	12.00
6	0.05	0.02	4.22	0.01	0.58	2.40	0.11	0.58	12.00

Table 2. Chemical composition of the sample $(BiYLu)_3(FeGa)_5O_{12}$. Metal atoms' content recalculated per 12 oxygen atoms (per formula unit).

	Ca	Fe	Ga	Y	Gd	Lu	Pt	Bi	O
1	0.02	3.52	1.45	0.88	0.01	1.30	0.04	0.78	12.00
2	0.02	3.51	1.45	0.87	0.02	1.32	0.02	0.78	12.00
3	0.02	3.51	1.47	0.84	0.01	1.32	0.03	0.78	12.00
4	0.03	3.50	1.45	0.86	0.01	1.32	0.03	0.79	12.00
5	0.03	3.49	1.47	0.87	0.02	1.31	0.03	0.78	12.00
6	0.03	3.51	1.45	0.86	0.02	1.31	0.03	0.79	12.00

As a first approach, we assumed that the chemical composition of the studied iron garnets contained 12 oxygen ions per formula unit. Using this assumption, we obtained the cation content per formula unit too (Tables 1 and 2). By combining these data together with known molecular field coefficients [15], we estimated the temperature of magnetic compensation and found a good agreement between molecular field theory and the experimental results.

To define the temperature of magnetic compensation, we used the temperature dependence of magnetization measured using vibrating magnetometry. Figure 1 gives the dependences obtained by using a Versa Lab setup (Quantum Design, Sao Paolo, Brazil) for both samples in a weak magnetic field of 10 mT applied perpendicular to the film plane. Both dependencies (Figure 1) had a qualitatively similar characteristic, which is typical for a ferrimagnetic material with a magnetic compensation temperature. The temperature of magnetic compensation is a temperature that results the deep minimum of the magnetization dependence below the Neel temperature. The absolute magnetization value is an unimportant feature in finding the temperature of magnetic compensation. This allowed us to ignore the normalization of the measured magnetic moment on the film volume and to present magnetization in the arbitrary units.

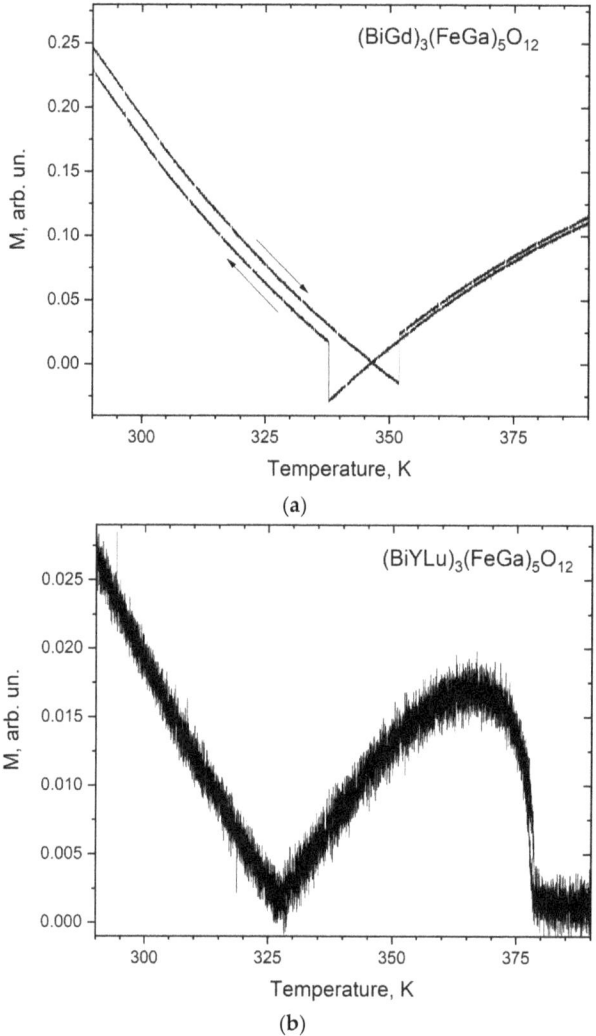

Figure 1. Temperature dependences of magnetization. External magnetic field of 0.01 T directed along film normal. The speed of temperature scanning was about 1 K/min. (a) $(BiGd)_3(FeGa)_5O_{12}$, (b) $(BiYLu)_3(FeGa)_5O_{12}$.

The special measurements of the substrate magnetization indicate the absence of any features that can affect the accuracy of finding the compensation temperature. The substrate paramagnetic magnetization decreases $1/T$ as the temperature T grows, and paramagnetic contribution is excluded from the samples' magnetization in Figure 1. As a result, we found a compensation temperature T_m = 346.5 K for the first type of sample (bismuth gadolinium iron garnet having three magnetic sublattices) and T_m = 327.5 K for the second type of sample (bismuth iron garnet having two magnetic sublattices).

3. Experiment

The experimental method used in this study is based on the gradient in-plane method proposed in [16,17] for observing phase transitions. In contrast to [16,17], we used a temperature gradient instead of the gradient of the composition in the film plane. The general scheme of the experiment is shown in Figure 2.

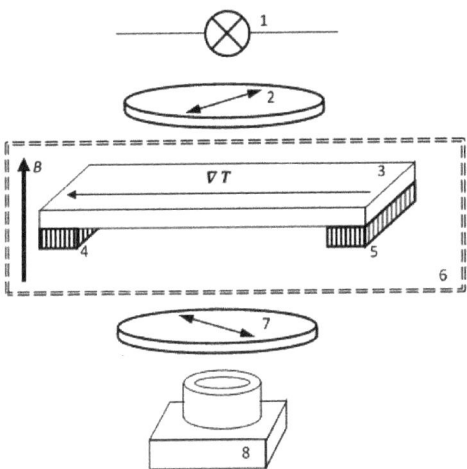

Figure 2. General scheme of the experiment. 1—unpolarized light source, 2—polarizer, 3—film sample, 4—heating Peltier element, 5—cooling Peltier element, 6—temperature-controlled stage, 7—analyzer, and 8—camera and microscope at 550× magnification, B—magnetic field, ∇T—lateral temperature gradient.

Polarized light was used to observe the magnetic phases. Passing through the magnetic film of the sample, the polarized light entered the digital camera through the analyzer. The angle between the polarizer and the analyzer was adjusted to achieve the maximum image. This provided some middle intensity of light if polarization remained unchanged between the polarizer and analyzer.

Due to the Faraday effect, polarization rotation is possible in magnetic materials that influence the intensity of the light passing through a film. The Faraday rotation in our samples provided the dominant contribution from the magnetic sublattice of the tetrahedral iron ions. The diamagnetic bismuth ions only enhanced this effect. The contribution of the octahedral sublattice was always less than that of the tetrahedral one. The angle of Faraday rotation depends on the projection of the tetrahedral magnetization onto the direction of the light beam and reaches the maximum value when magnetization is collinear to the beam, while the orthogonal direction results in unchanged polarization. As a result, the sample areas having different tetrahedral magnetization along the film normal should demonstrate different brightness in the image obtained with the digital camera.

In the experiments, a digital video camera was used together with a 550× magnification microscope. The observed area of the sample was a circle with a diameter of about 3 mm. A special holder made of non-magnetic materials was made for both the video camera and the microscope.

To create a lateral temperature gradient ∇T in the sample plane, two Peltier elements connected in opposite directions were used. Both Peltier elements, model TB–7–06–08 KryoTherm (St. Petersburg, Rusia) with a size of 4 × 4 mm, were located at the edges of the observation area at a distance of 3 mm. A temperature-controlled sample holder with Peltier elements was used to stabilize the temperature during measurements. The sample temperature monitoring and control system also included service elements, such as a temperature sensor and a differential thermocouple for monitoring the temperature gradient, not shown in Figure 2. The specially designed holder that held the optical system together, described above, temperature control equipment, and the test sample were placed in the working area of a superconducting magnet. The magnet provided a field B up to 10 T directed along the normal to the film plane. The cryogenic system for the superconducting magnet was a GFSG-510-2K-SCM10T-VTI29 (CryoTrade Engineering LLC, Moscow, Russia).

To automate the experiment, a computer, a temperature control system, measuring instruments, and a magnet power supply were networked. A computer managed the network utilizing the specially developed software. The developed system automatically stored into a file the measured values of magnetic field, temperature, and temperature gradient and displayed them on a monitor.

The magnetization redistribution due to the changing magnetic field resulted in the changing image registered with polarized light. The dynamically changing image together with the above-mentioned parameters were recorded as a video file. The proposed optical system along with the temperature and magnetic field management systems allowed the visualization of the magnetization behavior during the temperature-phase transition in high-quality, homogeneous films.

4. Results and Discussion

To identify the conditions for the existence of a non-collinear phase, measurements were carried out in magnetic fields ranging from 0 to 10 T. The sample temperature was maintained in the compensation temperature region.

Visually, both types of samples qualitatively demonstrated the same behavior, differing only in numerical values. The images for the $(BiYLu)_3(FeGa)_5O_{12}$ sample looked more contrasted than those for the first type sample, because the concentration of Bi ions in the second type of sample was significantly higher (Tables 1 and 2). The experimental results for the second type of sample are given in Figures 3 and 4.

In the absence of an external magnetic field, a large-block domain structure was observed at room temperature, and the temperature variations changed this structure. In magnetic fields exceeding the coercivity threshold, the sample was in saturation, and a monodomain state was observed.

To conduct experiments near the compensation point, we started from room temperature and gradually heated the sample up to the compensation point. The rate of temperature changes did not exceed 1 degree per minute.

Near the compensation temperature, the observed pattern consists of two areas of different brightness when the magnetic field is below the first critical value H_{c1} (Figure 3a). We attribute these areas to collinear phases of different magnetization directions perpendicular to the film plane. The observed behavior agreed with expectations because the theoretical temperature range for the canted phase existence is sufficient in low fields [2,10]. The other argument is that the "easy-axis" magnetic anisotropy crushes the canted phase in low fields. This anisotropy also responds to the temperature hysteresis in the magnetization near the compensation point (Figure 1a). Similar hysteresis took place for the second type of sample in the 1 mT field.

Let us note that the visualized magnetic phases differed from one other in the direction of the magnetization of tetrahedral iron, while the direction of the total magnetization was the same for both phases and laid along the magnetic field. A boundary between areas of the different phases is the 180-degree compensation wall. This compensation wall is similar to the compensation wall that has been reported in [17,18]. Unlike the temperature gradient reported here, the authors of [17,18] used samples that had a compensation point gradient due to the gradient of the chemical composition in a sample.

Starting from the first critical value H_{c1}, the field increase brought the extra features into the observed pattern. A transition region of intermediate brightness appears between regions that represent collinear phases (Figure 3b). The field growth up to the second critical value H_{c2} increased the width of the transition region.

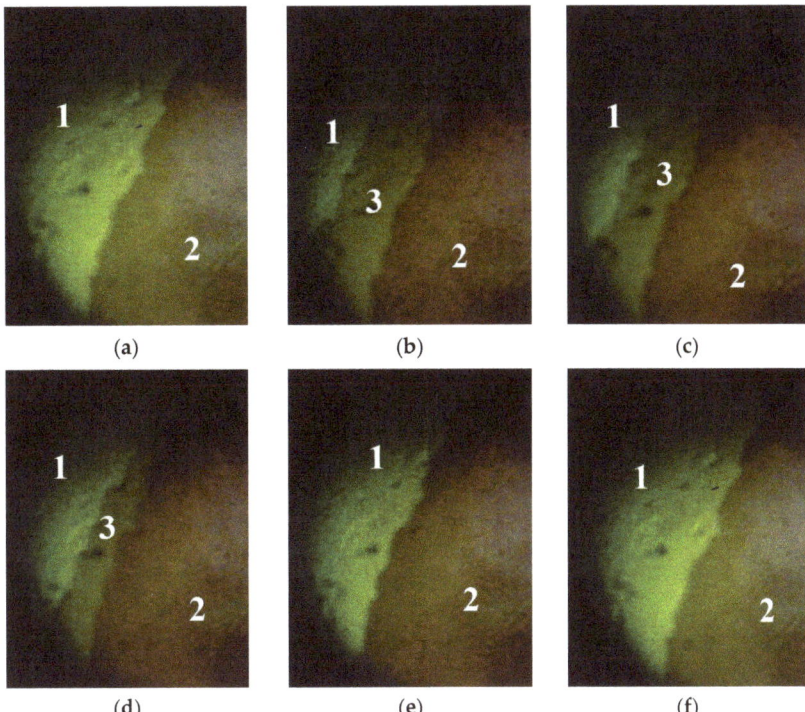

Figure 3. Magneto-optical visualization of phase states in the $(BiYLu)_3(FeGa)_5O_{12}$ sample at the temperature T = 327.5 K (the region of magnetic compensation) in an external magnetic field B = 0.5 T (**a**), B = 1.0 T (**b**), B = 1.03 T (**c**), B = 1.05 T (**d**), B = 1.1 T (**e**), and B = 1.2 T (**f**). (1), (2) Collinear phases of opposite magnetization direction; (3) non-collinear (canted) magnetic phase.

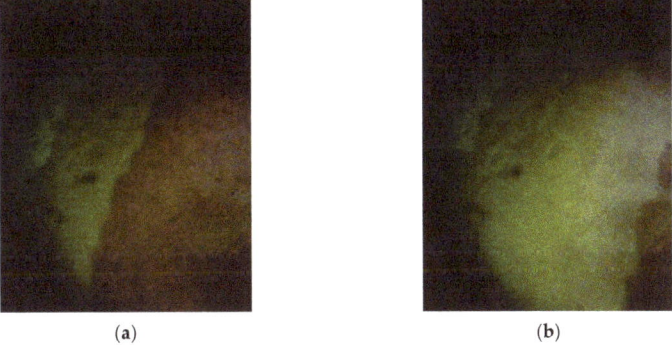

Figure 4. The temperature gradient influence on the non-collinear phase in the $(BiYLu)_3(FeGa)_5O_{12}$ sample. The lateral gradient for image (**a**) is approximately twice as small as that in image (**b**).

The new region appeared due to the canted phase, which gave a weaker Faraday rotation. In reality, the directions of the sublattices magnetization do not match the external magnetic field in the non-collinear (canted) phase. At the temperature of magnetic compensation, the sublattices' magnetizations are practically orthogonal to the external field and, as a result, provide a vanishingly small contribution to the Faraday rotation. The temperature offset from the compensation point results in out-of-plane magnetization. The angle between the magnetization and a perpendicular to the film plane should change smoothly in the canted phase [2].

This theoretical expectation contradicts the experimental results, which showed a sharp boundary between the canted and collinear phases. However, the sharp boundary occurred in the region of fields and temperatures where the coexistence of both collinear and canted phases was possible. A similar pattern was observed by magneto-optical method in experimental work [17].

With an increase in the magnetic field strength, the area corresponding to the non-collinear phase initially increased, and then, starting from the second critical field H_{c2}, it decreased. It can be seen that the canted phase region in the external field of 1.03 T (Figure 3c) was smaller than in the field of 1.0 T (Figure 3b). At external magnetic field strengths above a certain critical value, H_{c3}, the non-collinear phase completely disappeared, and two opposite collinear phases with maximum contrast were observed again (Figure 3e). A further increase in the intensity of the external magnetic field did not change anything in the observed picture (Figure 3f).

This experimental result conflicts with known theories [2,10]. According to theory, an increase in a magnetic field should only increase the width of the temperature region where a canted phase exists. The other problem relates to the maximum width of the temperature range for the canted phase. This region is too small, contrary to expectations.

On the other hand, we obtained preliminary results in the scope of molecular field theory. We found an extra magnetization in the collinear phase due to the paraprocess. This can narrow the temperature region of the canted phase and contribute to a decrease in the collinear phase energy in strong magnetic fields. As mentioned above, for a fixed magnetic field, the magnetization direction depends on the temperature in the canted phase. The temperature gradient induces an inhomogeneity in the magnetization direction for the canted phase. As a result, this can contribute to an increase in canted phase energy. These results will be discussed in detail in the next publication.

For the temperature gradient used for the magneto-optical visualization, we could tune it in the range from 0 K per cm to 12 K per cm. The most effective influence on the magnetic phase images was exerted by a temperature gradient in the range from 1.5 K per cm to 2.5 K per cm. Further gradient increases had little effect on the visualization quality but had a general temperature effect on the sample, moving the phase boundary out of the observation area. The data in Figure 4 demonstrate the influence of the lateral temperature gradient. The lateral gradient for the image in Figure 4b is approximately twice as large as that for the image in Figure 4a, while the sample temperature and external magnetic field are approximately the same for both cases.

The experiments showed that the value of the third critical field H_{C3} = 3.2 T for a three-sublattice bismuth gadolinium iron garnet having Neel temperature T_N = 450 K was almost three times higher than that for a two-sublattice $(BiYLu)_3(FeGa)_5O_{12}$ garnet that had a Neel temperature T_N = 380 K and H_{C3} = 1.1 T.

Determined experimentally, phase diagrams in the coordinate system of magnetic field vs. temperature for samples of both types are given in Figure 5 and Table 3. The arrows (Figure 5) schematically represent the sublattices' magnetizations.

Below the compensation temperature in the second type of sample (Figure 5b) dominated the octahedral magnetization, and it was directed along the magnetic field in the collinear phase. Above the compensation temperature, vice versa, it was dominated by the tetrahedral sublattice. In the canted phase, the deviation from the antiparallel magnetization arrangement was small, in the order of several degrees (our estimations using molecular field theory). A change in temperature primarily changed the angle between the magnetizations and the field, so that the magnetizations remained almost antiparallel to one other. Although the bismuth gadolinium garnet had three magnetic sublattices, the net iron magnetization could be considered as a single one. As a result, we could consider the first type of sample as a ferrimagnet of two magnetic sublattices, where one sublattice was gadolinium, and the other sublattice was iron (Figure 5a).

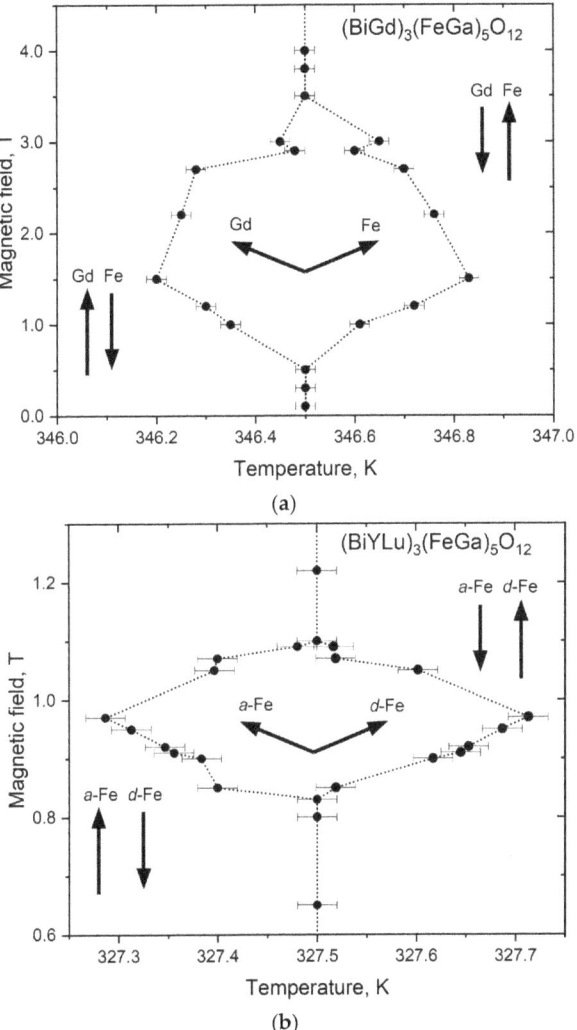

Figure 5. Phase diagrams in the vicinity of the compensation point in samples: (**a**) $(BiGd)_3(FeGa)_5O_{12}$; (**b**) $(BiYLu)_3(FeGa)_5O_{12}$. Arrows schematically represent sublattices magnetizations for Gd, total (Fe), octahedral (a-Fe), and tetrahedral (d-Fe) iron.

Table 3. Main characteristics of samples and their phase diagrams.

N	Composition	T_m, K	T_N, K	H_{c1}, T	H_{c2}, T	H_{c3}, T	dH, T	dT, K
1	$(BiGd)_3(FeGa)_5O_{12}$	346.5	450	0.5	1.5	3.2	2.7	0.6
2	$(BiYLu)_3(FeGa)_5O_{12}$	327.5	380	0.65	0.95	1.1	0.45	0.4

It can be seen that the magnetic field interval $dH = H_{C3} - H_{C1}$ and temperature interval $dT = T_{max} - T_{min}$, that define the area of a non-collinear phase near the compensation point had different values for the studied samples. The field interval $dH_1 = 3$ T and temperature range $dT_1 = 1$ K for the three-sublattice garnets were visibly larger than $dH_2 = 0.5$ T, $dT_2 = 0.7$ K for the two-sublattice garnet. The difference in behavior between the two types of structures is related to the difference in exchange interactions between them, as well as to the possible influence of magnetic anisotropy, the role of which in the formation of phase diagrams for iron garnet films near the compensation point requires more detailed theoretical consideration.

5. Conclusions

We experimentally demonstrated that the lateral temperature gradient permits the observation of three different magnetic phases (two collinear and one canted) in the same sample in the same magnetic field. The result was achieved for the cation-substituted iron garnet by means of the Faraday effect.

Phase T–H diagrams of the magnetic states in two- and three-sublattice ferrimagnetic structures in the vicinity of the compensation point were experimentally determined using magneto-optical methods. Below the critical field H_{c1}, two collinear magnetic phases separated by a compensational domain boundary were observed. In addition to collinear magnetic phases, a region of non-collinear magnetic phase was experimentally observed above this value. Starting from a certain field H_{c2}, the temperature range of the existence of the non-collinear phase decreased and, at H_{c3}, a phase transition to the collinear phase occurred.

Further theoretical and experimental studies should clarify the features of the magnetic phase T–H diagrams of rare-earth iron garnets near the spin-reorientation phase transition point.

Supplementary Materials: The following supporting information can be downloaded at: https://www.mdpi.com/article/10.3390/cryst13091297/s1, Table S1. Main LPE growth parameters [19]; Table S2. Melt composition in molar percent [mol%]; Figure S1. Rocking curve of the $(BiGd)_3(FeGa)_5O_{12}$ (888) peak.

Author Contributions: Conceptualization, P.M.V.; methodology, P.M.V. and S.N.P.; validation, S.N.P. and V.N.B.; formal analysis, V.N.B.; investigation, D.A.S., P.M.V. and S.V.T.; resources, A.V.M.; data curation, S.N.P.; writing—original draft preparation, D.A.S., P.M.V. and V.N.B.; writing—review and editing, S.N.P.; visualization, D.A.S. and P.M.V.; supervision, V.G.S.; project administration, P.M.V. All authors have read and agreed to the published version of the manuscript.

Funding: The research was carried out with the support of the Russian Science Foundation, project No. 22-22-00754.

Data Availability Statement: Not applicable.

Conflicts of Interest: The authors declare no conflict of interest.

References

1. Belov, K.P. Ferrimagnets with 'weak' magnetic sublattice. *Physics–Uspekhi* **1996**, *6*, 623–634. [CrossRef]
2. Clark, A.E.; Callen, E. Néel Ferrimagnets in Large Magnetic Fields. *J. Appl. Phys.* **1968**, *39*, 5972–5982. [CrossRef]
3. González, J.A.; Andrés, J.P.; Antón, R.L. Applied Trends in Magnetic Rare Earth/Transition Metal Alloys and Multilayers. *Sensors* **2021**, *21*, 5615. [CrossRef] [PubMed]
4. He, Y.; Lenne, S.; Gercsi, Z.; Atcheson, G.; O'Brien, J.; Fruchart, D.; Rode, K.; Coey, J.M.D. Noncollinear ferrimagnetism and anomalous Hall effects in Mn4N thin films. *Phys. Rev. B* **2022**, *106*, L060409. [CrossRef]
5. Xu, C.; Weng, J.; Li, H.; Xiong, W. The study of ultrafast magnetization reversal across magnetization compensation temperature in GdFeCo film induced by femtosecond laser pulses. *J. Magn. Magn. Mater.* **2014**, *352*, 25–29. [CrossRef]
6. Deb, M.; Molho, P.; Barbara, B.; Bigot, J.-Y. Controlling laser-induced magnetization reversal dynamics in a rare-earth iron garnet across the magnetization compensation point. *Phys. Rev. B* **2018**, *97*, 134419. [CrossRef]
7. Mashkovich, E.A.; Grishunin, K.A.; Zvezdin, A.K.; Blank, T.G.H.; Zavyalov, A.G.; van Loosdrecht, P.H.M.; Kalashnikova, A.M.; Kimel, A.V. Terahertz-driven magnetization dynamics of bismuth-substituted yttrium iron-gallium garnet thin film near a compensation point. *Phys. Rev. B* **2022**, *106*, 184425. [CrossRef]
8. Logunov, M.; Safonov, S.; Fedorov, A.; Danilova, A.; Moiseev, N.; Safin, T.; Nikitov, S.; Kirilyuk, A. Domain Wall Motion Across Magnetic and Spin Compensation Points in Magnetic Garnets. *Phys. Rev. Appl.* **2021**, *15*, 064024. [CrossRef]
9. Zvezdin, A.K.; Matveev, V.M. Some Features of the Physical Properties of Rare-Earth Iron Garnets Near the Compensation Temperature. *Sov. Phys. JETP* **1971**, *35*, 140–145.
10. Bernasconi, J.; Kuse, D. Canted Spin Phase in Gadolinium Iron Garnet. *Phys. Rev. B* **1971**, *3*, 811–815. [CrossRef]

11. Vetoshko, P.M.; Berzhansky, V.N.; Poluliach, S.N.; Suslov, D.A.; Mashirov, A.V.; Shavrov, V.G.; Pavliuk, E.V. Magneto-optical visualization of magnetic phases in an epitaxial garnet ferrite film near the compensation point. In Proceedings of the ICEEE-2022, RF, Alushta, Ukraine, 27 September–1 October 2022.
12. Prokopov, A.R.; Vetoshko, P.M.; Shumilov, A.G.; Shaposhnikov, A.N.; Kuz, A.N.; Koshlyakova, N.N.; Berzhansky, V.N.; Zvezdin, A.K.; Belotelov, V.I. Epitaxial Bi-Gd-Sc iron-garnet films for magnetophotonic applications. *J. Alloys Compd.* **2016**, *671*, 403–407. [CrossRef]
13. Zvezdin, A.K.; Kotov, V.A. *Modern Magnetooptics and Magnetooptical Materials*; CRC Press: Boca Raton, FL, USA, 1997; Volume 404, ISBN 9780367579494.
14. Shaposhnikov, A.N.; Berzhansky, V.N.; Prokopov, A.R.; Milyukova, E.T.; Karavaynikov, A.V. Interface properties single crystal films bismuth-constituted garnet—GGG substrate. *Scientific Notes of Taurida V. I. Vernadsky Univ. Ser. Phys.* **2009**, *22*, 127–141.
15. Hansen, P.; Röschmann, P.; Tolksdorf, W. Saturation magnetization of gallium-substituted yttrium iron garnet. *J. Appl. Phys.* **1974**, *45*, 2728–2732. [CrossRef]
16. Lisovskii, F.V.; Shapovalov, V.I. Noncollinearity of sublattices and existence of a domain structure in $Dy_3Fe_5O_{12}$ near the magnetic-compensation point in strong magnetization fields. *JETP Lett.* **1974**, *20*, 128–131.
17. Lisovskii, F.V.; Mansvetova, E.G.; Shapovalov, V.I. Phase diagram and domain-boundary structure in a uniaxial ferrimagnet near the compensation point. *Sov. Phys. JETP* **1976**, *44*, 755–760.
18. Hansen, P.; Krumme, J.-P. The compensation wall. *Philips Tech. Rev.* **1974**, *34*, 96–102.
19. Blank, S.L.; Nielsen, J.W. The growth of magnetic garnets by liquid phase epitaxy. *J. Cryst. Growth* **1972**, *17*, 302–311. [CrossRef]

Disclaimer/Publisher's Note: The statements, opinions and data contained in all publications are solely those of the individual author(s) and contributor(s) and not of MDPI and/or the editor(s). MDPI and/or the editor(s) disclaim responsibility for any injury to people or property resulting from any ideas, methods, instructions or products referred to in the content.

Article

First-Principles Study of the Effect of Sn Content on the Structural, Elastic, and Electronic Properties of Cu–Sn Alloys

Lingzhi Zhang [1], Yongkun Li [1,2,*], Rongfeng Zhou [1,2,*], Xiao Wang [1], Qiansi Wang [1], Lingzhi Xie [1], Zhaoqiang Li [1] and Bin Xu [3]

1. Faculty of Material Science and Engineering, Kunming University of Science and Technology, Kunming 650093, China; zlz2694@163.com (L.Z.); wang_xiao@kust.edu.cn (X.W.); wqs5248@163.com (Q.W.); 18315338611@163.com (L.X.); 18087147862@139.com (Z.L.)
2. City College, Kunming University of Science and Technology, Kunming 650093, China;
3. Chengdu Tonglin Casting Industrial Co., Ltd., Chengdu 610000, China; xubinyyy@126.com
* Correspondence: liyongkun@kust.edu.cn (Y.L.); zhourfchina@hotmail.com (R.Z.)

Abstract: In order to explore the mechanism of the influence of Sn contents on the relevant properties of Cu–Sn alloys, the structure, elasticity, electronic, and thermal properties of Cu–Sn alloys doped with different proportions of Sn (3.125 at%, 6.25 at%, and 9.375 at%) were established using the first-principles calculation based on density functional theory. Firstly, their lattice constants and Sn concentration comply with Vegard's Law. From the mixing enthalpy, it can be seen that Sn atoms can be firmly dissolved in the Cu matrix, and the structure is most stable when the Sn content is 3.125 at%. In addition, the introduction of mismatch strain characterized their solid solution strengthening effect. The elastic and electronic properties showed that when the Sn content is 6.25 at%, the Cu–Sn alloy has the best plasticity and the highest elastic anisotropy; when the Sn content is 3.125 at%, the Cu–Sn alloy is the most stable and has stronger bulk and shear modulus, which was mainly due to a stronger Cu-Cu covalent bond. Finally, the Debye temperature, thermal conductivity, and melting point were calculated. It is estimated that the thermal conductivity of Cu–Sn alloy is relatively good when the Sn content is low.

Keywords: first-principles; Cu–Sn alloys; solution strengthening; elastic properties; electronic properties

1. Introduction

As one of the important engineering structural materials, Cu–Sn (bronze) alloys have important application values in the fields of aerospace, marine, electrical appliances, and other fields due to their excellent wear resistance, corrosion resistance, thermal conductivity, and electrical conductivity, as well as sufficient strength and ductility [1–4]. The mechanical properties of Cu–Sn alloys are closely related to the content of the alloying element Sn. Due to different phase compositions, Cu–Sn alloys with different Sn content have different applications. When the Sn content is between 3 and 4 wt.%, it is mainly used for elastic components, wear-resistant parts, and antimagnetic parts; when the Sn content is between 5 and 11 wt.%, it is mainly used for bearings, shaft sleeves, turbines, etc. [5]. Therefore, Sn content is a key factor affecting the mechanical properties of Cu–Sn alloys.

In recent years, the first-principles calculations have become a powerful complement to solve the difficulties in the production and preparation process of Cu alloys and conduct extensive development and prediction of new Cu alloys, bridging the gap between theory and experiment. Wen et al. [6] studied the energy, elasticity, and electronic properties of Fe–Cu disordered solid solution alloys (Cu doping ratios of 25 at%, 37.5 at%, and 50 at%, respectively), and found that the elastic stability of Fe–Cu disordered solid solution was positively correlated with the Cu content. Zhou et al. [7] calculated key physical parameters such as elastic constants, bulk modulus, heat capacity, Debye temperature, and volumetric thermal expansion coefficient of Cu_6Sn_5 and Cu_5Zn_8 alloy phases. The calculated results were in agreement with experimental data, indicating that both Cu_6Sn_5 and Cu_5Zn_8 alloy phases were elastic anisotropic, and that Cu_6Sn_5 had a low bulk modulus. Rong et al. [8] calculated the elastic properties and anisotropy of Cu_3Sn, indicating that in-depth discussion of the anisotropy of intermetallic compounds with preferential growth and large volume fraction in the joint will be of great significance for accurately characterizing the mechanical behavior of the entire joint. In summary, the first-principles method can accurately study and predict the mechanical properties of copper alloys. However, currently, the calculation of Cu–Sn alloys mainly focuses on the specific phase structure of copper alloys, and there are few studies on the effect of Sn content on the properties and properties of copper alloy disordered solid solutions. Therefore, under the premise of ensuring the basic stability of the fcc structure, it is necessary to establish a model of Cu–Sn disordered solid solution to study the effect of tin solute on its related properties.

In this study, the phase stability, mechanical properties, and electronic properties of Cu–Sn alloys with Sn content of 3.125 at%, 6.25 at%, and 9.375 at% have been systematically studied using a first-principles calculation method. The lattice constant, mixing enthalpy, yield stress, elastic constant, elastic modulus, density of state, differential charge density, and Debye temperature were calculated. This provides a theoretical basis for the subsequent research on Cu–Sn alloys and the design, development, and wide application of new copper alloys. It is worth mentioning that, according to the Cu–Sn phase diagram, the solid solution limit of Sn in Cu matrix is 15.8 wt.% (9.2 at%) [9]. When the Sn content is greater than 15.8 wt.%, in addition to solid solution, the δ-phase ($Cu_{41}Sn_{11}$) occurs, which adversely affects the properties and applications of the material [10,11]. The generation of the δ-phase should be avoided or reduced as much as possible in the practical production applications. Therefore, the δ-phase was not discussed in this study.

2. Calculation Method and Details

The calculations were all performed using the CASTEP (Cambridge Serial Total-Energy Package) [12–14] code, which is based on the first-principles plane-wave pseudopotential method of density functional theory (DFT) [15] to perform quantum mechanical calculations. The ultrasoft pseudopotential (USPPs) was used to evaluate the interaction between valence electrons and ions. In this case, the valence electron configurations of Cu and Sn are $3p^63d^{10}4s^1$ and $4d^{10}5s^25p^2$, respectively. In addition, the generalized gradient approximation (GGA) of Perdew–Burke–Ernzerhof (PBE) is used to approximate the effect of the exchange–correlation energy on the calculated results [16]. In this calculation, $2 \times 2 \times 2$ supercells based on fcc structure were established by Perl Script enumeration of alloy structures. According to the Lowest Energy Principle, the stability models ($Cu_{31}Sn$, $Cu_{30}Sn_2$, and $Cu_{29}Sn_3$) of Cu–Sn alloys with Sn contents of 3.125 at%, 6.250 at%, and 9.375 at% were screened out, respectively, as shown in Figure 1 and Table 1. The detailed modelling methodology is shown in Appendix A.1.

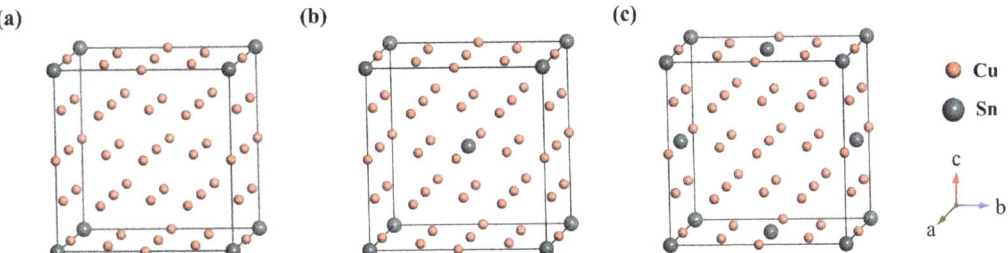

Figure 1. Crystal structure of Cu–Sn alloys (**a**) $Cu_{31}Sn$, (**b**) $Cu_{30}Sn_2$, and (**c**) $Cu_{29}Sn_3$.

Table 1. Cu–Sn alloys' model components.

Number of Sn Atoms	Structure	Mass Ratio of Sn Contents (wt.%)	Atomic Ratio of Sn Contents (at%)
0	Cu	0	0
1	$Cu_{31}Sn$	5.864	3.125
2	$Cu_{30}Sn_2$	11.075	6.250
3	$Cu_{29}Sn_3$	16.195	9.375

The BFGS (Broyden–Fletcher–Goldfarb–Shanno) [17] minimization algorithm was then chosen to optimize these structures by full relaxation to bring the system to a more stable state. Then, after convergence tests, the maximum truncation energy of the plane wave basal energy was set to 450 eV, and the k-point sampling network in the Brillouin zone was generated based on the Monkhorst–Pack scheme and set to 5 × 5 × 5. In the geometric optimization and electronic property calculations, the convergence tolerances for the total energy, maximum force, maximum stress, and maximum displacement were set to 1×10^{-5} eV/atom, 0.03 eV/Å, 0.05 GPa, and 0.001 Å. For the calculation of elastic properties, the convergence tolerances for total energy, maximum force, and maximum displacement were set to 2×10^{-6} eV/atom, 0.006 eV/Å, and 2×10^{-4} Å, respectively, and the number of steps and maximum strain amplitude for each strain were set to 4 and 0.003.

3. Results and Discussion

3.1. Lattice Constant

The lattice constant can reflect the structure of the crystal and its internal composition, which is the basic parameter of the crystal structure and the basis for the study of the material structure [18]. The optimized lattice constants of Cu, $Cu_{31}Sn$, $Cu_{30}Sn_2$, and $Cu_{29}Sn_3$ are shown in Table 2. To verify the accuracy of the calculation results, the lattice constants of the pure copper model were compared with the experimental result [19] reported in other literature, which showed a difference of 0.387%. In general, the difference of the lattice constant is within 1%, which means that the obtained pseudopotential can be considered as a good pseudopotential [20], thus indicating that the model, conditions, and parameters are more reasonable. Figure 2 shows the calculated values of the lattice constant as a function of solute concentration, which was fitted linearly to obtain the following equation:

$$a (\text{Å}) = 3.629 + 1.144c \text{ with } R = 0.99995 \tag{1}$$

Table 2. Experimental and theoretical lattice parameters (a, b, c and α, β, γ), mixing enthalpy ΔH (kJ/mol) for Cu–Sn alloys.

Structure	Source	a (Å)	b (Å)	c (Å)	α (deg)	β (deg)	γ (deg)	ΔH (kJ/mol)
	Exp. at 25 °C	3.615	-	-	90	-	-	[20]
Cu	Present	3.629	-	-	90	-	-	
	Error	0.387%	-	-	-	-	-	-
$Cu_{31}Sn$	Present	3.664	-	-	90	-	-	−3.25
$Cu_{30}Sn_2$	Present	3.700	-	-	90	-	-	−2.69
$Cu_{29}Sn_3$	Present	3.736	-	-	90	-	-	−1.79

The result from E. Sidot [21] is also reported in Figure 2, where the same linear regression calculation was performed on these data. The relevant equation is as follows:

$$a (\text{Å}) = 3.615 + 1.054c \text{ with } R = 0.9997 \tag{2}$$

The results show that the lattice constants of Cu–Sn alloys are proportional to the solute concentration, in full compliance with Vegard's law. As can be seen from Figure 2, the calculated data agreed well with the slope of the experimental data, although the calculated results do not fully agree with the experimental results in terms of intercept (equal to the lattice constant of pure Cu). The focus of this study is on the trend of lattice constant change, rather than the absolute value of lattice constant. The discrepancies between the calculated and experimental values of the pure Cu lattice constant are mainly due to thermal expansion and the limitations of the GGA [20]. Therefore, the optimized lattice constants can be used for subsequent calculations.

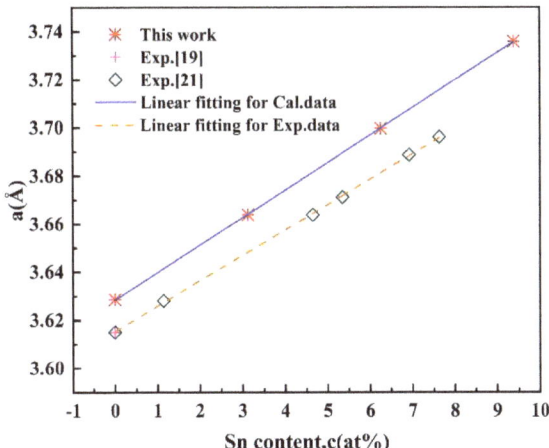

Figure 2. Relationship between the lattice parameter and the atomic Sn concentration of Cu–Sn alloys.

3.2. Enthalpy of Mixing

From the energy point of view, the mixing enthalpy ΔH_{mixing} is usually introduced to describe the dissolution of the solute atom Sn in the Cu matrix, which expresses the relationship between the energies of two binary alloys with the same structure. However, the most stable structure of the element Sn in the ground state is not the fcc structure. The Birch–Murnaghan equation reveals the internal structure and properties of solids by investigating their rate of change of volume and modulus of elasticity at different pressures. The energy–volume (E-V) curve can be obtained by fitting this equation to obtain the total static energy of the pure element, which can then be substituted to obtain the enthalpy of mixing. In this calculation, in order to obtain the equilibrium volume V_0 and the static energy E_0 of the element Sn in the fcc structure, the energy–volume (E-V) curve in the ground state was fitted by the Birch–Murnaghan equation of state with the following empirical equation [22]:

$$E(V) = E_0 + \frac{9V_0 B_0}{16} \left\{ \left[\left(\frac{V_0}{V}\right)^{\frac{2}{3}} - 1 \right]^3 B'_0 + \left[\left(\frac{V_0}{V}\right)^{\frac{2}{3}} - 1 \right]^2 \left[6 - 4\left(\frac{V_0}{V}\right)^{\frac{2}{3}} \right] \right\} \quad (3)$$

where E_0 and V_0 are the static energy and equilibrium volume of each atom at steady state, respectively, while B_0 and B'_0 are the first-order derivatives of the bulk modulus and bulk modulus with respect to the pressure, respectively.

The enthalpy of mixing can then be calculated by the following equation:

$$\Delta H_{mixing}(Cu_x Sn_y) = \frac{E_{total}(Cu_x Sn_y) - xE_{atom}(Cu) - yE_{atom}(Sn)}{x+y} \quad (4)$$

where $\Delta H_{mixing}(Cu_x Sn_y)$, $E_{total}(Cu_x Sn_y)$, $E_{atom}(Cu)$, and $E_{atom}(Sn)$ represent the mixing enthalpy, the static total energy of the Cu–Sn alloy, the static total energy of pure Cu and solute atom Sn, respectively, and x and y are the quantities of pure Cu and solute atom Sn, respectively.

Some scholars [23] pointed out that when evaluating the solid solubility of elements from the mixing enthalpy, if the contribution of entropy after heating is considered, there may be an uncertainty of approximately 0.05 eV/atom. It is shown that when the difference in radius between solute and solvent atoms $|\Delta R| < 15\%$, a solid solution with larger solid solution will be formed when other conditions are similar; conversely, when $|\Delta R|$ 15%, the larger the $|\Delta R|$, and the smaller the solid solution. As for the electronegativity, if the difference in electronegativity between the group elements is small, a larger solid solution degree will be formed; if the difference is large, it is easier to form stable intermetallic compounds, and even if a solid solution can be formed, its solid solution degree is not large. In this study, the radius difference $|\Delta R|$ between Cu and Sn is about 23.44%, which indicates a small solid solubility, while the electronegativity difference between the two is 0.06, indicating a large solid solubility. Therefore, the magnitude of solid solubility should be the result of a combination of multiple factors, which is related to the crystal structure, electron concentration, and temperature, in addition to the atomic size and electronegativity [24]. The more negative the mixing enthalpy, the stronger the chemical bond and the better the stability. As can be seen from Table 2, the mixing enthalpies of $Cu_{31}Sn$, $Cu_{30}Sn_2$, and $Cu_{29}Sn_3$ are all negative, and the negative value of the mixing enthalpy of $Cu_{31}Sn$ is the largest, which is -3.25 kJ/mol, indicating that 3.125 at%, 6.25 at%, and 9.375 at% Sn atoms can be solid-soluble in the Cu matrix, and $Cu_{31}Sn$ (3.125 at%) has the strongest chemical bond and the most stable structure.

3.3. Solid Solution Strengthening

Substitution of some atoms in the copper-based solid solutions by solute atoms will cause lattice distortion. At this point, a strain field is formed around the solute atoms, which hinders the movement of dislocations, leading to solid solution strengthening [18]. Several mechanisms have been proposed to describe the interaction between mobile dislocations and solute atoms, including the size effect [25], modulus effect [26], Suzuki effect [27], and electrostatic interaction [28]. Among them, the size effect and the modulus effect are of more importance since the effect of solid solution strengthening of copper substrates is difficult to present in a quantitative form using conventional experimental methods. Therefore, in this study, based on first principles, we introduce the parameter mismatch strain, which is the local lattice distortion around the solute atom and the size effect mentioned earlier, as a measure of the strength of the solute atom strengthening by the characteristic strain generated by the size difference between some solute atoms represented by elastic inclusions and the pores of the host material in an elastic continuous medium model. The mismatch strain is defined as follows:

$$\varepsilon = \frac{d - d_0}{d_0} \tag{5}$$

where d is the distance between the host atom (Cu) and the first nearest neighbor of the solute atom (Sn), and d_0 is the distance between the host atom (Cu) and the host atom (Cu). At zero pressure, the lattice constant is optimized to a_0, when the atomic positions are relaxed, and the distance d within the cell is measured. Then, the lattice constant is fixed to a_0, the atomic positions are fixed to the ideal fcc lattice position, and the distance d_0 within the cell is measured. For the first nearest neighbor solvent atom, the relationship between d_0 and the lattice constant is $d_0 = a_0/\sqrt{2}$. The mismatch strain ε for the first nearest neighbor in the Cu–Sn alloys is shown in Table 3. This parameter is determined based on the average value of the distance between the first nearest neighbor atoms in $Cu_{31}Sn$, $Cu_{30}Sn_2$, and $Cu_{29}Sn_3$ [20]. According to the Cottrell model [25], the maximum interaction force F_m between solute atoms and edge dislocations is:

$$F_m = \frac{\sqrt{3}}{2}\left(\frac{1+\nu}{1-\nu}\right)Gb^2|\varepsilon| \tag{6}$$

where v is the Poisson's ratio, G is the shear modulus, b is the Burns vector, and ε is the mismatch strain. In Friedel's theory [29], the interaction forces between atoms generate a critical decomposition shear stress $\Delta\tau_s$, which is defined as

$$\Delta\tau_s = \frac{\sqrt{2}F_m^{3/2}}{b^3}\sqrt{\frac{c}{G}} \tag{7}$$

where c is the concentration of solute atoms. Substituting F_m in Equation (6) into Equation (7) and then using the Taylor factor M, one obtains the yield stress $\Delta\sigma_s$ increased by solid solution strengthening of the polycrystalline alloy with the following relation:

$$\Delta\sigma_s = M\frac{3^{\frac{3}{4}}}{2}\left(\frac{1+v}{1-v}\right)^{3/2}G|\varepsilon|^{3/2}\sqrt{c} \tag{8}$$

In Cu–Sn alloys, the value of M is 3.06 [30]. Table 3 summarizes the misfit strain ε induced by solid solution of Sn atom into the Cu matrix and the contribution of solid solution strengthening to the yield stress of Cu–Sn alloys. The Poisson's ratio v and shear modulus G in Equation (8) are obtained from Table 4; Table 5 below.

Figure 3 shows the relationship between the Sn content and the yield stress values in Cu–Sn alloys, comparing the calculated results with the analytical results of some experimental results [31–33]. The experimental values are distributed on both sides of the calculated results. In fact, the yield stress values depend on two major factors. On the one hand, it depends on the intrinsic factors of the material, including the bonding bonds and the influence of the microstructure dominated by four major strengthening mechanisms: solid solution strengthening, strain strengthening, dispersion strengthening, and grain size strengthening; on the other hand, the yield stress values are also affected by some extrinsic factors such as the temperature, the strain rate, and the state of stress [31]. As shown in Table 3, different processing methods and heat treatment conditions lead to different yield stress results when the solute atomic concentration is the same. The first-principles calculations used in this study simulate the relevant properties of the material at a nearly ideal 0 K condition, which differs from the experimental conditions. Hence, the yield strength values are somewhat deviated, but observing the overall trend in Figure 3, the yield stress values of the Cu–Sn alloys increase with the increase of the Sn content. The mismatch strain data obtained from this calculation, to a certain extent, can provide theoretical guidance for the solid solution strengthening effect of Cu–Sn alloys, which is of reference value for the development of new copper alloys with very high yield strength.

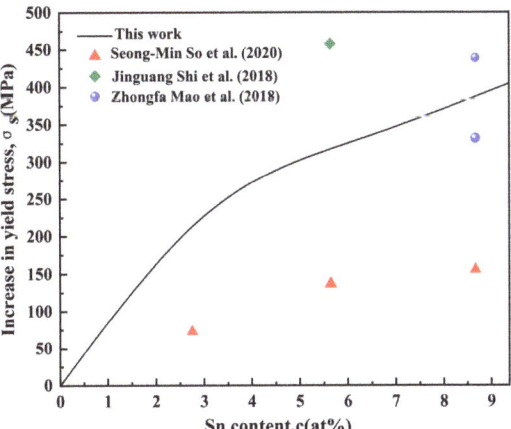

Figure 3. The relationship between the Sn content and the yield stress values in Cu–Sn alloys. This includes calculation results and experimental results [31–33].

Table 3. Theoretical results of mismatch strain caused by Sn atoms in Cu and the contribution of solution strengthening to the yield stress of Cu–Sn alloys. It also includes some solid solution strengthening experimental values.

	c (at%)	ε (%)	$\Delta\sigma_s$ (MPa)	Notes	
Present	3.125		233.52		
	6.25	1.97	330.24		
	9.375		404.46		
Exp.	2.753		68.5 ± 4.8	As-cast	[31]
	5.639		133.4 ± 3.5	As-cast	[31]
			458	SLM	[32]
	8.668		139.2 ± 16.6	As-cast	[31]
			436 ± 3	SLM	[33]
			328 ± 4	SLM + annealing	[33]

3.4. Elastic Properties

The elastic constants of metallic materials usually express their mechanical properties, especially the stability and stiffness of the material [8]. They express the stress condition required to maintain a certain deformation.

In this study, the elastic constants will be obtained by the "stress-strain" method [34], and for cubic crystal, the system has three independent elastic constants: C_{11}, C_{12}, and C_{44} [35].

Table 4 summarizes the elastic constants obtained from this calculation and compares the single-crystal elastic constants C_{ij} of pure Cu with the experimentally reported and previously calculated values. As can be seen from Table 4, the C_{ij} of $Cu_{31}Sn$, $Cu_{30}Sn_2$, and $Cu_{29}Sn_3$ do not satisfy the cubic crystal structure relationship because the number of independent elastic constants will increase after geometric optimization of the model obtained with supercell disordered modeling, whose crystal structure symmetry is slightly broken due to the quasi-random distribution of solute atoms. Therefore, in the present study, we used the symmetry-based projection (SBP) technique [36,37] to correct the elastic tensor of $Cu_{31}Sn$, $Cu_{30}Sn_2$, and $Cu_{29}Sn_3$. We usually take the average of the relevant elastic parameters to obtain the elastic constants of these quasi-random systems [35]. The relation is as follows:

$$\overline{C_{11}} = \frac{(C_{11} + C_{22} + C_{33})}{3} \quad (9)$$

$$\overline{C_{12}} = \frac{(C_{12} + C_{13} + C_{23})}{3} \quad (10)$$

$$\overline{C_{44}} = \frac{(C_{44} + C_{55} + C_{66})}{3} \quad (11)$$

The average values of the relevant elastic parameters calculated for $Cu_{31}Sn$, $Cu_{30}Sn_2$, and $Cu_{29}Sn_3$ are shown in Table 4.

For stable structures, the elastic constants C_{ij} should satisfy the corresponding Born stability criterion [38]. For the cubic crystal system, the elastic constants should satisfy the following criteria: $C_{11} - C_{12} > 0$, $C_{11} + 2C_{12} > 0$ and $C_{44} > 0$. Observing Table 4, it can be found that the calculated elastic constants of the alloys satisfy the stability criterion, indicating that the Cu–Sn alloys are stable at 0 K. These results are consistent with the actual situation and correspond to the previously calculated mixing enthalpy results.

From the elastic constants, the corresponding bulk modulus B, shear modulus G, Young's modulus E, and Poisson's ratio v can be obtained using the Voigt–Reuss–Hill approximation [39]. The Voigt, Reuss, and Hill approximations of the elastic modulus are denoted by the subscripts V, R, and H, respectively. For cubic structures, the modulus of elasticity can be defined as:

$$B_V = B_R = (C_{11} + 2C_{12})/3 \quad (12)$$

$$G_V = (C_{11} - C_{12} + 3C_{44})/5 \tag{13}$$

$$G_R = 5(C_{11} - C_{12})C_{44}/[4C_{44} + 3(C_{11} - C_{12})] \tag{14}$$

In the V-R-H model, B and G in the Hill model are obtained by taking the average of B or G in the Voigt and Reuss models,

$$B_H = \frac{1}{2}(B_V + B_R) \tag{15}$$

$$G_H = \frac{1}{2}(G_V + G_R) \tag{16}$$

Meanwhile, the relationship between Young's modulus E and Poisson's ratio v can be obtained,

$$E = \frac{9BG}{3B + G} \tag{17}$$

$$v = \frac{3B - 2G}{6B + 2G} \tag{18}$$

Calculated values of elastic parameters for Cu and Cu–Sn alloys ($Cu_{31}Sn$, $Cu_{30}Sn_2$, and $Cu_{29}Sn_3$) are presented in Table 5. In order to verify the reliability of the calculated results, the calculated values of the elastic parameters for copper in Table 4; Table 5 were compared with the previously reported experimental values [40–42] and theoretical values [43,44]. The elastic parameters obtained in this study are in better agreement with the reference values, indicating that the calculated parameters and method have high reliability and certain reference values.

Table 4. Elastic constants C_{ij} of Cu–Sn alloys. The present calculation results are compared with experimentally reported and other theoretical values.

Structure	Source	Elastic Constants of Crystals (GPa)								
		C_{11}	C_{12}	C_{13}	C_{22}	C_{23}	C_{33}	C_{44}	C_{55}	C_{66}
Cu	Present	184.5	116.7					77.1		
	Exp.at 4.2 K [a]	176.2	124.9					81.8		
	Exp.at RT [b]	170	122.5					75.8		
	Exp.at RT [c]	168.1	121.5					75.1		
	Cal. [d]	176	118.2					81.9		
	Cal. [e]	183.5	125.9					80.9		
$Cu_{31}Sn$	Present	182.98	109.69	109.69	183.01	109.71	182.80	78.13	78.13	78.13
	Present (SBP)	182.93	109.73					78.13		
$Cu_{30}Sn_2$	Present	158.44	130.77	128.96	159.97	129.74	156.39	61.33	61.35	61.35
	Present (SBP)	158.27	129.82					61.34		
$Cu_{29}Sn_3$	Present	160.51	107.29	106.40	161.68	106.97	159.88	75.63	75.62	75.62
	Present (SBP)	160.69	106.89					75.62		

[a] Experimental data reported in Ref. [40]. [b] Experimental data reported in Ref. [41]. [c] Experimental data reported in Ref. [42]. [d] Calculated data reported in Ref. [43]. [e] Calculated data reported in Ref. [44].

Figure 4a shows the changes in B_H, G_H, and E of Cu, $Cu_{31}Sn$, $Cu_{30}Sn_2$, and $Cu_{29}Sn_3$ as the content of Sn increases. In general, the bulk modulus B_H is used to characterize the incompressibility of a material. The higher the B_H value, the less likely the material is to compress under external forces. The shear modulus G_H is defined as the ability of a material to resist shear deformation. If the shear modulus G_H is larger, it indicates that the directional bonding between atoms is more significant. The Young's modulus E is a physical quantity used to describe the stiffness of a material. As the Young's modulus E increases, the hardness of the material also increases.

Table 5. The calculated bulk modulus B_H (GPa), shear modulus G_H (GPa), Young's modulus E (GPa), Poisson's ratio υ, Pugh's ratio G_H/B_H, Cauchy pressure $C_{12}-C_{44}$, dislocation strain energy W, and universal elastic anisotropy A^U for Cu–Sn alloys. The present calculation results are compared with experimentally reported and previously computed values.

Structure	Source	Modulus			υ	G_H/B_H	$C_{12}-C_{44}$	$W/J\cdot m^{-1}$	A^U	
		B_H (GPa)	G_H (GPa)	E (GPa)						
Cu	Present	139.7	55.7	147.5	0.32	0.40	39.6	0.367	0.83	
	Exp.at 4.2 K	142	51.5	137.8	0.34	0.36			1.80	[40]
	Exp.at RT	138.3	47.7	128.3	0.35	0.35			1.81	[41]
	Exp.at RT	137.0	47.1	126.7	0.35	0.34			1.84	[42]
	Cal.	137.4	54.0	143.3	0.33	0.39			1.42	[43]
	Cal.	145.1	53.5	142.9	0.34	0.37			1.40	[44]
$Cu_{31}Sn$	Present	134.13	57.63	151.23	0.31	0.43	31.60	0.387	0.72	
$Cu_{30}Sn_2$	Present	139.01	34.43	95.42	0.39	0.25	68.48	0.236	3.06	
$Cu_{29}Sn_3$	Present	124.81	49.99	132.31	0.32	0.40	31.27	0.349	1.40	

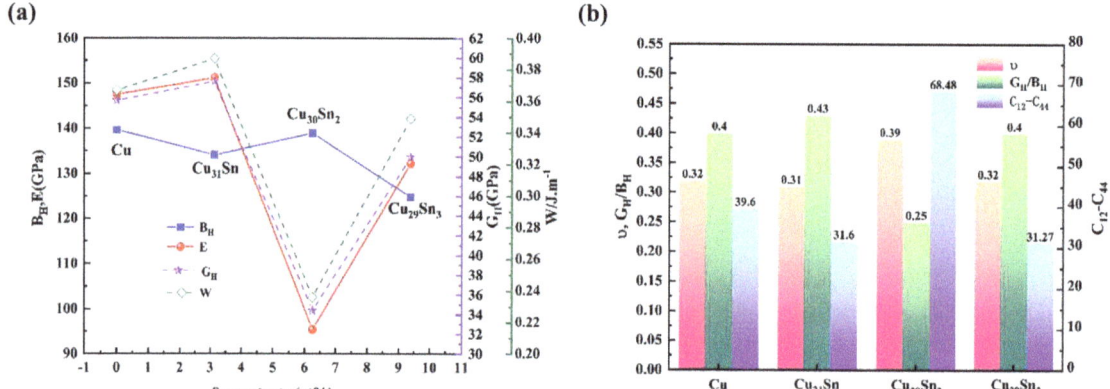

Figure 4. Variations in B_H, G_H, E, and W (a), υ, G_H/B_H, and $C_{12}-C_{44}$ (b) for Cu–Sn alloys.

As can be seen from Figure 5a, the order of values for GH and E is: $Cu_{31}Sn > Cu > Cu_{29}Sn_3 > Cu_{30}Sn_2$. $Cu_{31}Sn$ has the highest G_H value (57.63 GPa) and the highest E value (151.23 GPa), while $Cu_{30}Sn_2$ has the lowest G_H value (34.43 GPa) and the lowest E value (95.42 GPa), indicating that among these Cu–Sn alloys, $Cu_{31}Sn$ has the most significant directional bonding, the strongest shear deformation resistance, and the highest hardness. On the contrary, $Cu_{30}Sn_2$ has the weakest shear deformation resistance and the highest plasticity. In addition, in Cu–Sn alloys, the bulk modulus presents a "downward-upward-downward" trend with the increase of Sn content. Compared with pure Cu, an increase in Sn content will reduce its incompressibility.

The lattice distortion will occur when Sn is solidly dissolved into the Cu matrix. The elastic stress field caused by this deformation increases the crystal energy, which is defined as the strain energy of the dislocation [45],

$$W \approx Gb^2 \tag{19}$$

where G is the shear modulus and b is the Burgers vector. For fcc crystals, $b^2 = 0.5a^2$. The greater the dislocation strain energy, the poorer its plastic deformation ability, and the higher its tensile strength. The dislocation strain energy of Cu–Sn alloys is shown in Figure 4. With the increase of Sn content, the dislocation strain energy presents a trend of first increasing, then decreasing, and then increasing, indicating that its plastic deformation ability first decreases, then increases, and then decreases. This trend is the same as that of shear modulus G and Young's modulus E, which can be explained by the dislocation motion theory.

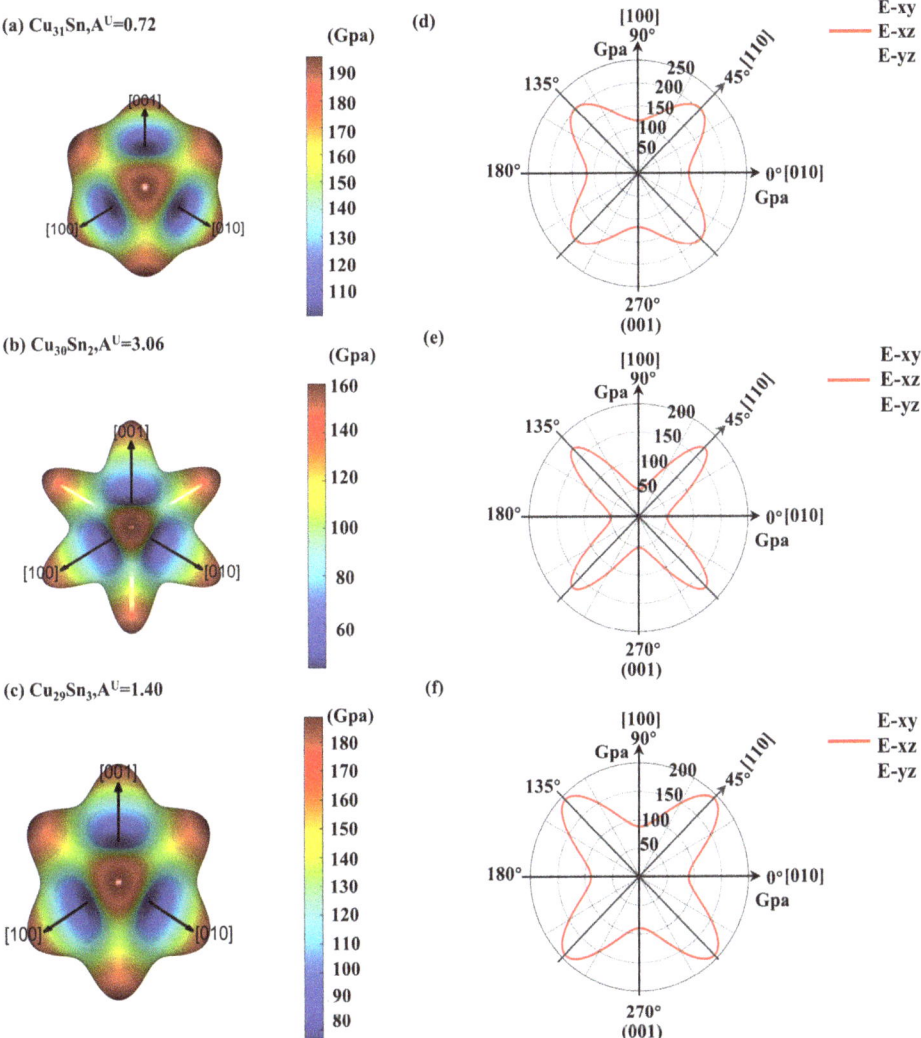

Figure 5. Three-dimensional surfaces (**a**–**c**) and planar projections (**d**–**f**) of the Young's modulus E for Cu–Sn alloys.

Elastic modulus and Poisson's ratio are important indicators that can reflect the mechanical properties of materials to a certain extent. However, to have a better understanding of their mechanical properties, in any service environment, it is also necessary to associate their bonding properties with toughness. Poisson's ratio υ, G_H/B_H [46], and Cauchy enact pressure on C_{12}-C_{44} [47] to evaluate the ductility trend of the material. According to Pettifor and Pugh criteria, ductile materials should meet: $\upsilon > 0.26$, $G_H/B_H < 0.57$, C_{12}-$C_{44} > 0$; conversely, brittle materials: $\upsilon < 0.26$, $G_H/B_H > 0.57$, C_{12}-$C_{44} < 0$. From Table 5 and Figure 4b, these Cu–Sn alloys meet the toughness criteria and have ductility. With the increase of Sn content, the ductility presents a "downward-upward-downward" trend, with $Cu_{30}Sn_2$ having the largest υ (0.39), C_{12}-C_{44} maximum (68.48), G_H/B_H minimum (0.25), indicating that $Cu_{30}Sn_2$ has the best ductility.

It is well known that elastic anisotropy is one of the causes that induce microcracking in materials [1]. Therefore, it is necessary to study its elastic anisotropy to evaluate the mechanical durability of Cu–Sn alloys. Among others, Ranganathan and Ostoja-Starzewski [48] improved the concept of the universal anisotropy index (A^U) to study the degree of anisotropy in different directions of bonding between atoms in different crystal planes, which can be expressed as

$$A^U = 5\frac{G_V}{G_R} + \frac{B_V}{B_R} - 6 \quad (20)$$

A^U takes into account the contributions of both shear and bulk modulus, where the deviation of A^U from 0 determines the degree of crystal anisotropy, and as can be seen from Table 5, $Cu_{30}Sn_2$ exhibits a higher degree of anisotropy compared to Cu, $Cu_{31}Sn$, and $Cu_{29}Sn_3$. Furthermore, this degree of anisotropy can be visually represented, as it is in Figure 5.

The Young's modulus E is not only color-coded in all directions by Elastic POST [49], but also its specific magnitude is shown in a two-dimensional plot. The Cu–Sn alloys examined in this study belong to the cubic crystal system, and the directional dependence of its Young's modulus can be obtained from the calculated flexibility constant [50], which can be expressed as

$$\frac{1}{E} = S_{11} - (2S_{11} - 2S_{12} - S_{44})\left(l_1^2 l_2^2 + l_2^2 l_3^2 + l_3^2 l_1^2\right) \quad (21)$$

where E is the Young's modulus, S_{ij} is the elastic flexibility coefficient, and l_1, l_2, and l_3 are the directional cosines.

Observing the three-dimensional diagram of Young's modulus anisotropy of Cu–Sn alloys in Figure 5a–c, the degree of elastic anisotropy of $Cu_{30}Sn_2$ can be described in more detail using the ratio of directional elastic modulus in Planar Projection, Figure 5d–f. The greater the deviation of this ratio from 1, the higher the elastic anisotropy of the surface [48]. For cubic crystal systems, the directional elastic modulus satisfies the following conditions: [100] = [010] = [001] ≠ [110]. $E_{(100)}/E_{(110)}$ represents the directional Young's modulus elastic anisotropy in the (110) plane. According to Figure 5d–f, the Young's moduli of $Cu_{31}Sn$, $Cu_{30}Sn_2$, and $Cu_{29}Sn_3$ in the <100> direction are 120 GPa, 50 GPa, and 85 GPa, respectively; the Young's moduli in the <110> direction are 205 GPa, 175 GPa, and 190 GPa, respectively. The deviation between $E_{(100)}/E_{(110)}$ and 1 for $Cu_{30}Sn_2$ is the largest (0.714), followed by the deviation between $E_{(100)}/E_{(110)}$ and 1 for $Cu_{29}Sn_3$ (0.553), and the deviation between $E_{(100)}/E_{(110)}$ and 1 for $Cu_{31}Sn$ (0.415) is the smallest. This indicates that the Young's modulus anisotropy of Cu–Sn alloys satisfies the following requirements: $Cu_{30}Sn_2$ > $Cu_{29}Sn_3$ > $Cu_{31}Sn$, which is the same as the order of A^U.

3.5. Electronic Properties

The electronic structure can explain the source of mechanical properties at a microscopic level. To further grasp the phase stability and bonding characteristics of the Cu–Sn alloys, the relevant electronic properties of the solid solution were investigated based on structural optimization. Figure 6 shows the total density of states (TDOS) and the partial density of stats (PDOS) of the Cu–Sn alloys in the energy range of −12 eV to 6 eV. From Figure 6, it can be seen visually that the distribution of density of states and their trends are relatively similar for the Cu–Sn alloys. First, the TDOS below the Fermi energy level (0 eV) is contributed mainly by the Cu-3d states, with partial contributions from the Sn-5s and Sn-5p states, while the TDOS above the Fermi energy level mainly originates from the Sn-5s and Sn-5p states, while partly from the Cu-3p states. It is well known that the DOS values ($N_{(EF)}$) at the Fermi energy level are related to the phase stability, where the smaller the $N_{(EF)}$, the more stable the corresponding phase is [51]. The $N_{(EF)}$ values of $Cu_{31}Sn$, $Cu_{30}Sn_2$, and $Cu_{29}Sn_3$ are 7.0843, 9.1978, and 8.6509 electrons/(eV·f.u.), respectively. The order of $N_{(EF)}$ values is $Cu_{31}Sn$ < $Cu_{29}Sn_3$ < $Cu_{30}Sn_2$. As discussed earlier, the enthalpy of mixing indicates that $Cu_{31}Sn$ is the most stable. Second, all Cu–Sn alloys have non-zero TDOS values at the Fermi energy level, which indicates the metallic character of these Cu–Sn alloys. Thirdly, the peak values of the Cu-3d state undergo splitting at −4 eV to −2 eV. As the Sn content increases, the three peaks gradually change from uniform to non-uniform in $Cu_{31}Sn$, $Cu_{30}Sn_2$, and $Cu_{29}Sn_3$, with a decrease at −4 eV and an increase at −2 eV. This is mainly attributed to the characteristics of the crystal structure and the symmetry of coordination, resulting in the crossing or overlapping of energy levels, which in turn affects the state and degree of peaks in DOS.

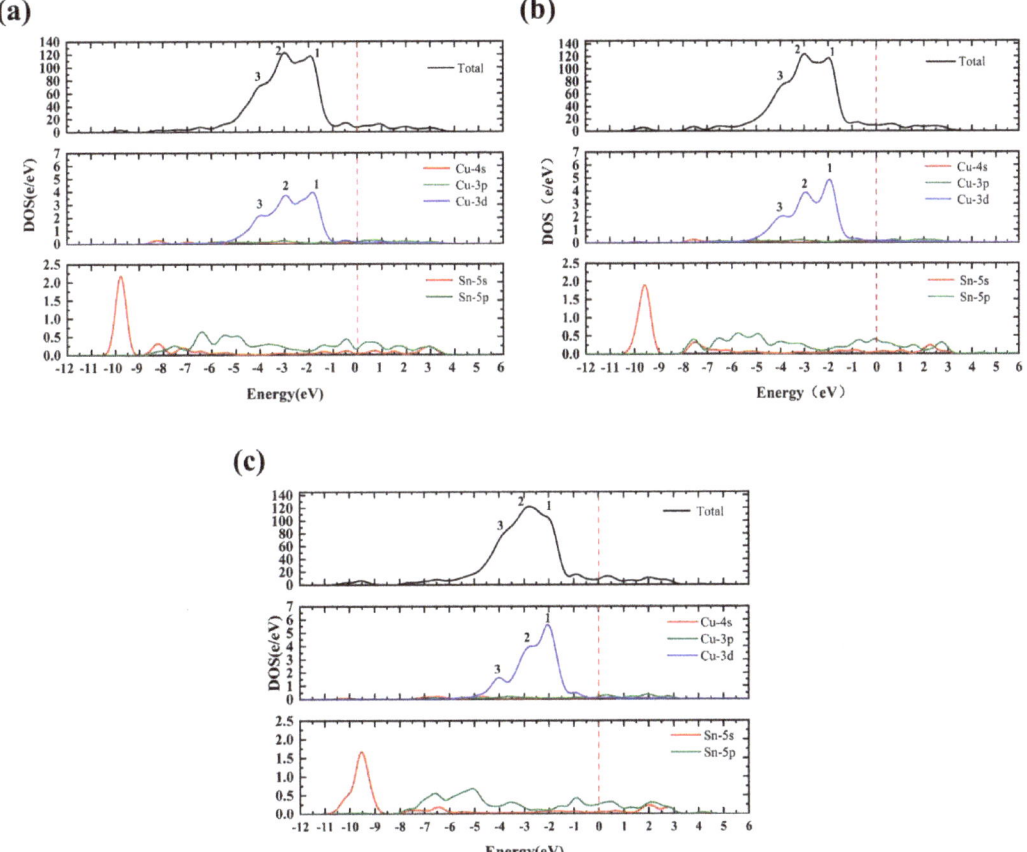

Figure 6. Total and partial electronic densities of states (TDOS and PDOS) near Fermi level of Cu–Sn alloys (**a**) $Cu_{31}Sn$, (**b**) $Cu_{30}Sn_2$, and (**c**) $Cu_{29}Sn_3$. The red dotted line indicates the Fermi level, and 1, 2, and 3 indicate the peak splitting of the Cu-3d state.

The differential charge density can directly characterize the nature of chemical bonding between different atoms and the electron gain and loss. Figure 7 shows the differential charge density diagram for Cu–Sn alloys in the range of −0.250 to 0.047 e/Å^3, where the red region indicates the accumulation of electrons and the blue region indicates the depletion of electrons. As shown in Figure 7, in the Cu–Sn alloys, a large number of electrons gather between the Cu and Sn atoms, and the Cu atoms have a significant loss of charge in the outer layers, which can be clearly observed as a "sea of electrons" phenomenon, thus indicating the existence of metallic bonds [52]. As shown in Figure 7, the distribution of electron clouds around Cu atoms is in the shape of petal, with directionality. The petal distributions are closely related to the shapes of d orbitals [53]. Moreover, the electron cloud is in the shape of petal, indicating that polarization is relatively severe, resulting in uneven distribution of electrons. It is speculated that there may be other bonds in the Cu matrix besides metal bonds. In addition, some electrons accumulate between Cu and adjacent Cu atoms, which indicates the presence of metallic bonds and Cu-Cu covalent bonds in Cu–Sn alloys. Among them, compared with Cu and adjacent Cu atoms, Cu and adjacent Sn atoms direction, the blue area around Cu is larger and dense, indicating a serious electron loss and the formation of stronger Cu–Sn covalent bonds. By the non-uniformity of the charge causes anisotropy in the relevant properties of the material (e.g., elastic properties). Observing Figure 7, it is found that the blue area around Cu in $Cu_{30}Sn_2$ is large and dense compared to $Cu_{31}Sn$ and $Cu_{29}Sn_3$, and the non-uniformity of the charge is more significant, thus its elastic anisotropy is the highest, reflecting the highest Young's modulus elastic anisotropy of $Cu_{30}Sn_2$ discussed earlier. With the addition of the alloying element Sn, the distribution of electron clouds around the atoms changes subsequently, and the electron cloud of the Sn element has a red sphere shape, indicating the accumulation of electrons in the alloying element.

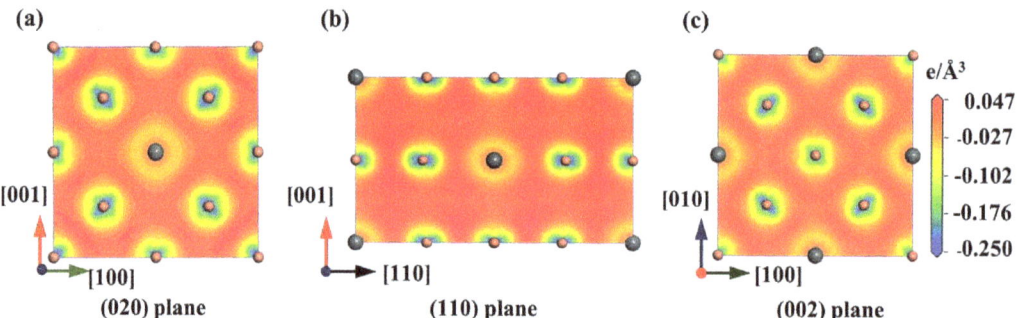

Figure 7. Charge density differences for Cu–Sn solid solutions (a) $Cu_{31}Sn$, (b) $Cu_{30}Sn_2$ and (c) $Cu_{29}Sn_3$.

On the other hand, the atomic Mulliken charge (AMC) can adequately describe the charge transfer between Cu and Sn atoms. If the atom has a negative AMC, it indicates that this atom gains charge; otherwise, this atom loses charge. In $Cu_{29}Sn_3$ and $Cu_{31}Sn$, most of the Cu atoms gain charge from Sn atoms or other Cu atoms, and some lose charge. In $Cu_{30}Sn_2$, most of the Cu atoms gain charge from Sn atoms or other Cu atoms, and a few have no gain or loss of electrons. Bond population (BP) and the bond length L are also important parameters to assess the bonding properties. In general, the shorter the bond length L and the larger bond population (BP), the stronger the bond, and a bond with a BP value of zero is a perfect ionic bond; otherwise, it is a covalent bond. A larger absolute BP value indicates a stronger covalent bond. Positive and negative BP values indicate bonding interactions and antibonding interactions in the bond, respectively [54,55]. As shown in Table 6, the BP values of Cu–Sn bonds and Cu-Cu bonds in these Cu–Sn alloys are much larger than zero, thus indicating the presence of Cu–Sn covalent bonds and Cu-Cu covalent bonds. Furthermore, it can be found that Sn-Sn bonds do not exist in these Cu–Sn alloys. Therefore, Sn atoms are prone to displacement and will first form vacancies at Sn sites [47]. It was shown that bond population (BP) is also an important indicator of the mechanical properties of the material. In general, the phase stability, shear modulus and hardness of Cu–Sn alloys are positively correlated with the strength of the covalent bond, and this relationship can be obtained by bond population (BP), and the stronger the covalent bond, the larger bond population (BP). The strong phase stability, shear modulus and hardness of $Cu_{31}Sn$ obtained in this study can be attributed to the formation of a stronger Cu-Cu covalent bond.

Table 6. Atomic Mulliken charge (AMC), bond population (BP) analysis and mean bond length (Å) for Cu–Sn alloys. The numbers in brackets for the atom represent the number of Cu or Sn ions, whereas the number in brackets for the bond represents the number of Cu–Cu and Cu–Sn bonds.

Species	Atom	Charge Number				AMC	Bond	BP	Length (Å)
		s	p	d	Total				
$Cu_{31}Sn$	Cu(1)	0.51	0.77	9.72	11.01	−0.01	Cu-Cu(12)	0.27	2.50443
	Cu(12)	0.51	0.78	9.72	11.01	−0.01	Cu-Cu(48)	0.23	2.5588
	Cu(3)	0.51	0.81	9.72	11.03	−0.03	Cu-Cu(12)	0.2	2.57028
	Cu(12)	0.53	0.81	9.73	11.06	−0.06	Cu-Cu(24)	0.21	2.57028
	Cu(3)	0.51	0.74	9.73	10.98	0.02	Cu-Cu(24)	0.19	2.59082
	Sn(1)	0.65	2.42	0	3.08	0.92	Cu-Cu(24)	0.21	2.59218
							Cu-Cu(12)	0.19	2.61121
							Cu-Cu(24)	0.16	2.67706
							Cu–Sn(12)	0.17	2.67706
$Cu_{30}sn_2$	Cu(24)	0.53	0.82	9.73	11.07	−0.07	Cu-Cu(8)	0.26	2.54638
	Cu(6)	0.51	0.77	9.73	11	0	Cu-Cu(16)	0.26	2.54639
	Sn(2)	0.71	2.42	0	3.13	0.87	Cu-Cu(44)	0.26	2.54828
							Cu-Cu(26)	0.2	2.61677
							Cu-Cu(22)	0.2	2.61678
							Cu–Sn(8)	0.18	2.68531
							Cu-Cu(45)	0.16	2.68531
							Cu-Cu(12)	0.18	2.68532
							Cu-Cu(4)	0.16	2.68532
							Cu–Sn(3)	0.18	2.68533
$Cu_{29}sn_3$	Cu(12)	0.54	0.85	9.74	11.13	−0.13	Cu-Cu(12)	0.24	2.56039
	Cu(10)	0.52	0.82	9.73	11.07	−0.07	Cu-Cu(24)	0.26	2.56039
	Cu(3)	0.5	0.71	9.74	10.96	0.04	Cu-Cu(12)	0.23	2.62491
	Cu(1)	0.5	0.66	9.75	10.92	0.08	Cu-Cu(59)	0.21	2.64287
	Sn(3)	0.79	2.46	0	3.25	0.75	Cu-Cu(48)	0.19	2.67471
							Cu–Sn(12)	0.14	2.72284

3.6. Debye Temperature

The Debye temperature (θ_D) is an important parameter of crystalline materials. On the one hand, it can reflect the thermal properties of the material, and on the other hand, it can be used as a link between the thermal and mechanical properties of the material. At low temperatures, the acoustic vibration is the only factor that triggers the vibration excitation, so at low temperatures, the Debye temperature calculated by the elastic constant is equivalent to the Debye temperature determined by the specific heat measurement. Thus, it can be calculated by the following equation [56,57],

$$\theta_D = \frac{h}{k}\left[\frac{3n}{4\pi}\left(\frac{N_A\rho}{M}\right)\right]^{\frac{1}{3}} v_m \qquad (22)$$

where h, k, and N_A are Planck's constant, Boltzmann's constant, and Avogadro's constant, respectively, n is the total number of atoms per unit cell, ρ is the density, M is the molecular weight, and v_m is the average speed of sound, which can be defined as [57,58],

$$v_m = \left[\frac{1}{3}\left(\frac{2}{v_t^3} + \frac{1}{v_l^3}\right)\right]^{-\frac{1}{3}} \qquad (23)$$

where v_l and v_t are the longitudinal and transverse sound velocities, respectively, and can be obtained from the shear modulus G, the bulk modulus B, and the density ρ, which are related as follows,

$$v_l = \left(\frac{B + \frac{4}{3}G}{\rho}\right)^{\frac{1}{2}} \qquad (24)$$

$$v_t = \left(\frac{G}{\rho}\right)^{\frac{1}{2}} \tag{25}$$

The calculated Debye temperatures, sound velocities and densities of pure Cu, and Cu–Sn alloys are shown in Table 7, which shows that the calculated sound velocities and Debye temperatures of Cu are in good agreement with the experimental values and previous calculations by scholars. In general, the higher the Debye temperature, the higher the melting point of the corresponding crystal and the stronger the covalent bond, the more stable the structure. As shown in Table 7 and Figure 8, among the Cu–Sn alloys, the Debye temperature of $Cu_{31}Sn$ is the highest, the corresponding covalent bond strength is the strongest, and the stability is the best, which is exactly in line with the results discussed in Table 2 and Figure 4a.

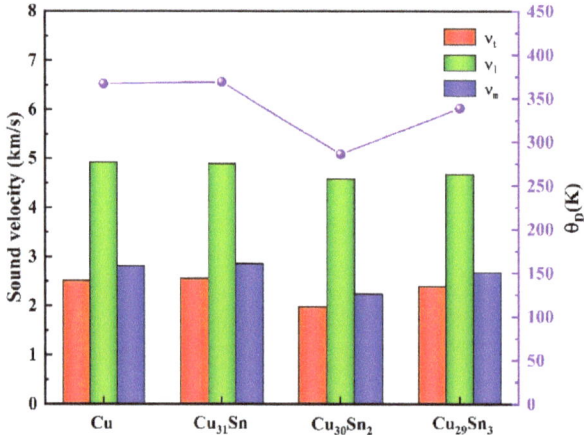

Figure 8. Variations in the sound velocity and Debye temperature for Cu–Sn alloys.

In addition, the lattice thermal conductivity κ_{ph} is also one of the most fundamental physical properties of the material, which characterizes the thermal conductivity of the material and is of great significance for exploring the application of the material at high temperatures. In general, we consider the minimum value of the lattice thermal conductivity k_{min} [59], which is related by the equation,

$$k_{min} = \frac{k_B}{2.48} n^{\frac{2}{3}} (2v_t + v_l) \tag{26}$$

where k_B is the Boltzmann constant, n is the number of atoms per unit volume, and v_l and v_t are the longitudinal and transverse velocities of sound, respectively.

Melting point is also an important parameter of the material and is currently a hot issue of research, playing a crucial role in predicting new intermetallic compounds for high-temperature applications. It can be obtained by the following empirical equation [60],

$$T_m = 354 + 4.5 \frac{2C_{11} + C_{33}}{3} \tag{27}$$

In addition to the speed of sound (v_l, v_t, v_m) and Debye temperature θ_D, the density ρ, the minimum value of lattice thermal conductivity k_{min}, and the melting point T_m of the Cu–Sn alloys are included in Table 7. It can be found that the minimum value of lattice thermal conductivity and melting point of Cu–Sn alloys follow the following pattern: $Cu_{31}Sn > Cu_{29}Sn_3 > Cu_{30}Sn_2$. The above calculation results indicate that Cu–Sn alloys are good thermally conductive materials, and it is tentatively predicted that the thermal conductivity of Cu–Sn alloy is relatively good when the Sn content is low. However, there are few reports on the experimental Debye temperature and other thermal properties of Cu–Sn alloys. Therefore, it is hoped that the results of the present calculations can provide a reference value for subsequent studies of Cu–Sn alloys.

Table 7. The calculated and experimental results of density (ρ), transverse, longitudinal, average sound velocity (v_t, v_l, v_m in m/s), Debye temperatures (θ_D, K), the minimum thermal conductivity (k_{min} in Wm^{-1}K^{-1}) and melting point (T_m in K) of Cu–Sn alloys.

Structure	Source	ρ	v_t	v_l	v_m	θ_D	κ_{min}	T_m	Refs.
Cu	Present	8.828	2512	4923	2815	367	1.06	1184	
	Exp.	8.937						1353	[61]
	Exp.					343			[62]
	Cal.	9.353	2639	5209	2958	391	1.16	1330	[63]
	Cal.	8.930	2277	4723	2560	335			[47]
Cu$_{31}$Sn	Present	8.815	2557	4892	2861	369	1.19	1177	
Cu$_{30}$Sn$_2$	Present	8.787	1980	4587	2237	286	0.88	1066	
Cu$_{29}$Sn$_3$	Present	8.754	2390	4677	2678	339	0.96	1077	

4. Conclusions

The lattice constants, phase stability, solution strengthening, elastic properties, electronic properties, and Debye temperature of Cu–Sn alloys with different Sn contents (Cu$_{31}$Sn, Cu$_{30}$Sn$_2$, Cu$_{29}$Sn$_3$) were studied by first principles. The relevant conclusions are as follows.

The calculated lattice constants are proportional to the solute concentration, consistent with the Vegard's law, and have a linear relationship across the entire Cu–Sn solid solution region; the Sn atoms of 3.125 at%, 6.25 at%, and 9.375 at% can be solidly dissolved in the Cu matrix. The negative mixing enthalpy of Cu$_{31}$Sn (3.125 at%) is the largest, indicating that its chemical bond is the strongest and its structure is the most stable. In the aspect of solution strengthening, the mismatch strain parameter is introduced to quantify the effect of solution strengthening. The calculated values can be used to predict the solution strengthening effect of Cu-based solid solutions, and are of great significance for developing copper alloys with ultra-high yield strength.

In Cu–Sn alloys, Cu$_{30}$Sn$_2$ has the smallest shear modulus and Young's modulus. Its variation trend is the same as that of dislocation strain energy (Cu$_{30}$Sn$_2$ has a minimum dislocation strain energy of 0.236 Jm^{-1}), indicating that when the Sn content is 6.25 at%, the plasticity of Cu–Sn alloys is the largest. In addition, Cu$_{30}$Sn$_2$ has the highest Young's modulus and elastic anisotropy.

The electronic structure and bonding properties of the Cu–Sn alloys have been calculated, and their relationship with the stability and mechanical properties of the alloys is analyzed and discussed. Three types of bonding existed in Cu–Sn alloys: Cu-Cu covalent bonds, Cu-Cu metallic bonds, and Cu-Sn covalent bonds, of which Cu$_{31}$Sn had the best stability and the highest shear modulus, which depended to a certain extent on the fact that it had stronger Cu-Cu covalent bonds.

The Debye temperature of the Cu–Sn alloys, the minimum lattice thermal conductivity, and the melting point all decrease sequentially along the order of Cu$_{31}$Sn, Cu$_{29}$Sn$_3$, and Cu$_{30}$Sn$_2$. This indicates that Cu–Sn alloys are good thermal conductivity materials. Additionally, it is tentatively predicted that the thermal conductivity of Cu–Sn alloy is relatively good when Sn content is low.

Author Contributions: L.Z., Y.L., X.W. and R.Z. designed most of the experiments, L.Z. analyzed the results and wrote this manuscript, Y.L. helped analyze the experiment data and gave some constructive suggestions about how to write this manuscript. L.Z., L.X., Q.W. and Z.L. performed most experiments. B.X. provided financial support. All authors have read and agreed to the published version of the manuscript.

Funding: This work was financially supported by the National Natural Science Foundation of China (No.52205373).

Data Availability Statement: Not applicable.

Acknowledgments: The authors gratefully acknowledge the National Natural Science Foundation of China (No.52205373) for the financial support of this research work.

Conflicts of Interest: The authors declare no conflict of interest.

Appendix A

Appendix A.1 Modeling Method:Script

1:
use strict;use Getopt::Long;
use MaterialsScript qw(:all);
my $disorderedStructure = $Documents{"$Cu_{31}Sn$.xsd"};
my $results = Tools->Disorder->StatisticalDisorder->GenerateSuperCells ($disorderedStructure,2,2,2);
my $table = $results->StudyTable;
print "Number of disorder configurations generated:".$results->NumIrreducibleConfigurations. "\n";

2:
use strict;
use Getopt::Long;
use MaterialsScript qw(:all);
my $disorderedStructure = $Documents{"$Cu_{30}Sn_2$.xsd"};
my $results = Tools->Disorder->StatisticalDisorder->GenerateSuperCells ($disorderedStructure,2,2,2);
my $table = $results->StudyTable;
print "Number of disorder configurations generated:".$results->NumIrreducibleConfigurations. "\n";

3:
use strict;
use Getopt::Long;
use MaterialsScript qw(:all);
my $disorderedStructure = $Documents{"$Cu_{29}Sn_3$.xsd"};
my $results = Tools->Disorder->StatisticalDisorder->GenerateSuperCells ($disorderedStructure,2,2,2);
my $table = $results->StudyTable;
print "Number of disorder configurations generated:".$results->NumIrreducibleConfigurations. "\n";

Note:
This script references the content of the following web site: https://zhuanlan.zhihu.com/p/50322042.

Table A1. Structures.

Structures		Weighting	Configuration	E (eV/atom)
$Cu_{31}Sn$	1	32	baaaaaaaaaaaaaaaaaaaaaaaaaaaaaaa	−45,867.723
$Cu_{30}Sn_2$	1	192	baaaaaaaaaaabaaaaaaaaaaaaaaaaaaa	−44,485.574
	2	192	baaaaaaabaaaaaaaaaaaaaaaaaaaaaaa	−44,484.631
	3	16	baaaaaabaaaaaaaaaaaaaaaaaaaaaaaa	−44,485.663
	4	48	baabaaaaaaaaaaaaaaaaaaaaaaaaaaaa	−44,485.603
	5	48	bbaaaaaaaaaaaaaaaaaaaaaaaaaaaaaa	−44,485.427
$Cu_{29}Sn_3$	1	256	baaaaaaaaaaabaaaaabaaaaaaaaaaaaa	−43,103.560
	2	768	baaaaaaabaaaaaaaaabaaaaaaaaaaaaa	−43,102.542
	3	768	baaaaaaabaaaaaaabaaaaaaaaaaaaaaa	−43,101.536
	4	256	baaaaaaabaaaaaabaaaaaaaaaaaaaaaa	−43,100.554
	5	384	baaaaaabbaaaaaaaaaaaaaaaaaaaaaaa	−43,102.624
	6	768	baabaaaaaaaaaaaabaaaaaaaaaaaaaaa	−43,102.558
	7	192	baabaaaaaaaabaaaaaaaaaaaaaaaaaaa	−43,103.594
	8	192	baabaaaabaaaaaaaaaaaaaaaaaaaaaaa	−43,101.576
	9	32	baababaaaaaaaaaaaaaaaaaaaaaaaaaa	−43,103.623
	10	384	bbaaaaaaaaaaaaaaaaaaaabaaaaaaaaa	−43,102.360
	11	384	bbaaaaaaaaabaaaaaaaaaaaaaaaaaaaa	−43,103.366
	12	384	bbaaaaaabaaaaaaaaaaaaaaaaaaaaaaa	−43,101.419
	13	96	bbaaaabaaaaaaaaaaaaaaaaaaaaaaaaa	−43,103.483
	14	96	bbbaaaaaaaaaaaaaaaaaaaaaaaaaaaaa	−43,103.230

Appendix A.2

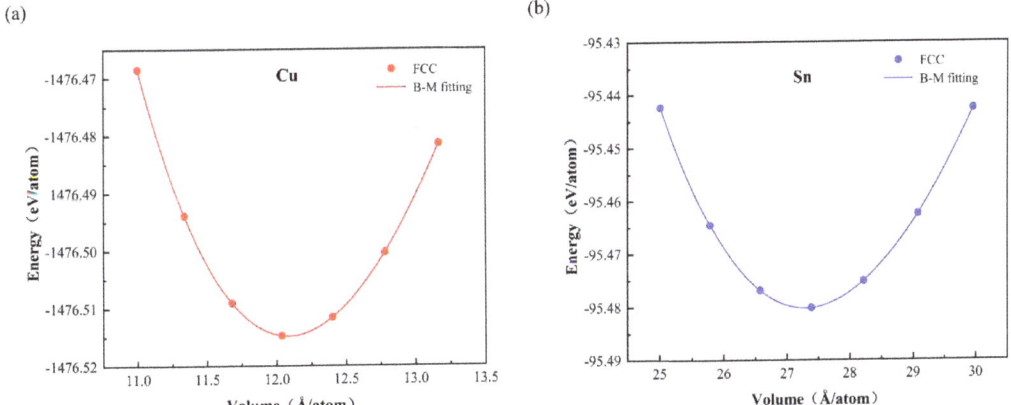

Figure A1. E-V fitting curves of pure elements (a) Cu, (b) Sn.

Appendix A.3

Table A2. Equilibrium volume V_0 (Å^3/atom), bulk modulus B_0 (GPa), first-order derivative of bulk modulus with respect to pressure B_0' and static energy E_0 (eV/atom) of Cu and Sn.

Element	Pure			
	V_0 (Å^3/atom)	B_0 (GPa)	B_0'	E_0 (eV/atom)
Cu	12.04	128.16	4.33	−1476.515
Sn	27.29	54.47	4.40	−95.480

Appendix A.4

Table A3. Volume V (Å^3/atom) and total energy E (eV/atom) of the Cu–Sn alloys.

Structure	V (Å^3/atom)	E (eV/atom)
Cu$_{31}$Sn	393.47	−45,868.47
Cu$_{30}$Sn$_2$	405.02	−44,487.20
Cu$_{29}$Sn$_3$	417.17	−43,105.81

References

1. Yang, L.; Kang, H. Measurements of mechanical properties of α-phase in Cu–Sn alloys by using instrumented nanoindentation. *J. Mater. Res.* **2012**, *27*, 192–196.
2. Sun, J.; Ming, T.Y. Electrochemical behaviors and electrodeposition of Single-Phase Cu-Sn Alloy coating in [BMIM]Cl. *Electrochim. Acta* **2018**, *297*, 87–93.
3. Liu, Y.; Wang, L. Electro-deposition preparation of self-standing Cu-Sn alloy anode electrode for lithium ion battery. *J. Alloys Compd.* **2019**, *775*, 818–825. [CrossRef]
4. Singh, J.B.; Cai, W. Dry sliding of Cu–15 wt%Ni–8 wt%Sn bronze: Wear behaviour and microstructures. *Wear* **2007**, *263*, 830–841. [CrossRef]
5. Zhu, S.Q.; Ringer, S.P. On the role of twinning and stacking faults on the crystal plasticity and grain refinement in magnesium alloys. *Acta Mater.* **2017**, *144*, 365–375. [CrossRef]
6. Wen, Y.F.; Sun, J. Elastic stability of face-centered cubic Fe-Cu random solid solution alloys based on special quasirandom structure model. *Chin. J. Nonferrous Met.* **2012**, *22*, 2522–2528.
7. Wei, Z.; Liu, L. Structural, electronic and thermo-elastic properties of Cu$_6$Sn$_5$ and Cu$_5$Zn$_8$ intermetallic compounds: First-principles investigation. *Intermetallics* **2010**, *18*, 922–928.
8. An, R.; Wang, C. Determination of the Elastic Properties of Cu$_3$Sn Through First-Principles Calculations. *J. Electron. Mater.* **2008**, *37*, 477–482.
9. Davis, J.R. *ASM Specialty Handbook: Copper and Copper Alloys*; ASM International: Almere, The Netherlands, 2001.
10. Scudino, S.; Unterdoerfer, C. Additive manufacturing of Cu-10Sn bronze. *Mater. Lett.* **2015**, *156*, 202–204. [CrossRef]
11. Ping, H.; Xiao, F.R. Influence of hot pressing temperature on the microstructure and mechanical properties of 75% Cu–25% Sn alloy. *Mater. Des.* **2014**, *53*, 38–42.
12. Payne, M.C.; Teter, M.P. Iterative minimization techniques for ab initio total-energy calculations: Molecular dynamics and conjugate gradients. *Rev. Mod. Phys.* **1992**, *64*, 1045. [CrossRef]
13. Ming, X.; Wang, X.L. First-principles study of pressure-induced magnetic transition in siderite FeCO$_3$. *J. Alloys Compd.* **2011**, *510*, L1–L4. [CrossRef]
14. Varadachari, C.; Ghosh, A. Theoretical Derivations of a Direct Band Gap Semiconductor of SiC Doped with Ge. *J. Electron. Mater.* **2014**, *44*, 167–176.
15. Ullrich, C.A.; Kohn, W. Degeneracy in Density Functional Theory: Topology in the ν and n Spaces. *Phys. Rev. Lett.* **2002**, *89*, 156401. [CrossRef] [PubMed]
16. Perdew, J.P.; Burke, K. Generalized Gradient Approximation Made Simple. *Phys. Rev. Lett.* **1996**, *77*, 3865. [CrossRef] [PubMed]
17. Head, J.D.; Zerner, M.C. A Broyden—Fletcher—Goldfarb—Shanno optimization procedure for molecular geometries. *Chem. Phys. Lett.* **1985**, *122*, 264–270. [CrossRef]
18. Liu, Y.; Liu, T. Research progress on intermetallic compounds and solid solutions of Mg alloys based on first-principlescalculation. *Chongqing Daxue Xuebao/J. Chongqing Univ.* **2018**, *41*, 30–44.
19. Straumanis, M.E.; Yu, L.S. Lattice parameters, densities, expansion coefficients and perfection of structure of Cu and of Cu–In α phase. *Acta Crystallogr.* **2014**, *25*, 676–682. [CrossRef]
20. Uesugi, T.; Higashi, K. First-principles studies on lattice constants and local lattice distortions in solid solution aluminum alloys. *Comput. Mater. Sci.* **2013**, *67*, 1–10. [CrossRef]
21. Sidot, E.; Kahn-Harari, A. The lattice parameter of α-bronzes as a function of solute content: Application to archaeological materials. *Mater. Sci. Eng. A* **2005**, *393*, 147–156. [CrossRef]
22. Liu, T.; Chong, X.Y. Changes of alloying elements on elasticity and solid solution strengthening of α-Ti alloys: A comprehensive high-throughput first-principles calculations. *Rare Met.* **2022**, *41*, 2719–2731. [CrossRef]
23. Mxw, A.; Hong, Z.B. Solid-solution strengthening effects in binary Ni-based alloys evaluated by high-throughput calculations. *Mater. Des.* **2020**, *198*, 109359.
24. Petrman, V.; Houska, J. Trends in formation energies and elastic moduli of ternary and quaternary transition metal nitrides. *J. Mater. Sci.* **2013**, *48*, 7642–7651. [CrossRef]
25. Cottrell, A. Effect of solute atoms on the behavior of dislocations. In *Report of a Conference on Strength of Solids*; The Physical Society London: London, UK, 1948; pp. 30–36.
26. Fleischer, R.L. Substitutional solution hardening. *Acta Metall.* **1963**, *11*, 203–209. [CrossRef]
27. Suzuki, H. Segregation of Solute Atoms to Stacking Faults. *J. Phys. Soc. Jpn.* **1962**, *17*, 322–325. [CrossRef]

28. Cottrell, A.H.; Hunter, S.C. CXI. Electrical interaction of a dislocation and a solute atom. *Philos. Mag.* **1953**, *44*, 1064–1067. [CrossRef]
29. Friedel, J. Hardness of a Crystal Containing Uniformly Distributed Impurities or Precipitates. *Dislocations* **1964**, *20*, 368–384.
30. Uesugi, T.; Takigawa, Y. Deformation Mechanism of Nanocrystalline Al-Fe Alloys by Analysis from Ab-Initio Calculations. *Mater. Sci. Forum* **2006**, *503–504*, 209–214. [CrossRef]
31. Sms, A.; Kyk, A. Effects of Sn content and hot deformation on microstructure and mechanical properties of binary high Sn content Cu–Sn alloys. *Mater. Sci. Eng. A* **2020**, *796*, 140054.
32. Shi, J.G.; Liu, P. Selective Laser Melting Experiment of $Cu_{10}Sn$ Alloy. *Ind. Technol. Innov.* **2018**, *5*, 7–11.
33. Mao, Z.; Zhang, D.Z. Processing optimisation, mechanical properties and microstructural evolution during selective laser melting of Cu-15Sn high-tin bronze. *Mater. Sci. Eng. A* **2018**, *721*, 125–134. [CrossRef]
34. Page, Y.L.; Saxe, P. Symmetry-General Least-Squares Extraction of Elastic Data for Strained Materials From ab Initio Calculations of Stress. *Phys. Rev. B* **2002**, *65*, 104104. [CrossRef]
35. Jy, A.; Po, A. First-Principles Study of the Effect of Aluminum Content on the Elastic Properties of Cu-Al Alloys. *Mater. Today Commun.* **2022**, *31*, 103399.
36. Browaeys, J.T.; Chevrot, S. Decomposition of the elastic tensor and geophysical applications. *Geophys. J. R. Astron. Soc.* **2010**, *159*, 667–678. [CrossRef]
37. Moakher, M.; Norris, A.N. The Closest Elastic Tensor of Arbitrary Symmetry to an Elasticity Tensor of Lower Symmetry. *J. Elast.* **2006**, *85*, 215–263. [CrossRef]
38. Waller, I. Dynamical Theory of Crystal Lattices by M. Born and K. Huang. *Acta Crystallogr.* **1956**, *9*, 837–838. [CrossRef]
39. Chung, D.H.; Buessem, W.R. The Voigt-Reuss-Hill Approximation and Elastic Moduli of Polycrystalline MgO, CaF2, β-ZnS, ZnSe, and CdTe. *J. Appl. Phys.* **1967**, *38*, 2535–2540. [CrossRef]
40. Overton, W.C.; Gaffney, J. Temperature Variation of the Elastic Constants of Cubic Elements. I. Copper. *Phys. Rev.* **1955**, *98*, 969–977. [CrossRef]
41. Chang, Y.A.; Himmel, L. Temperature Dependence of the Elastic Constants of Cu, Ag, and Au above Room Temperature. *J. Appl. Phys.* **1966**, *37*, 3567–3572. [CrossRef]
42. Schmunk, R.E.; Smith, C.S. Elastic constants of copper-nickel alloys. *Acta Metall.* **1960**, *8*, 396–401. [CrossRef]
43. Cheng, L.; Shuai, Z. Insights into structural and thermodynamic properties of the intermetallic compound in ternary Mg–Zn–Cu alloy under high pressure and high temperature. *J. Alloys Compd.* **2014**, *597*, 119–123. [CrossRef]
44. Zhou, W.; Liu, L. Structural, Elastic, and Electronic Properties of Al-Cu Intermetallics from First-Principles Calculations. *J. Electron. Mater.* **2009**, *38*, 356–364. [CrossRef]
45. Li, F.; Chen, Y. First-Principles Calculations on the Enhancing Effect of Zr on the Mechanical and Thermodynamic Properties of Ir-Rh Alloys. *Trans. Indian Inst. Met.* **2023**, *76*, 1809–1817. [CrossRef]
46. Pugh, S.F. XCII. Relations between the elastic moduli and the plastic properties of polycrystalline pure metals. *Philos. Mag.* **2009**, *45*, 823–843. [CrossRef]
47. Qu, D.; Li, C. Structural, electronic, and elastic properties of orthorhombic, hexagonal, and cubic Cu_3Sn intermetallic compounds in Sn–Cu lead-free solder. *J. Phys. Chem. Solids* **2019**, *138*, 109253. [CrossRef]
48. Ranganathan, S.I.; Ostoja-Starzewski, M. Universal Elastic Anisotropy Index. *Phys. Rev. Lett.* **2008**, *101*, 055504. [CrossRef] [PubMed]
49. Liao, M.; Yong, L. Alloying effect on phase stability, elastic and thermodynamic properties of Nb-Ti-V-Zr high entropy alloy. *Intermetallics* **2018**, *101*, 152–164. [CrossRef]
50. Shuvalov, L.A. *Electrical Properties of Crystals*; Springer: Berlin/Heidelberg, Germany, 1988.
51. Ma, L.; Duan, Y. Phase stability, anisotropic elastic properties and electronic structures of C15-type Laves phases ZrM2 (M = Cr, Mo and W) from first-principles calculations. *Philos. Mag.* **2017**, *97*, 2406–2424. [CrossRef]
52. Zhu, Y.D.; Yan, M.F. First-principles investigation of structural, mechanical and electronic properties for Cu–Ti intermetallics. *Comput. Mater. Sci.* **2016**, *123*, 70–78. [CrossRef]
53. Wei, Y.A.; Yz, A. Investigation on elastic properties and electronic structure of dilute Ir-based alloys by first-principles calculations. *J. Alloys Compd.* **2021**, *850*, 156548.
54. Segall, M.; Shah, R. Population analysis of plane-wave electronic structure calculations of bulk materials. *Phys. Rev. B Condens. Matter* **1996**, *54*, 16317. [CrossRef] [PubMed]
55. Liu, D.; Duan, Y. Structural properties, electronic structures and optical properties of WB2 with different structures: A theoretical investigation. *Ceram. Int.* **2018**, *44*, 11438–11447. [CrossRef]
56. Anderson, O. A simplified method for calculating the debye temperature from elastic constants. *J. Phys. Chem. Solids* **1963**, *24*, 909–917. [CrossRef]
57. Reffas, M.; Bouhemadou, A. Ab initio study of structural, elastic, electronic and optical properties of spinel $SnMg_2O_4$. *Phys. B Condens. Matter* **2010**, *405*, 4079–4085. [CrossRef]
58. Schreiber, E.; Anderson, O.L.; Soga, N.; Bell, J.F. Elastic Constants and Their Measurement. *J. Appl. Mech.* **1975**, *42*, 747–748. [CrossRef]
59. Cahill, D.G.; Pohl, R.O. Heat flow and lattice vibrations in glasses. *Solid State Commun.* **1989**, *70*, 927–930. [CrossRef]
60. Fine, M.E. Elastic constants versus melting temperature in metals. *Scr. Metall.* **1984**, *18*, 951–956. [CrossRef]

61. Murray, J.L. The CuTi (Copper-Titanium) system. *J. Phase Equilibria* **1983**, *4*, 81–95.
62. Lebedev-Stepanov, P.V. Plasma frequency approach to estimate the Debye temperature of the ionic crystals and metal alloys. *J. Phys. Chem. Solids* **2014**, *75*, 903–910. [CrossRef]
63. Li, Y.; Ma, X.J. First-principles calculations of the structural, elastic and thermodynamic properties of tetragonal copper-titanium intermetallic compounds. *J. Alloys Compd.* **2016**, *687*, 984–989. [CrossRef]

Disclaimer/Publisher's Note: The statements, opinions and data contained in all publications are solely those of the individual author(s) and contributor(s) and not of MDPI and/or the editor(s). MDPI and/or the editor(s) disclaim responsibility for any injury to people or property resulting from any ideas, methods, instructions or products referred to in the content.

Article

Optical, Dielectric, and Electrical Properties of Tungsten-Based Materials with the Formula $Li_{(2-x)}Na_xWO_4$ (x = 0, 0.5, and 1.5)

Moufida Krimi [1], Mohammed H. Al-Harbi [2], Abdulelah H. Alsulami [3], Karim Karoui [1], Mohamed Khitouni [4,*] and Abdallah Ben Rhaiem [1]

[1] Laboratory LaSCOM, Faculty of Sciences of Sfax, University of Sfax, BP1171, Sfax 3000, Tunisia; krimi.fayda@gmail.com (M.K.); karouikarim36@yahoo.com (K.K.); abdallahrhaiem@yahoo.fr (A.B.R.)
[2] Department of Science, King Abdulaziz Military Academy, Riadh 11538, Saudi Arabia
[3] Chemistry Department, Faculty of Science and Arts in Baljurashi, Al-Baha University, Al-Baha 65431, Saudi Arabia; aalsulami@bu.edu.sa
[4] Department of Chemistry, College of Science, Qassim University, Buraidah 51452, Saudi Arabia
* Correspondence: kh.mohamed@qu.edu.sa

Abstract: In the present study, three chemical compounds, Li_2WO_4, $Li_{0.5}Na_{1.5}WO_4$, and $Li_{1.5}Na_{0.5}WO_4$, are produced using the solid–solid method. Unlike the compound $Li_{0.5}Na_{1.5}WO_4$, which crystallizes in the orthorhombic system with the space group Pmmm, both compounds Li_2WO_4 and $Li_{1.5}Na_{0.5}WO_4$ crystallize in the monoclinic system with the space group P2/m. A morphological analysis reveals that all three compounds have a compact structure with some porosity present. An EDX analysis confirms the chemical composition of the three samples. The optical measurements provide information on the optical gaps and Urbach energies of the materials under consideration. Their dielectric characteristics are investigated in a frequency range of 100–106 Hz and at temperatures ranging from 300 to 600 K. Moreover, this research enables us to determine the ferroelectric transition as well as the type of dielectric material. In this study, an investigation of electrical conductivity was conducted for well-defined temperature and frequency values; which provided us with information about the mechanism of conduction and charge carrier transport models.

Keywords: tungsten-based materials; physical chemistry; optical properties; electrical conductivity; mechanism of conduction

1. Introduction

In an effort to find new materials with a wide range of commercial uses, several ceramics, single crystals, and thin films of various structural families have been synthesized and described during the past few years, utilizing a variety of experimental techniques [1–24]. In this regard, research projects carried out in the last several years have shown a great deal of interest in tungsten based materials [3–6]. There is constant innovation in this subject because tungsten has so many attractive characteristics. In fact, the use of tungsten-based materials is popular in a variety of fields, including photoluminescence [1,2], magnetic characteristics [3], supercapacitors [4,5], laser hosts [6,7], gas sensing [8,9], catalysts [10,11], photocatalysts [12–14], scintillator materials [15,16], humidity sensors [17,18], microwave applications [19,20], fiber optics [21,22], lithium batteries [23,24], etc. Because of their high specific capacity, good operating voltage, large reserves, and environmental friendliness, these materials have proven beneficial for electrochemical energy storage. With a melting temperature of roughly 3380 °C and a boiling point of roughly 5900 °C, tungsten is a thermally stable substance. Furthermore, because of its excellent coloration efficiency, strong reversibility ratio, rapid colorization bleaching speed, and lengthy cyclic stability, it is recyclable and not considered a health hazard [25,26]. These characteristics help explain tungsten's remarkable success in a variety of applications. In fact, the literature mentions a number of tungsten-based compounds such as $ABWO_4$ and A_2WO_4 (A, B = Li, Na, K,

etc.), which were discovered to have a variety of structures [27,28], substantial phase transitions [29,30], and considerable electrical [31,32] and ferroelectric properties [33]. Li_2WO_4 and Na_2WO_4 compounds, for instance, have been the topic of numerous prior investigations employing various techniques of production. Both compounds show a ferroelectric transition, at Tc = 278 K for Li_2WO_4 and at Tc = 300 K for Na_2WO_4 [5,6]. Ferroelectric and ferroelastic transitions are mainly found in tungsten forms containing either a monovalent element alone or a monovalent element coupled with a bivalent element. Moreover, the samples containing mainly bivalent elements are unlikely to exhibit ferroelectric transitions. The XRD, EDF, and SEM analyses of $Li_{2-x}Na_xWO_4$ (x = 0, 0.5, and 1.5), which have been realized and published by Krimi et al. confirm the purity and the stoichiometry of these materials [6,34,35]. The samples Li_2WO_4 and $Li_{1.5}Na_{0.5}WO_4$ crystallize in the monoclinic system (space group: P2/m), while the other compounds, Na_2WO_4 and $Li_{0.5}Na_{1.5}WO_4$, crystallize in the orthorhombic system (space groups: Pbca for Na_2WO_4 and Pmmm for $Li_{0.5}Na_{1.5}WO_4$) [6,34,35]. In fact, a simple comparison of the obtained results leads us to the conclusion that a rise in the sodium rate is accompanied by an increase in the volume of the unit cell, which is expected given that the radius of the sodium atom is larger than that of the lithium atom.

The purpose of the present research is to examine the thermal, optical, and electrical characteristics of the compound $Li_{2-x}Na_xWO_4$, as well as the effect of substituting lithium for sodium. Such findings are evaluated in light of the parent chemicals' characteristics.

2. Materials and Methods

2.1. Synthesis

The solid–solid method was used to synthesize three compounds: Li_2WO_4, $Li_{0.5}Na_{1.5}WO_4$, and $Li_{1.5}Na_{0.5}WO_4$. For this, well-calculated stoichiometric quantities of the precursors Li_2CO_3, Na_2CO_3, and WO_3 were mixed and then pulverized. They were initially annealed at 723 K to ensure the discharge of undesired chemicals, particularly carbon. They were then ground again, pelletized, and reheated at 823 K to ensure the compactness of the precursors and reduce the size of the grains.

2.2. Equipment

The purity of the samples was determined using the X-ray diffraction powder "Siemens D5000" with CuKα radiation (λ = 1.5406 Å). The powder morphology was examined using scanning electron microscopy (SEM) in secondary-electron mode at a voltage of 15 kV, the microscope used was the DSM960A ZEISS type (Zeiss, Oberkochen, Germany). The SEM was coupled with a Vega&Tescan energy dispersive X-ray spectrometry (EDS) analyzer (Zeiss, Oberkochen, Germany).

To determine the optical band gap and Urbach energy, a UV-Vis measurement was made at room temperature using a spectrophotometer (Shimadzu, Kyoto, Japan, UV-3101PC), while an electric measurement was carried out as a function of frequency in the 10–160 kHz range at temperatures ranging from 458 K to 623 K using a TEGAM 3550 impedance analyzer driven by a microcomputer.

The phase transition in different samples was first detected using a calorimetric study. This was later carried out in a Mettler–Toledo DSC822 apparatus under an Argon atmosphere, at temperatures ranging from 300 to 450 K and at a heating rate of 10 K/min. The annealing of different samples was also performed in the DSC under a 20 mL/min Ar flow.

3. Results

3.1. Structural and Morphological Studies

Figure 1 depicts the XRD patterns of the synthesized compounds: $Li_{0.5}Na_{1.5}WO_4$, $Li_{1.5}Na_{0.5}WO_4$, and Li_2WO_4. Each diffractogram shows the following: the experimental points (black circles), the calculated curve (red line), the difference between the experimental and calculated profiles (blue line), and the Bragg positions (green line). We have been careful to follow the known sequence of steps in order to release different crystallographic parameters during the structural refining process of the three samples [36]. This confirms the refinement's stability when all the parameters become available. As shown, there is a good level of agreement between the calculated and experimental diagrams. The structural study demonstrates that the compound $Li_{0.5}Na_{1.5}WO_4$ crystallizes in the orthorhombic system with the space group Pmmm (a = 22.42 Å, b = 15.21 Å, c = 7.18 Å) [35]. As long as the compounds $Li_{1.5}Na_{0.5}WO_4$ and Li_2WO_4 crystallize in the monoclinic system with the space group P2/m, the lattice parameters of the compound $Li_{1.5}Na_{0.5}WO_4$ are as follows: a = 15.92 Å, b = 7.19 Å, c = 7.18 Å, $\alpha = \gamma = 90°$, and $\beta = 114.31°$, while those of the compound Li_2WO_4 are a = 7.49 Å, b = 8.31 Å, c = 5.92 Å, $\alpha = \gamma = 90°$, and $\beta = 96.81°$ [6,34]. It may be observed that the volume of the elementary unit of the crystal lattice increases as the Na proportion rises.

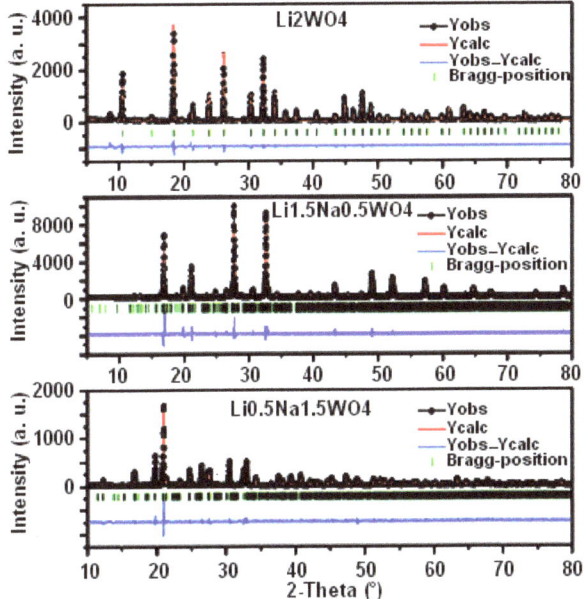

Figure 1. XRD patterns and Rietveld refinements for samples of $Li_{0.5}Na_{1.5}WO_4$, $Li_{1.5}Na_{0.5}WO_4$, and Li_2WO_4 [6,34,35].

On the other hand, several methods have been developed to separate the microstructural parameters including the Williamson–Hall [37] and Halder–Wagner [38] methods. Crystallite size and lattice distortions yield to line broadening. Thus, the total line broadening β can be expressed as [39]

$$\beta = \frac{k\lambda}{D}\frac{1}{\cos\theta} + 4\varepsilon \tan\theta \tag{1}$$

where D is the crystallite size and ε is the lattice strain. The first term in the right-hand part of Equation (1) is the size contribution and the second term is the lattice distortion. If the peak broadening is due solely to a finite crystallite size, it is assumed that $\beta = k\lambda/(<D>\cos\theta)$, where k is close to 1 [40]; this is known as the Scherrer equation [41].

Using the model of Equation (1), the average crystallite size for each studied sample was determined in the current study. This model contains a distinct component for predicting peak broadening related to the crystallite microstrain and uses diffraction peak broadening from at least four diffraction peaks as a foundation for calculating the crystallite size. Equation (1) is a linear equation that considers the isotropic character of the crystals. The inverse of the average crystallite size $<D>$ can be found from the intercept, and the lattice microstrain (ε) can be found from the slope straight line of the plot made with ($4.sin\theta$) along the x-axis and ($\beta.cos\ \theta$) along the y-axis for the three samples. The average particle sizes for $Li_{0.5}Na_{1.5}WO_4$, $Li_{1.5}Na_{0.5}WO_4$, and Li_2WO_4 samples were found to be approximately 200, 150, and 100 nm, respectively. The size of a crystallite, which is not always the same as the particle size, is thought to be the size of a coherently diffracting domain. The samples $Li_{0.5}Na_{1.5}WO_4$, $Li_{1.5}Na_{0.5}WO_4$, and Li_2WO_4 are expected to have lattice strain values of approximately 0.017, 0.035, and 0.043%, respectively. The lattice contraction or expansion in the crystallites, which is the source of the lattice microstrain, is mostly caused by the arrangement of atoms within the crystal lattice. On the other hand, many structural defects (point defects like vacancies, stacking faults, grain boundaries, etc.) are also created in the lattice structure as a result of size refinement and internal–external stresses that result in lattice strain [42,43]. The theoretical estimations of crystallite size derived from the Rietveld refinement for the samples $Li_{0.5}Na_{1.5}WO_4$, $Li_{1.5}Na_{0.5}WO_4$, and Li_2WO_4 were roughly 27% less than the estimated average crystallite size for the three samples.

Figure 2 gives the SEM images, with resolutions of 9 μm and 40 μm, and the EDX spectra of the three studied compounds. We were able to verify the presence of Na, O, and W elements, as indicated in the EDX spectra. Lithium is the only component that is lacking, and this is because of its low Z = 1 value, which prevents it from emitting X-ray radiation. The SEM images clearly show that the three compounds have a compact structure with just a few pores and a tendency for clustering together because of the humid environment that lithium creates. Also, these images demonstrate a homogeneous distribution of grains with sizes around 5 [36], 2.6, and 4 μm [35] for the three compositions $Li_{0.5}Na_{1.5}WO_4$, $Li_{1.5}Na_{0.5}WO_4$, and Li_2WO_4, respectively.

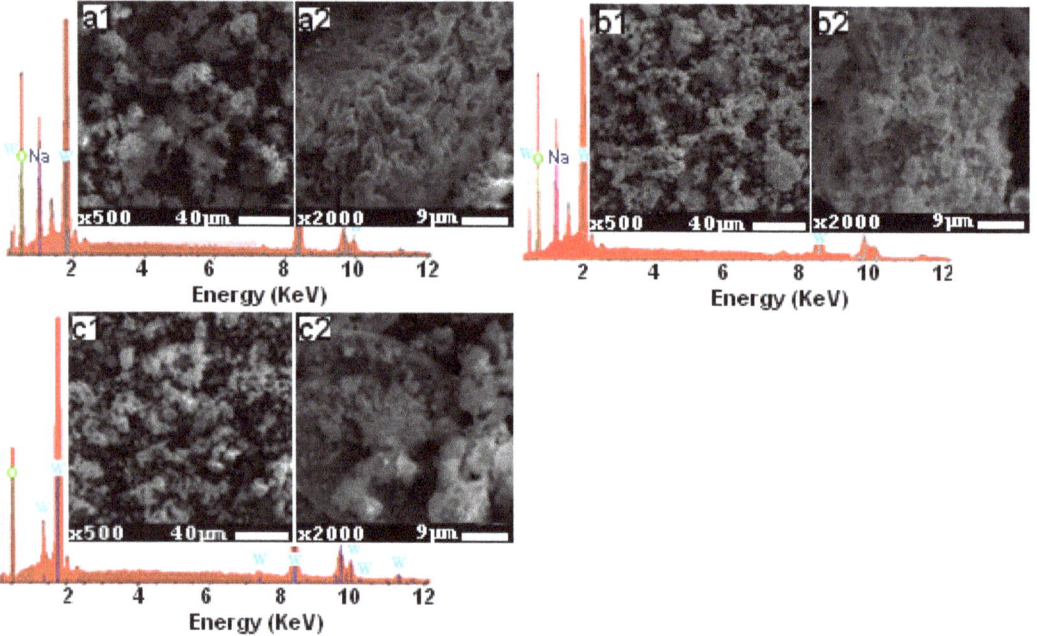

Figure 2. SEM micrographs for the studied samples: (**a1,a2**) $Li_{0.5}Na_{1.5}WO_4$, (**b1,b2**) $Li_{1.5}Na_{0.5}WO_4$, and (**c1,c2**) Li_2WO_4, and the corresponding EDX spectra.

3.2. Optical Characterization

Considering the significance of optical characteristics in establishing light efficiency, the gap energy was examined using UV-Vis spectroscopy. The gap energy (E_g) is a crucial parameter for the description of materials in the solid state because it links disordered materials to either a direct or indirect transition through an optical forbidden band and the absorption coefficient (α). It takes note of the presence of broad bands rather than peaks in the spectrum. This demonstrates that these molecules contain energetically realized transitions. The investigation of such a spectrum leads to the calculation of the greatest absorption, and thus, based on the literature, the attribution of these bands can be determined. In the spectra presented in Figure 3, one can observe a broad band around 244 nm for the molecule Li_2WO_4 and a broad band around 246 nm for the complex $Li_{1.5}Na_{0.5}WO_4$. These bands are attributed to the electronic transition from the 2p state of oxygen to the 5d state of tungsten in the $(WO)^{-2}$ group [8]. The compound $Li_{0.5}Na_{1.5}WO_4$ shows a similar transition at a wavelength of roughly 300 nm. It can be seen that the structural change reported for this molecule may be the cause of the wavelength change. Indeed, a change in symmetry (from monoclinic to orthorhombic) results in a shift in the atoms' locations and sites.

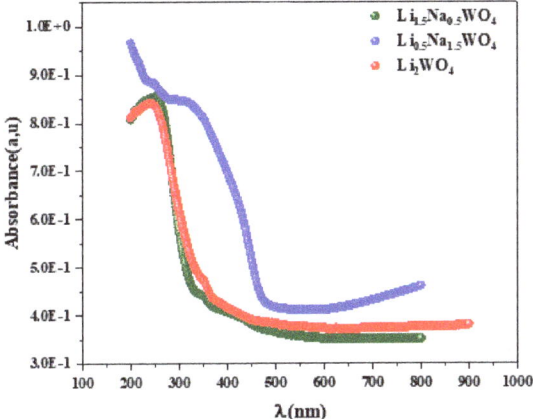

Figure 3. Absorbance spectra for the studied compounds.

The Kubelka–Munk method is used to calculate the Eg on the basis of Equation (1) [44,45]:

$$\frac{F(R)}{e} = \frac{(1-R)^2}{2R} \qquad (2)$$

where e is the compound's thickness ($e = 1$ mm), R is its reflectance, and $(F(R))/e$ is proportional to the absorption coefficient (α). Thus, a modified Kubelka–Munk equation was employed to determine the Eg by multiplying $F(R)$ by $h\nu$ and applying the corresponding coefficient (n), which is linked to an electronic transition, as shown by [46]

$$\left(\frac{F(R)}{e} h\nu\right)^n = f(h\nu) \qquad (3)$$

in which $n = 1/2$ represents a direct permitted transition (plotted as $\alpha(h\nu)^2$ versus E) and $n = 2$ represents an indirect permitted transition (plotted as $\alpha(h\nu)^{1/2}$ versus E).

For our compound, the Eg was calculated for both types of transitions (Figure 4). To determine the band gap (shown by the red line), we actually extrapolate the curves in the linear section to zero. The band gap values for both the directly permitted transition and the indirectly permitted transition are shown in Table 1. These values clearly show that increased Li levels are associated with rising direct and indirect Eg, given the variations in the absorption band wavelength and symmetry.

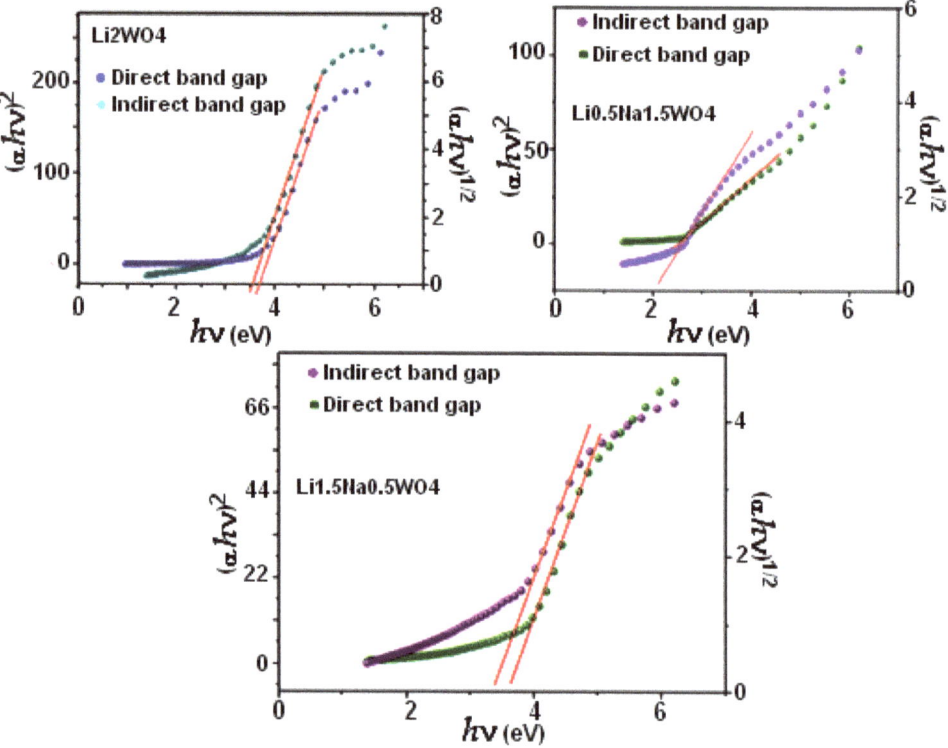

Figure 4. Direct and indirect band gaps for the studied compounds.

Table 1. Calculated values of gap energies for the studied compounds.

Sample	Direct Gap (eV)	Indirect Gap (eV)
Li_2WO_4	4.1	3.5
$Li_{1.5}Na_{0.5}WO_4$	3.71	3.4
$Li_{0.5}Na_{1.5}WO_4$	2.6	2.1

Thus, in order to further investigate this pattern, we computed the Urbach Eu energy, which is associated with the changes from the valence band's stretched states to the conduction band's restricted states. On the other hand, the Urbach Eu energy causes the formation of localized states at the tail of the band, near the forbidden band's limits, bounded by the valence and conduction bands. It also describes the compound's disorder [47]. Then, the following empirical formula [46] expresses the relationship between the Urbach energy and the disorder:

$$\alpha = \alpha_0 e^{\frac{h\gamma}{E_u}} \quad (4)$$

Furthermore, when plotting the variation in $\ln(\alpha)$ as a function of energy ($h\nu$), we notice that the following expression governs this curve:

$$\ln(\alpha) = \ln(\alpha_0) + \frac{h\gamma}{E_u} \quad (5)$$

where α_0 is a constant and Eu is the Urbach energy, which is equal to 0.9 eV, 1.36 eV, and 0.35 eV for the Li_2WO_4, $Li_{1.5}Na_{0.5}WO_4$, and $Li_{0.5}Na_{1.5}WO_4$ compounds, respectively. In fact, as shown in Figure 5, these values are actually determined by taking the inverse of the curves' slopes. Then, Table 2 shows the computed Urbach energy values. Therefore, based on the above-mentioned results, we can conclude that the compound $Li_{1.5}Na_{0.5}WO_4$ is the most disordered of the tested compounds. Moreover, an examination of the values of the Urbach energy in relation to the gap one revealed the existence of a harmony in variation, which indicates that the larger the disorder, the wider the gap will be. This suggests that the latter is affected by disorder in our materials.

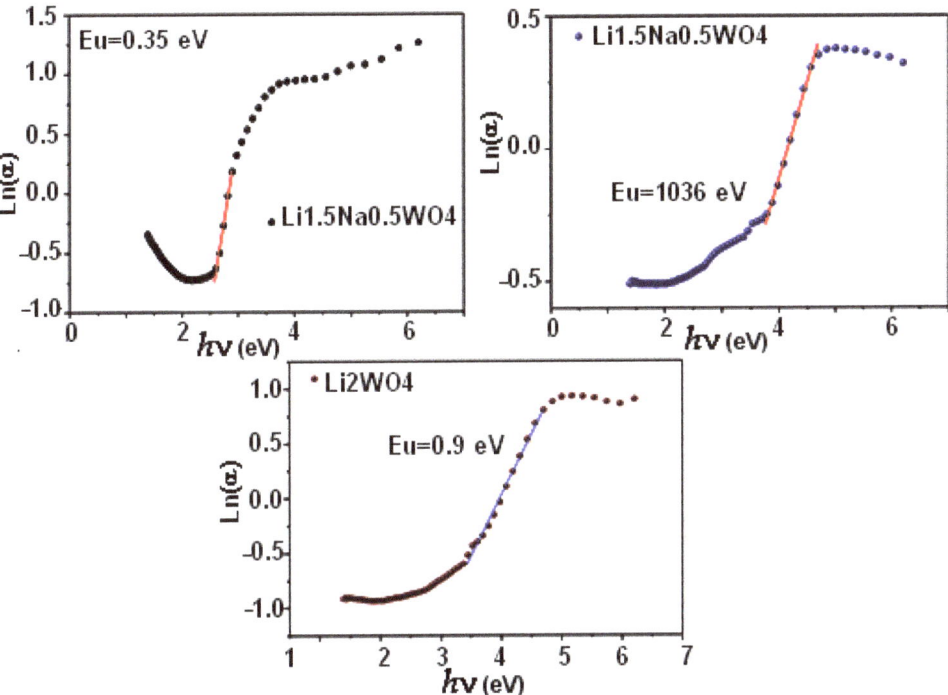

Figure 5. Urbach energy for the three compounds.

Table 2. Densities of localized states and the values of Urbach energy for the studied compounds.

Sample	$Li_{0.5}Na_{1.5}WO_4$	$Li_{1.5}Na_{0.5}WO_4$	Li_2WO_4
Density of localized states	0.6×10^{20}–6×10^{20}	1.10^{22}–2.10^{24}	1.10^{22}–9.10^{23}
Value of Urbach energy	0.35 (eV)	1.36 (eV)	0.9 (eV)

3.3. Thermal Analysis

A differential scanning calorimetric analysis is commonly used to analyze a material's thermal change, such as melting, glass transition, crystallization, and so on. Therefore, to avoid reactivity with the atmosphere, this analysis was carried out in the presence of an argon atmosphere. The thermal investigations were carried out in a temperature range of ambient temperature to 450 K with a scan rate of 5 C°/min. The obtained thermograms are shown in Figure 6. As shown, the compound $Li_{1.5}Na_{0.5}WO_4$ has three endothermic peaks, at T = 365 K, T = 379 K, and T = 397 K (Figure 6a) [6], whereas $Li_{0.5}Na_{1.5}WO_4$ has just two endothermic peaks, at T = 379 K and T = 397 K (Figure 6b) [35].

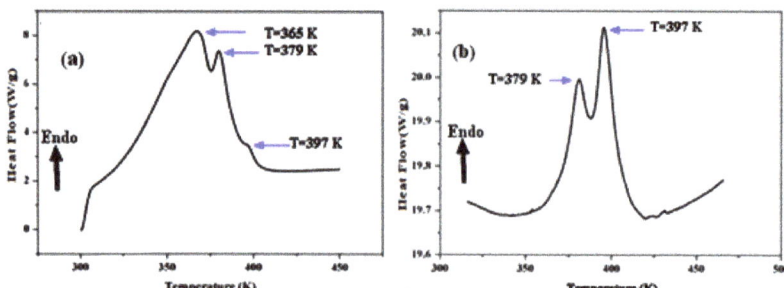

Figure 6. Differential scanning calorimetric of (**a**) Li$_{1.5}$Na$_{0.5}$WO$_4$ and (**b**) Li$_{0.5}$Na$_{1.5}$WO$_4$ [36,38].

The thermal examination of the parent compounds Li$_2$WO$_4$ and Na$_2$WO$_4$ revealed the presence of a peak at 373 K, which was attributed to the ferroelectric/paraelectric transition and confirmed by the dielectric research [5,6]. The broad peak at 365 K may be due to the progressive release of H$_2$O absorbed by the material at ambient temperature, which shows that this material has a hydroscopic character. Based on Ref. [6] and the impact of the composition difference, we can credit the peak found at T = 397 K for both compounds to the ferroelectric–paraelectric transition. In addition, to verify this result we investigated the dielectric characteristics of the compounds Li$_{1.5}$Na$_{0.5}$WO$_4$ and Li$_{0.5}$Na$_{1.5}$WO$_4$.

3.4. Dielectric Studies

Figures 7 and 8 show the temperature dependency of the dielectric constant for the compounds Li$_{0.5}$Na$_{1.5}$WO$_4$ and Li$_{1.5}$Na$_{0.5}$WO$_4$. The study of dielectric characteristics is, therefore, essential in order to detect the changes in the phases, the properties, and the behavior of the examined substance when an electric field is applied. This investigation is undertaken in the present case as a function of temperature in the range of 300–650 K, and in a frequency range of 100 Hz to 1 MHz. In fact, Figures 7a,b and 8a,b show the temperature dependence of the relative dielectric constant with a fixed frequency for the compounds Li$_{0.5}$Na$_{1.5}$WO$_4$ and Li$_{1.5}$Na$_{0.5}$WO$_4$ [34], respectively.

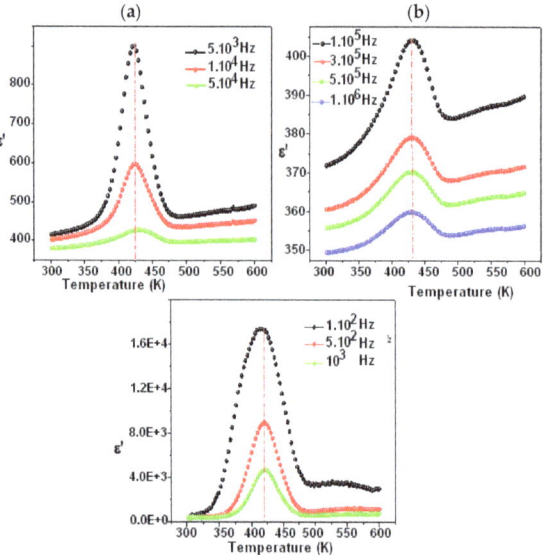

Figure 7. Variation in ε' as a function of temperature for Li$_{1.5}$Na$_{0.5}$WO$_4$ ((**a**) frequency from 100 Hz to 10,000 Hz and (**b**) frequency from 100.000 to 1000,000 Hz) [36].

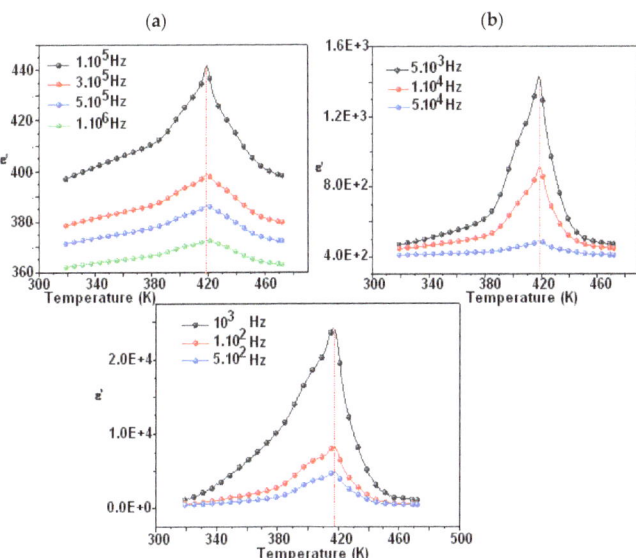

Figure 8. Variation in ε' as function temperature of $Li_{0.5}Na_{1.5}WO_4$ ((**a**) frequency from 100 Hz to 10,000 Hz and (**b**) frequency from 100.000 to 1000,000 Hz).

We can see a dielectric relaxation of about 421 K for the compound $Li_{1.5}Na_{0.5}WO_4$ and 418 K for the compound $Li_{1.5}Na_{0.5}WO_4$. Therefore, a para–ferroelectric phase transition may be related to this large dielectric peak at T_c. Its maximum does not change (T_c remains constant at different frequencies), but its magnitude diminishes as the frequency increases. This relaxation is also observed for the parent molecule at about 373 K [5,6]. This shift in transition temperature could be attributed to a change in symmetry caused by the composition change. Actually, a qualitative study of these curves revealed that T_c remained constant when the frequency was varied, confirming the examined compound's traditional characteristic [48]. Hence, in order to verify the nature of classic or relaxer ferroelectrics, Uchino and Nomura [49] proposed a more general expression of the Curie–Weiss law by introducing the degree of relaxation γ:

$$\frac{1}{\varepsilon'} - \frac{1}{\varepsilon'_{max}} = \frac{(T - T_c)^\gamma}{C} \tag{6}$$

where C is the Curie–Weiss constant, ε'_{max} is the real dielectric constant at $T = T_C$, and γ ($1 < \gamma < 2$) denotes the degree of relaxation. The diffuse nature of the transition is translated by the component γ. However, the γ value is close to 1 for classical ferroelectrics and should equal 2 for perfect relaxer ferroelectrics [50,51].

Figures 9 and 10 illustrate the logarithmic charts of Equation (5) at various frequencies (the curve fitted to the modified Curie–Weiss law is indicated by the solid red lines).

The obtained values for the $Li_{1.5}Na_{0.5}WO_4$ and $Li_{0.5}Na_{1.5}WO_4$ compounds are around 1.3 [34] and 1.5, respectively. This finding demonstrates that our sample is a typical ferroelectric compound. The dielectric measurements in the paraelectric state match the Curie–Weiss law rather well, as stated by the following relation:

$$\varepsilon' = \frac{C}{T - T_0} \tag{7}$$

where C and T_0 are the Curie–Weiss constant and temperature, respectively.

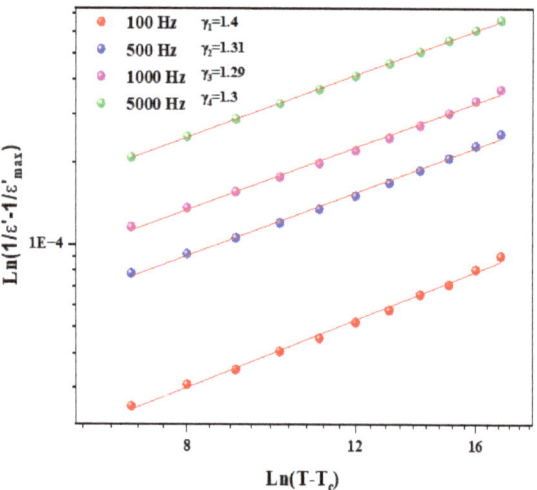

Figure 9. $\ln\left(\frac{1}{\varepsilon'} - \frac{1}{\varepsilon'_{max}}\right)$ as function of $\ln(T - Tc)$ for different frequencies for $Li_{1.5}Na_{0.5}WO_4$ compound.

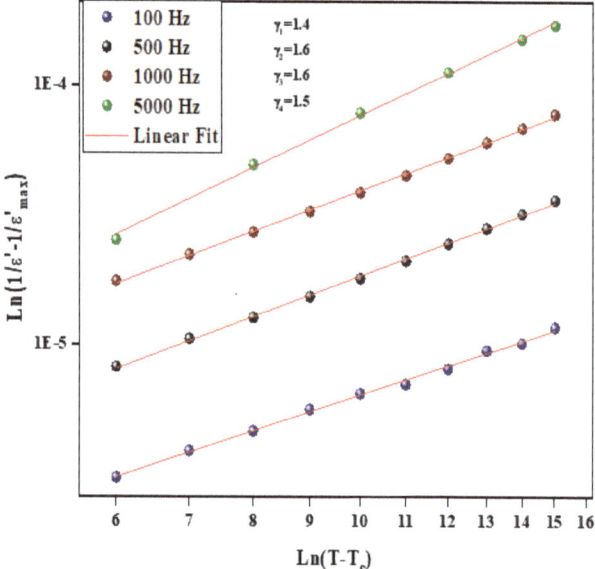

Figure 10. $\ln\left(\frac{1}{\varepsilon'} - \frac{1}{\varepsilon'_{max}}\right)$ as function of $\ln(T - Tc)$ for different frequency for $Li_{0.5}Na_{1.5}WO_4$ compound.

Figure 11 depicts the temperature dependency of the dielectric reciprocal $(1/\varepsilon')$ at 500 Hz. Indeed, for $Li_{0.5}Na_{1.5}WO_4$ and $Li_{1.5}Na_{0.5}WO_4$, these plots display straight lines, with the x-axis intercepts at $T_0 = 418$ K and 421 K [34]. In classical ferroelectrics, the value of T_0 indicates the order of the para–ferroelectric phase transition. In the case of $T_0 \neq T_C$, the phase transition is first order. If $T_0 = T_C$, then this transition is of second order [52]. According to our observations, the value of T_0 is comparable to T_C, suggesting that this transition is of second order [53]. Beginning with the concept of ε', which is connected to capacitance and the alignment of the dipole that is observed, we may deduce that a rise in frequency corresponds to an increase in disorder, which causes ε' to drop. It is evident from comparing the ε' values of the studied compounds that $Li_{0.5}Na_{1.5}WO_4$ has higher values than $Li_{1.5}Na_{0.5}WO_4$. We could, therefore, conclude that the latter is the most disordered.

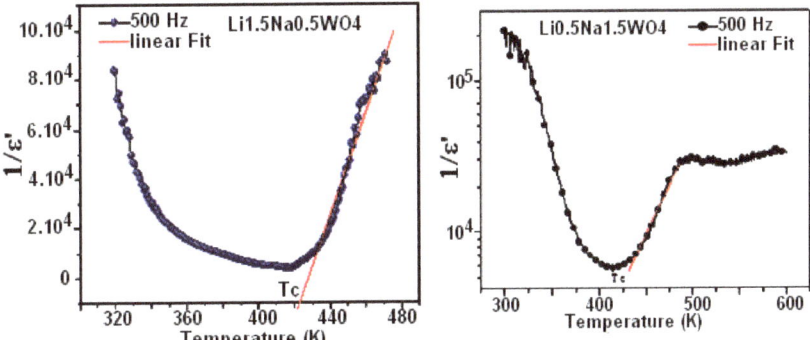

Figure 11. Temperature variation in $(1/\varepsilon')$ at 500 Hz.

3.5. Conductivity Analysis

For the three compounds under investigation, we have calculated the conductivity fluctuation as a function of the frequency of the various temperature values [6,34,35]. We also observed that the experimental variation in conductivity is described by Jonscher's universal law. The experimental results were adopted using the equation

$$\sigma_{AC}(W) = \sigma_{dC} + A\omega^S \tag{8}$$

In fact, as shown in Figure 12, and based on this adjustment, we have plotted the curves $Ln(\sigma_{dC}) = f\left(\frac{1000}{T}\right)$ which are adjusted by the following expression:

$$\sigma_{dC} = \sigma_0 \exp\left(\frac{-E_a}{K_B T}\right) \tag{9}$$

where σ_0 is the pre-exponential factor given by the following expression:

$$\sigma_0 = \left(\frac{e_a^2 a_h^2 \gamma_0}{6k}\right) N(T) \exp\left(\frac{S_\mu}{k}\right) \tag{10}$$

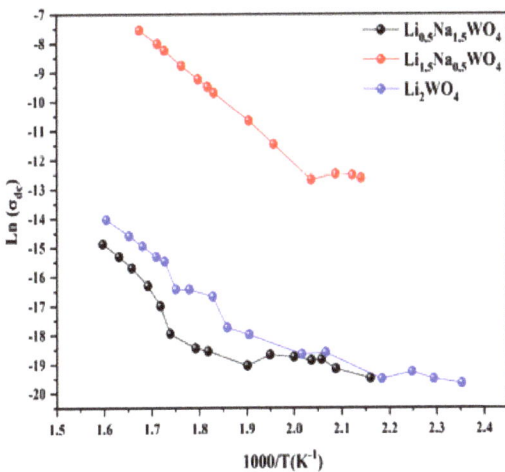

Figure 12. Variation in conductivity for the three studied compounds.

The number of charge carriers that contribute to the conduction N and the entropy S are both proportional to this factor. Since the activation energies of the three materials are of the same order of magnitude, the high conductivity of the $Li_{1.5}Na_{0.5}WO_4$ combination can be explained by other factors (Table 2). Moreover, given that the number of charge carriers is related to the density of the localized states (N = kTN (EF)) [54], the number in $Li_{1.5}Na_{0.5}WO_4$ is greater than those in Li_2WO_4 and $Li_{0.5}Na_{1.5}WO_4$. The enormous value of the Urbach energy for the composite $Li_{1.5}Na_{0.5}WO_4$ compared to the other two compounds indicates that the disorder (translated by the entropy S) in this material is more important, and thus: S(x = 0.5) > S(x = 1) > S(x = 1.5). Actually, a comparative study revealed that the conductivity of the compound $Li_{1.5}Na_{0.5}WO_4$ is greater than that of the other two compounds, which can be explained by the fact that the disorder (entropy represented by the Urbach energy) and the number of charge carriers (represented by the number of localized states) is greater for this compound than for the other two.

4. Conclusions

The obtained results allow us to deduce that the compounds Li_2WO_4 and $Li_{1.5}Na_{0.5}WO_4$ crystallize in the monoclinic system while the compound $Li_{0.5}Na_{1.5}WO_4$ crystallizes in the orthorhombic system. The values obtained from the estimation of the crystallite sizes of these compounds range between 0.2 and 0.1 μm, which is smaller than the grain sizes that were determined from the SEM images, which range from 3 to 5 μm. The SEM images clearly show that the compounds' structures are compact, with a small number of pores resulting from the compounds' tendency to form agglomerations. The EDX analysis verifies that the chemical elements are present in the compositions of the samples. Additionally, we can state that the optical analysis carried out with a UV-Vis spectrophotometer showed that the two compounds $Li_{1.5}Na_{0.5}WO_4$ and Li_2WO_4 each had a single absorption band present at 245 nm, while $Li_{0.5}Na_{1.5}WO_4$ had a single absorption band at 300 nm. In fact, this difference could be explained by the shift in symmetry from orthorhombic to monoclinic. Then, the optical gap energy was calculated using the Kubelka–Munk equation. In addition, the Urbach energy calculation revealed that the compound $Li_{1.5}Na_{0.5}WO_4$ is the most disordered of the three investigated. The dielectric studies of the two compounds also revealed the presence of ferro–paraelectric transitions at 421 K and 418 K for $Li_{0.5}Na_{1.5}WO_4$ and $Li_{1.5}Na_{0.5}WO_4$. Thermal research detected this transition (DSC), which occurs in the parent compounds at about 373 K; where the temperature offset may be due to symmetry differences. Moreover, a comparison study revealed that the conductivity of the compound $Li_{1.5}Na_{0.5}WO_4$ is greater than that of the other two compounds, which may be due to the fact that the disorder and the charge number are higher in the case of $Li_{1.5}Na_{0.5}WO_4$.

Author Contributions: Conceptualization, M.K. (Moufida Krimi) and A.B.R.; methodology, K.K.; software, M.K. (Moufida Krimi) and M.H.A.-H.; validation, M.K., A.B.R. and A.H.A.; formal analysis, M.K. (Moufida Krimi); investigation, A.B.R.; data curation, K.K.; writing—original draft preparation, M.K. (Moufida Krimi) and A.B.R.; writing—review and editing, A.B.R., M.K. (Mohamed Khitouni) and K.K.; supervision, A.B.R. All authors have read and agreed to the published version of the manuscript.

Funding: This research received no external funding.

Data Availability Statement: Data can be requested from the authors.

Acknowledgments: Researchers would like to thank the Deanship of Scientific Research, Qassim University for the funding publication of this project.

Conflicts of Interest: The authors declare no conflict of interest.

References

1. Deb, K.K. Pyroelectric characteristics of $(Pb_{0.9}Sm_{0.1})TiO_3$ ceramics. *Ferroelectrics* **1988**, *82*, 45–53. [CrossRef]
2. Tandon, R.P.; Singh, R.; Singh, V.; Swani, N.H.; Hans, V.K. Ferroelectric properties of lead titanate/polymer composite and its application in hydrophones. *J. Mater. Sci. Lett.* **1992**, *11*, 883–885. [CrossRef]
3. Alvarez-Vega, M.; Rodriguez-Carvajal, J.; Reyes-Cardenas, J.G.; Fuentes, A.F.; Amador, U. A Topological Analysis of Void Spaces in Tungstate Frameworks: Assessing Storage Properties for the Environmentally Important Guest Molecules and Ions: CO_2, UO_2, PuO_2, U, Pu, Sr^{2+}, Cs^+, CH_4, and H_2. *J. Chem. Mater.* **2001**, *13*, 3871–3875.
4. Wong, C.P.P.; Lai, C.W.; Lee, K.M. Tungsten-Based Materials for Supercapacitors. In *Book Inorganic Nanomaterials for Supercapacitor Design*, 1st ed.; CRC Press: Boca Raton, FL, USA, 2020; Volume 11. [CrossRef]
5. Isupov, V.A. Ferroelectric and Ferroelastic Phase Transitions in Molybdates and Tungstates of Monovalent and Bivalent Elements. *Ferroelectrics* **2005**, *322*, 83–114. [CrossRef]
6. Sharma, S.; Choudhary, R.N.P. Phase transition in Li_2WO_4. *Ferroelectrics* **1999**, *4*, 129–137. [CrossRef]
7. Besozzi, E.; Dellasega, D.; Pezzoli, A.; Conti, C.; Passoni, M.; Beghi, M.G. Amorphous, ultra-nano- and nano-crystalline tungsten-based coatings grown by Pulsed Laser Deposition: Mechanical characterization by Surface Brillouin Spectroscopy. *Mater. Des.* **2016**, *106*, 14–21. [CrossRef]
8. Sharma, S.; Choudhary, R.N.P.; Shanigrahi, S.R. Structural and electrical properties of Na_2WO_4 ceramics. *J. Mater. Lett.* **1999**, *40*, 134–139. [CrossRef]
9. Bazarova, Z.G.; Arkhincheyeva, S.I.; Batuyeva, I.S.; Bazarov, B.G.; Tushinova, Y.I.; Bazarova, S.T.; Fyodorov, K.N. Complex Oxide Compounds of Polyvalent Metals: Synthesis, Structure and Properties. *Chem. Sustain. Dev.* **2000**, *8*, 135–139.
10. Nagasaki, T.; Kok, K.; Yahaya, A.H.; Igawa, N.; Noda, K.; Ohno, H. Phase identification and electrical conductivity of Li_2WO_4 Reviewed. *Solid State Ion.* **1997**, *96*, 61–74. [CrossRef]
11. Busey, R.H.; Keller, O.L., Jr. Structure of the Aqueous Pertechnetate Ion by Raman and Infrared Spectroscopy. Raman and Infrared Spectra of Crystalline $KTcO_4$, $KReO_4$, Na_2MoO_4, Na_2WO_4, $Na_2MoO_4 \cdot 2H_2O$, and $Na_2WO_4 \cdot 2H_2O$. *J. Chem. Phys.* **1964**, *41*, 215–225. [CrossRef]
12. Pal, I.; Agarwal, A.; Sanghi, S.; Shearan, A.; Ahlawat, N. Conductivity and dielectric relaxation in sodium borosulfate glasses. *J. Alloys Compd.* **2009**, *472*, 40–45. [CrossRef]
13. Dyre, J.C. The random free-energy barrier model for ac conduction in disordered solids. *J. Appl. Phys.* **1988**, *64*, 2456–2468. [CrossRef]
14. Karoui, K.; Rhaiem, A.B.; Jomni, F.; Moneger, J.L.; Bulou, A.; Guidara, K. Characterization of phase transitions of $[N(CH_3)_4]_2ZnCl_2Br_2$ mixed crystals. *J. Mol. Struct.* **2013**, *1048*, 287–294. [CrossRef]
15. Huebner, J.S.; Dillenburg, R.G. Impedance spectra of hot, dry silicate minerals and rock: Qualitative interpretation of spectra. *J. Am. Mineral.* **1995**, *80*, 46–64. [CrossRef]
16. Imran, M.M.A.; Lafi, O.A. Electrical conductivity, density of states and optical band gap in $Se_{90}Te_6Sn_4$ glassy semiconductor. *Physica B* **2013**, *410*, 201–205. [CrossRef]
17. Bhowmik, R.N.; Vijayasri, G. Study of microstructure and semiconductor to metallic conductivity transition in solid state sintered $Li_{0.5}Mn_{0.5}Fe_2O_4 - \delta$ spinel ferrite. *J. Appl. Phys.* **2013**, *114*, 223701. [CrossRef]
18. Bhowmik, R.N.; Aneesh Kumar, K.S. Role of pH value during material synthesis and grain-grain boundary contribution on the observed semiconductor to metal like conductivity transition in $Ni_{1.5}Fe_{1.5}O_4$ spinel ferrite. *Mater. Chem. Phys.* **2016**, *177*, 417–428. [CrossRef]
19. Saraswat, V.K.; Singh, K.; Saxena, N.S.; Kishore, V.; Sharma, T.P.; Saraswat, P.K. Composition dependence of the electrical conductivity of $Se_{85-x}Te_{15}Sb_x$ ($x = 2, 4, 6, 8$ and 10) glass at room temperature. *Curr. Appl. Phys.* **2006**, *6*, 14–18. [CrossRef]
20. Pradhan, S.K.; Kalidoss, J.; Barik, R.; Sivaiah, B.; Dhar, A.; Bajpai, S. Development of high density tungsten based scandate by Spark Plasma Sintering for the application in microwave tube devices. *Int. J. Refract. Metals Hard Mater.* **2016**, *61*, 215–224. [CrossRef]
21. Namikawa, H. Characterization of the diffusion process in oxide glasses based on the correlation between electric conduction and dielectric relaxation. *J. Non-Cryst. Solids* **1975**, *18*, 173–195. [CrossRef]
22. Gudmundsson, J.T.; Svavarsson, H.G.; Gudjonsson, S.; Gislason, H.P. Frequency-dependent conductivity in lithium-diffused and annealed GaAs. *Phys. B Condens. Matter* **2003**, *340–342*, 324–328. [CrossRef]
23. van den Berg, A.J.; Tuinstra, F.; Warczewski, J. Modulated structures of some alkali molybdates and tungstates. *Acta Cryst. B* **1973**, *29*, 586–589. [CrossRef]
24. Mollah, S.; Som, K.K.; Chaudri, K.B. AC conductivity in $Bi_4Sr_3Ca_3Cu_yO_x$ (y=0–5) and $Bi_4Sr_3Ca_{3-z}Li_zCu4O_x$ (z=0.1–1.0) semiconducting oxide glasses. *J. Appl. Phys.* **1993**, *74*, 931–937. [CrossRef]
25. Hayashi, T.; Okada, J.; Toda, E.; Kuzuo, R.; Matsuda, Y.; Kuwata, N.; Kawamura, J. Electrochemical effect of lithium tungsten oxide modification on $LiCoO_2$ thin film electrode. *J. Power Sources* **2015**, *285*, 559–567. [CrossRef]
26. Radzikhovskaya, M.A.; Garkushin, I.K.; Danilushikina, E.G. Ternary systems $LiBr-Li_2MoO_4-Li_2WO_4$ and $LiF-Li_2MoO_4-Li_2WO_4$. *Russ. J. Inorg. Chem.* **2012**, *57*, 1616–1620. [CrossRef]
27. Barinova, O.; Sadovskiy, A.; Ermochenkov, I.; Kirsanova, S.; Khomyakov, A.; Zykova, M.; Kuchuk, Z.; Avetissov, I. Solid solution $Li_2MoO_4 - Li_2WO_4$ crystal growth and characterization. *Cryst. Growth* **2017**, *468*, 365–368. [CrossRef]

28. Dkhilalli, F.; Megdiche, S.; Guidara, K.; Rasheed, M.; Barillé, R.; Megdiche, M. AC conductivity evolution in bulk and grain boundary response of sodium tungstate Na$_2$WO$_4$. *Ionics* **2017**, *24*, 169–180. [CrossRef]
29. Johan, M.R.; Han, T.K.; Arof, A.K. Growth and sintering effects of hydrated polycrystalline Li$_2$WO$_4$. *Ionics* **2010**, *16*, 323–333. [CrossRef]
30. Luz Lima, C.; Saraiva, G.D.; Freire, P.T.C.; Maczka, M.; Paraguassu, W.; de Sousa, F.F.; Filho, J.M. Temperature-induced phase transformations in Na$_2$WO$_4$ and Na$_2$MoO$_4$ crystals. *Raman Spectrosc.* **2011**, *42*, 799–802. [CrossRef]
31. Zhou, D.; Randall, C.A.; Pang, L.-X.; Wang, H.; Guo, J.; Zhang, G.-Q.; Wu, X.-G.; Shui, L.; Yao, X. Microwave Dielectric Properties of Li$_2$WO$_4$ Ceramic with Ultra-Low Sintering Temperature. *J. Am. Ceram. Soc.* **2011**, *94*, 348–350. [CrossRef]
32. Bárbara, F.; Joséa, E. Antiferromagnetic and Ferroelectric Phase Transitions and Instabilities in PFW-PT Multiferroic Solid Solution Characterized by Anelastic Measurement. *Ferroelectrics* **2013**, *448*, 86–95.
33. Urusova, M.A.; Valyashko, V.M. High-temperature equilibria and critical phenomena in the Na$_2$CO$_3$-NaCl-H$_2$O and Na$_2$CO$_3$-Na$_2$WO$_4$-H$_2$O systems. *Russ. J. Inorg. Chem.* **2011**, *56*, 430–441. [CrossRef]
34. Krimi, M.; Karoui, K.; Sunol, J.J.; Rhaiem, A.B. Phase transition, impedance spectroscopy and conduction mechanism of Li$_{0.5}$Na$_{1.5}$WO$_4$ material. *Phys. E Low-Dimens. Syst. Nanostruct.* **2018**, *102*, 137–145. [CrossRef]
35. Krimi, M.; Karoui, K.; Suñol, J.J.; Rhaiem, A.B. Optical and electrical properties of Li$_2$WO$_4$ compound. *Phase Trans.* **2019**, *92*, 737–754. [CrossRef]
36. Mallah, A.; Al-Thuwayb, F.; Khitouni, M.; Alsawi, A.; Suñol, J.J.; Greneche, J.-M.; Almoneef, M.M. Synthesis, Structural and Magnetic Characterization of Superparamagnetic Ni$_{0.3}$Zn$_{0.7}$Cr$_{2-x}$Fe$_x$O$_4$ Oxides Obtained by Sol-Gel Method. *Crystals* **2023**, *13*, 894. [CrossRef]
37. Warren, B.E. *X-ray Diffraction*; Dover: New York, NY, USA, 1990; pp. 251–275.
38. Williamson, G.K.; Hall, W.H. X-ray line broadening from filed aluminium and wolfram. *Acta Metall.* **1953**, *1*, 22–31. [CrossRef]
39. Kuschke, W.M.; Keller, R.M.; Grahle, P.; Mason, R.; Arzt, E. Mechanisms of Powder Milling Investigated by X-ray Diffraction and Quantitative Metallography. *Z. Metallkd.* **1995**, *86*, 804–813. [CrossRef]
40. Aleksandrov, I.V.; Valiev, R.Z. Studies of Nanocrystalline Materials by X-ray Diffraction Techniques. *Phys. Met. Metallogr.* **1994**, *77*, 623–629.
41. Halder, N.C.; Wagner, C.N. Analysis of the Broadening of Powder Pattern Peaks Using Variance, Integral Breadth, and Fourier Coefficients of the Line Profile. *Adv. X-ray Anal.* **1966**, *9*, 91–102.
42. Daly, R.; Khitouni, M.; Kolsi, A.W.; Njah, N. The studies of crystallite size and microstrains in aluminum powder prepared by mechanical milling. *Phys. Stat. Solidi C* **2006**, *3*, 3325–3331. [CrossRef]
43. Khitouni, M.; Kolsi, A.W.; Njah, N. The effects of boron additions on the disordering and crystallite refinement of NI$_3$Al powders during mechanical milling. *Ann. Chim. Sci. Matériaux* **2003**, *28*, 17–29. [CrossRef]
44. Barhoumi, A.; Leroy, G.; Duponchel, B.; Gest, J.; Yang, L.; Waldhoff, N.; Guermazi, S. Aluminum doped ZnO thin films deposited by direct current sputtering: Structural and optical properties. *Superlattices Microstruct.* **2015**, *82*, 483–498. [CrossRef]
45. Kubelka, P.; Munk, F. An article on optics of paint layers. *Z. Tech. Phys.* **1931**, *12*, 259–274.
46. Bougrine, A.; El Hichou, A.; Addou, M.; Ebothé, J.; Kachouane, A.; Troyon, M. Structural, optical and cathodoluminescence characteristics of undoped and tin-doped ZnO thin films prepared by spray pyrolysis. *J. Mater. Chem. Phys.* **2003**, *80*, 438–445. [CrossRef]
47. Enneffati, M.; Maaloul, N.K.; Louati, B.; Guidara, K. Synthesis, vibrational and UV–visible studies of sodium cadmium orthophosphate. *Opt. Quantum Electron.* **2017**, *49*, 331. [CrossRef]
48. Ben Nasr, W.; Karoui, K.; Bulou, A.; Ben Rhaiem, A. Li$_{1.5}$Rb$_{0.5}$MoO$_4$: Ferroelectric properties and characterization of phase transitions by Raman spectroscopy. *Phys. E Low-Dimens. Syst. Nanostruct.* **2017**, *93*, 339–344. [CrossRef]
49. Uchino, K.; Nomura, S. Critical Exponents of the Dielectric Constants in Diffused-Phase-Transition Crystals. *Ferroelectr. Lett. Sect.* **1982**, *44*, 55–61. [CrossRef]
50. Viehland, D.; Wuttig, M.; Cross, L.E. The glassy behavior of relaxor ferroelectrics. *Ferroelectrics* **2011**, *120*, 71–77. [CrossRef]
51. Tan, Y.-Q.; Yuan, Y.; Hao, Y.-M.; Dong, S.-Y.; Yang, Y.-W. Structure and dielectric properties of Ba$_5$NdCu$_{1.5}$Nb$_{8.5}$O$_{30-\delta}$ tungsten bronze ceramics. *Mater. Res. Bull.* **2013**, *48*, 1934–1938. [CrossRef]
52. Rhaiem, A.; Jomni, F.; Karoui, K.; Guidara, K. Ferroelectric properties of the [N(CH$_3$)$_4$]$_2$CoCl$_2$Br$_2$ compound. *J. Mol. Struct.* **2013**, *1035*, 140–144. [CrossRef]
53. Hajji, R.; Oueslati, A.; Hajlaoui, F.; Bulou, A.; Hlel, F. Structural characterization, thermal, ac conductivity and dielectric properties of (C$_7$H$_{12}$N$_2$)$_2$[SnCl$_6$]Cl$_2$·1.5H$_2$O. *Phase Trans.* **2016**, *89*, 523–542. [CrossRef]
54. Long, A.R. Frequency-Dependent Loss in Amorphous Semiconductors. *J. Adv. Phys.* **1982**, *31*, 553–637. [CrossRef]

Disclaimer/Publisher's Note: The statements, opinions and data contained in all publications are solely those of the individual author(s) and contributor(s) and not of MDPI and/or the editor(s). MDPI and/or the editor(s) disclaim responsibility for any injury to people or property resulting from any ideas, methods, instructions or products referred to in the content.

Article

Theoretical Investigations of the Structural, Dynamical, Electronic, Magnetic, and Thermoelectric Properties of Co*M*RhSi (*M* = Cr, Mn) Quaternary Heusler Alloys

Abdullah Hzzazi [1,2], Hind Alqurashi [3,4], Eesha Andharia [1,3,*], Bothina Hamad [1,5,*] and M. O. Manasreh [6]

1. Department of Physics, University of Arkansas, Fayetteville, AR 72701, USA; ahhzzazi@uark.edu or abdullahhzzazi@gmail.com
2. Basic Sciences Department, King Saud Bin Abdulaziz University for Health Sciences, Riyadh 14611, Saudi Arabia
3. Materials Science and Engineering, University of Arkansas, Fayetteville, AR 72701, USA; halquras@uark.edu
4. Physics Department, College of Science, Al-Baha University, Alaqiq 65779, Saudi Arabia
5. Physics Department, The University of Jordan, Amman 11942, Jordan
6. Department of Electrical Engineering, University of Arkansas, Fayetteville, AR 72701, USA; manasreh@uark.edu
* Correspondence: esandhar@uark.edu (E.A.); bothinah@uark.edu (B.H.)

Abstract: The structural, dynamical, electrical, magnetic, and thermoelectric properties of Co*M*RhSi (*M* = Cr, Mn) quaternary Heusler alloys (QHAs) were investigated using density functional theory (DFT). The Y-type-II crystal structure was found to be the most stable configuration for these QHAs. Both CoCrRhSi and CoMnRhSi alloys possess a half-metallic behavior with a 100% spin-polarization as the majority spin channel is metallic. On the other hand, the minority spin channel is semiconducting with narrow indirect band gaps of 0.54 eV and 0.57 eV, respectively, along the $\Gamma - X$ high symmetry line. In addition, both CoCrRhSi and CoMnRhSi alloys possess a ferromagnetic structure with total magnetic moments of 4 μ_B and 5 μ_B, respectively, which are prominent for spintronics applications. The thermoelectric properties of the subject QHAs were calculated by using Boltzmann transport theory within the constant relaxation time approximation. The lattice thermal conductivities were also evaluated by Slack's equation. The predicted values of the figure-of-merit (ZT) for CoCrRhSi and CoMnRhSi were found to be 0.84 and 2.04 at 800 K, respectively, making them ideal candidates for thermoelectric applications.

Keywords: ab initio investigations; quaternary Heusler; transport coefficients; Slack's equation; ferromagnetic; half-metallic

Citation: Hzzazi, A.; Alqurashi, H.; Andharia, E.; Hamad, B.; Manasreh, M.O. Theoretical Investigations of the Structural, Dynamical, Electronic, Magnetic, and Thermoelectric Properties of Co*M*RhSi (*M* = Cr, Mn) Quaternary Heusler Alloys. *Crystals* 2024, 14, 33. https://doi.org/10.3390/cryst14010033

Academic Editors: Dmitri Donetski, Thomas M. Klapötke and Raphaël P. Hermann

Received: 8 November 2023
Revised: 18 December 2023
Accepted: 25 December 2023
Published: 27 December 2023

Copyright: © 2023 by the authors. Licensee MDPI, Basel, Switzerland. This article is an open access article distributed under the terms and conditions of the Creative Commons Attribution (CC BY) license (https://creativecommons.org/licenses/by/4.0/).

1. Introduction

The shortage of fossil fuels and increased global warming have created a demand for renewable energy sources. Thermoelectric materials, which directly convert waste heat or temperature gradient into electricity, have garnered a great deal of attention in recent years. Based on the range of temperature of operation, conventional thermoelectric materials are classified into three categories: (1) Bi_2Te_3 compounds with for T < 150 °C, (2) TAGS [$(AgSbTe_2)_{1-x}(GeTe)_x$] and PbTe-based compounds for 150 °C < T < 500 °C, and (3) SiGe for T > 500 °C [1]. Over time, half-Heusler alloys have also been investigated for their thermoelectric applications [2]. At the other end of the spectrum, Heusler alloys have also found wide applications in spintronics, a branch of nanoelectronics that relies on the spin of an electron rather than its charge. These devices operate on the principle of quantum tunneling and spin-transfer torque for MRAM applications [3]. Hence, Heusler alloys represent an important class of material for thermoelectric and spintronics applications.

Half-metallic ferromagnetic materials have attracted considerable attention in the last few years owing to their unique physical characteristics. These materials display a metallic behavior near the Fermi energy surface in one spin channel and a semiconducting behavior near the Fermi energy level in the other spin channel, which yields a perfect spin polarization of 100% [4]. As a result, these materials could produce highly spin-polarized currents that could potentially improve the efficiency of spintronics devices. Numerous materials, for instance Heusler alloys, perovskites, and chalcogenides, have been reported to have a half-metallic behavior [5–8]. Several Heusler alloys (HAs) have been investigated theoretically for their half-metallicity and high Curie temperatures [9–11] that were further confirmed experimentally [12,13]. There are three groups of HAs that can be classified based on their chemical structures: full-Heusler alloys (FHAs), half-Heusler alloys (HHAs), and quaternary-Heusler alloys (QHAs). The FHAs have a space group of $Fm\bar{3}m$ with prototype structure, such as Cu_2MnAl, and four interpenetrating cubic lattices. The chemical formula of these alloys is X_2YZ, where X and Y correspond to transition metal atoms and Z refers to an s-p atom [14]. The structure of HHAs is similar to that of FHAs, except for a missing an X atom that leads to a $C1_b$ structure with an XYZ chemical formula and an $F\bar{4}3m$ space group [15]. The chemical formula of the QHAs is XX'YZ with a Y-type structure (LiMgPdSb prototype structure) and a space group of $F\bar{4}3m$ (no. 216) [16,17]. The QHAs have applications in spin-dependent electronics, including spin filters, spin valves, and thermoelectric devices [18,19].

Several studies have indicated that QHAs possess significant thermoelectric properties, such as high transport coefficients (Seebeck coefficient and electrical conductivity), variable lattice thermal conductivity, and a good thermoelectric performance [20–24]. This leads to promising figure of merit (ZT) values where ZT is defined as follows [25]:

$$ZT = \frac{S^2 \sigma T}{(\kappa_e + \kappa_L)}. \quad (1)$$

In this equation, S, σ, T, κ_e, and κ_L refer to the Seebeck coefficient, electrical conductivity, absolute temperature, electronic thermal conductivity, and lattice thermal conductivity, respectively. To obtain a high value of figure of merit ZT, alloys should have a high S and σ values, whereas κ_L should be as low as possible [26].

There are several studies on the subject of the thermoelectric properties of QHAs. For instance, Alqurashi and Hamad predicted ZT values of 1.13, 0.62, and 0.92 for VTiRhSi, VTiRhGe, and VTiRhSn alloys, respectively, at 800 K [24]. Previous computations predicted maximum S values of 44.3 and 53.44 µV/K for CoRuMnAs and CoRhMnAs alloys, respectively [11]. Promising power factor values of 07.56×10^{-5} and 21.05×10^{-5} $Wm^{-1}K^{-2}s^{-1}$ at 300 K were predicted for LaCoCrAl and LaCoCrGa QHAs, respectively [27]. Other calculations predicted ZT values of 0.61 and 0.71 for CoFeTiGe and CoFeCrGe, respectively [28]. Furthermore, the PdZrTiAl alloy was found to be a half-metallic ferromagnet with a 100% spin polarization and a 3 μ_B total magnetic moment [29]. Alqurashi et al. [30] predicted a similar half-metallic ferromagnetic behavior for VTiRhGa and VTiRhIn QHAs with a 100% spin polarization and a 3 μ_B total magnetic moment.

In this work, we report on ab initio investigations based on DFT to calculate the structural, dynamical, electronic, magnetic, and thermoelectric properties of novel CoMRhSi (M = Cr, Mn) quaternary Heusler alloys. To the best of our knowledge, this study investigates and reports the formation of these alloys along with spintronic and thermoelectric properties for the first time. The paper is organized as follows: in Section 2 the computational methodology is described, Section 3 presents the results and discussions, and Section 4 is devoted to the conclusions.

2. Computational Methodology

The structural optimization and energetic calculations are performed using the DFT method as implemented in VASP code [31]. The cut-off energy is selected to be 520 eV, while the total energy tolerance is 10^{-8} eV. The total energy calculations are performed using a $22 \times 22 \times 22$ k-mesh. The formation energy is calculated based on the results of the total energy. The phonon calculations using the Phonopy package are considered to investigate the dynamical stability of the QHAs [32]. For these calculations, a supercell of 4 × 4 × 4 is used with a 4 × 4 × 4 k-mesh and cut-off energy of 500 eV. The supercell of QHAs unit-cell contains 256 atoms. The electronic and magnetic properties were computed within the full-potential linearized augmented plane wave (FP-LAPW) method as implemented in WIEN2k code [33]. The generalized gradient approximation (GGA) within Perdew–Burke–Ernzerhof (PBE) formalism was utilized to treat the exchange–correlation potential [34]. The $K_{max} \times R_{MT}$ value was selected to be 9, where K_{max} is the highest reciprocal lattice vector of the plane wave expansion and R_{MT} is the smallest atomic muffin tin radius. The R_{MT} values were selected as 2.3, 2.2, 2.1 and 1.8 atomic units (a.u.) for Co, M (M = Cr, Mn), Rh, and Si atoms, respectively. The total energy and force tolerances were set to 10^{-4} Ry and 1 mRy/au, respectively. We adopted the VASP pseudopotential code for the structural optimization and WIEN2K full-potential code for the electronic and magnetic properties. For instance, in the case of structural optimization, VASP uses a smaller basis set which leads to faster calculations [35]; whereas, for electronic and magnetic properties, all electron calculations (WIEN2k) give more precise results [36].

The transport coefficients are considered by applying the Boltzmann transport theory as implemented in BoltzTraP code [37,38]. The S, σ, and κ_e parameters are estimated using the following equations [24,39]:

$$S_{\alpha\beta}(T,\mu) = \frac{1}{eT\Omega\sigma_{\alpha\beta}(T,\mu)} \int \overline{\sigma}_{\alpha\beta}(\varepsilon)(\varepsilon-\mu)\left[-\frac{\partial f_0(T,\varepsilon,\mu)}{\partial \varepsilon}\right]d\varepsilon \quad (2)$$

$$\sigma_{\alpha\beta}(T,\mu) = \frac{1}{\Omega} \int \overline{\sigma}_{\alpha\beta}(\varepsilon)\left[-\frac{\partial f_0(T,\varepsilon,\mu)}{\partial \varepsilon}\right]d\varepsilon \quad (3)$$

$$\kappa^0_{\alpha\beta}(T,\mu) = \frac{1}{e^2T\Omega} \int \overline{\sigma}_{\alpha\beta}(\varepsilon)(\varepsilon-\mu)^2\left[-\frac{\partial f_0(T,\varepsilon,\mu)}{\partial \varepsilon}\right]d\varepsilon \quad (4)$$

Here α and β are tensor indices; μ, Ω, and f_0 are the chemical potential, unit cell volume, and the Fermi–Dirac distribution function, respectively. The thermoelectric properties are calculated by using 36 × 36 × 36 centered k-mesh. The σ and κ_e were computed within the constant relaxation time (τ) approximation, which was selected to be 0.5×10^{-15} s. This value was utilized for comparable structures for instance VTiRhZ (Z = Si, Sn, In) and FeRhCrZ (Z = Si, Ge) [24,40]. Slack's formula was used to compute the lattice thermal conductivity (κ_l) as follows [41–43]:

$$\kappa_l = A \frac{\overline{M}\Theta_D^3 V^{1/3}}{\gamma^2 n^{2/3} T}, \quad (5)$$

where, $A = \frac{2.43 \times 10^{-6}}{1 - \frac{0.514}{\gamma} + \frac{0.228}{\gamma^2}}$, and $\overline{M}$, Θ_D, V, γ, n, and T are the average atomic mass, Debye temperature, volume per atom, Grüneisen parameter, number of atoms in the primitive unit cell, and temperature, respectively. The Θ_D and γ are estimated by calculating the elastic constants of CoMRhSi (M = Cr, Mn) QHAs. It is important to note that the Debye temperature can also be calculated from the vibrational density of states spectra, which could lead to a different value of Θ_D. However, there are reports as presented in reference [44] that the results of these two methods give a comparable output.

To compute the elastic constant, the IBRON was selected to be 6 in order to calculate the fourth-order elastic moduli tensor. The above IBRON number uses finite differences to calculate second derivatives of the Hessian matrix (second order derivative of energy with respect to atomic positions) by performing a total of six distortions to the crystal structure. The shear (G) and bulk (B) moduli were calculated based on the Voigt–Reuss–Hill approximations [45,46] that are defined as

Voigt average:

$$G_V = \frac{1}{5}[(C_{11} - C_{12}) + 3C_{44}], \text{ and } B_V = \frac{1}{3}(C_{11} + 2C_{12}), \qquad (6)$$

Reuss average:

$$G_R = \frac{5C_{44}(C_{11} - C_{12})}{3(C_{11} - C_{12}) + 4C_{44}}, \text{ and } B_R = \frac{(C_{11} + C_{12})C_{11} - 2C_{12}^2}{3(C_{11} - C_{12})}, \qquad (7)$$

Hill average:

$$G = \frac{1}{2}(G_V + G_R), \text{ and } B = \frac{1}{2}(B_V + B_R), \qquad (8)$$

Using G and B from Equation (8), Young's modulus (E) and Poisson's ratio (v) are calculated by [45,46] such that

$$E = \left(\frac{9BG}{3B + G}\right), \qquad (9)$$

$$v = \left(\frac{3B - 2G}{2(3B + G)}\right), \qquad (10)$$

Moreover, the Θ_D and (γ) are estimated as follows [45,46]:

$$\Theta_D = \frac{h}{k_B}\left(\frac{3n\rho N_A}{4\pi M}\right)^{1/3} v_m, \qquad (11)$$

$$\gamma = \frac{9 - 12(v_t/v_l)^2}{2 + 4(v_t/v_l)^2}. \qquad (12)$$

The parameters h, ρ, N_A, k_B, and M refer to the Planck constant, density, Avogadro's number, the Boltzmann constant, and the molecular weight, respectively. In addition, v_m, v_l and v_t are the average, transverse, and longitudinal sound velocities, which are given by [47,48] as follows:

$$v_t = \sqrt{\frac{G}{\rho}} \qquad (13)$$

$$v_l = \sqrt{\frac{Y(1 - v)}{\rho(1 + v)(1 - 2v)}} \qquad (14)$$

$$v_a = \left[\frac{1}{3}\left(\frac{2}{v_t^3} + \frac{1}{v_l^3}\right)\right]^{-\frac{1}{3}} \qquad (15)$$

The lattice thermal conductivity (κ_l) values are computed by substituting Θ_D and γ values into Equation (5).

3. Results and Discussions

3.1. Structural Properties

The CoMRhSi (M = Cr, Mn) QHAs have a face-centered cubic LiMgPdSn (Y-type) configuration with (1:1:1:1) stoichiometry and a space group of $F\bar{4}3m$ (no. 216). Figure 1 illustrates the atomic configurations of these QHAs structures, Y-type-I, Y-type-II, and Y-type-III. The Wyckoff positions of the atoms in the three types are 4a (0,0,0), 4c (1/4, 1/4, 1/4), 4b (1/2, 1/2, 1/2), and 4d (3/4, 3/4, 3/4), see Table 1. The values of convex hull for both compounds, as predicted using OQMD, are -1.864 eV/f.u. for CoCrRhSi and -2.08 eV/f.u. for CoMnRhSi. The total energy calculations were performed for both QHAs in their three configuration types. These calculations predicted the Y-type-II configuration as the most stable structure of CoMRhSi (M = Cr, Mn) QHAs, see Table 2. These findings are consistent with prior calculations [49]. To determine if these QHAs can be synthesized experimentally and to confirm their thermodynamical stability, the formation enthalpy per formula unit is calculated using the following equation [50]:

$$E_{form} = E_{tot} - \left(E_{Co}^{bulk} + E_{Rh}^{bulk} + E_{Mn,Cr}^{bulk} + E_{Si}^{bulk}\right), \quad (16)$$

where E_{tot} is the equilibrium total energy per formula unit of the CoCrRhSi and CoMnRhSi alloys, and E_{Co}^{bulk}, E_{Rh}^{bulk}, E_{Y}^{bulk}, and $E_{Z\,=\,Si}^{bulk}$ are the total energies per atom in the bulk structure. In general, the negative value of the formation enthalpy indicates the feasibility of synthesizing the QHAs. The formation enthalpies of the CoCrRhSi and CoMnRhSi alloys are -1.32 and -1.91 eV/f.u., as shown in Table 3. These values agree with other previous calculations of similar structures such as CoNbMnSi (-1.74 eV/f.u.) and CoMoMnSi (-1.90 eV/f.u.) [48]. Moreover, the lattice parameters are found to be 5.78 Å and 5.83 Å for CoCrRhSi and CoMnRhSi as indicated in Table 3. These findings are in good agreement with the previously reported theoretical calculations [51,52].

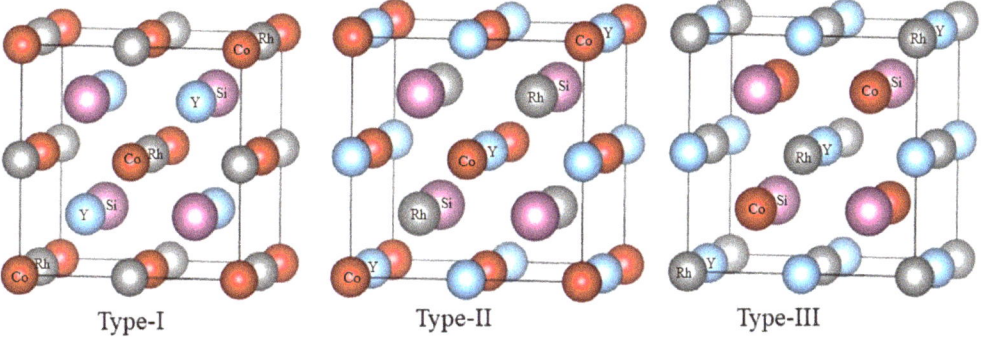

Figure 1. The three different types of the quaternary Heusler alloy primitive cells.

Table 1. The Wyckoff positions (4a, 4c, 4b, and 4d) of the elements in CoMRhSi (M = Cr, Mn) QHAs for three types of configurations.

Y	4a (0,0,0)	4b (1/2, 1/2, 1/2)	4c (1/4, 1/4, 1/4)	4d (3/4, 3/4, 3/4)
Type-I	Co	Rh	Y	Z
Type-II	Co	Y	Rh	Z
Type-III	Rh	Co	Y	Z

Table 2. The total energy in eV of CoMRhSi (M = Cr, Mn) QHAs in the three types of configurations.

Alloys	Type-I	Type-II	Type-III
CoCrRhSi	−29.719	−30.633	−30.145
CoMnRhSi	−29.187	−30.749	−29.918

Table 3. The formation energy E_{form} (eV/f.u.), lattice constant a (Å), elastic constants C_{ij} (GPa), bulk modulus B (GPa), Young's modulus E (GPa), isotropic shear modulus G (GPa), anisotropy factor A, Pugh's ratio B/G, Cauchy pressure C_p (GPa), anisotropy factor A, and the melting temperature T_{melt} (K) of CoMRhSi (M = Mn, Cr) QHAs.

Physical Parameter	CoCrRhSi	CoMnRhSi
E_{form}	−1.32	−1.91
a	5.78	5.83
C_{11}	335.89	334.40
C_{12}	145.23	141.94
C_{44}	58.91	72.14
B	216.88 238.85 [a)]	211.45
E	185.23	216.13
G	68.35	81.29

Table 3. Cont.

Physical Parameter	CoCrRhSi	CoMnRhSi
B/G	3.17 3.39 [a)]	2.60
C_p	86.31	69.05
A	0.61	0.74
T_{melt}	2624	2574

[a)] Ref [43].

3.2. Phonon Calculation

The analysis of the dynamical phonon properties of the CoMRhSi (M = Cr, Mn) alloys is presented here. The phonon dispersion curves (PDCs) provide information about the dynamical stability of the system. These phonon dispersion curves show only positive frequencies, indicating that both alloys are dynamically stable as indicated in Figure 2. The phonon dispersion curves of CoMRhSi (M = Mn, Cr) alloys exhibit twelve phonon branches since the primitive cell consists of four atoms. There are three acoustic phonon modes (one longitudinal (LA) and two transversal acoustic (TA)), and nine optical branches (three longitudinal (LO) and six transverse optical (TO)). These results are in good agreement with previous investigations on CoFeCrGe and CoFeTiGe QHAs [28].

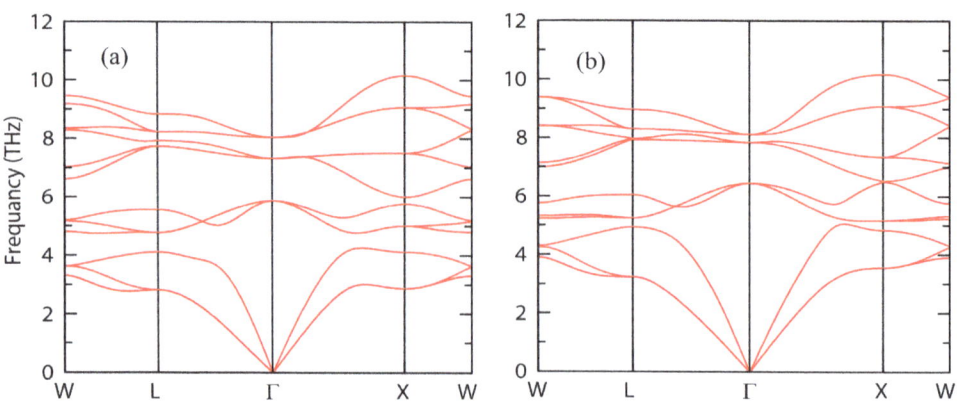

Figure 2. Phonon dispersion relation of (**a**) CoRhCrSi and (**b**) CoRhMnSi QHAs.

3.3. Mechanical Properties

The mechanical properties of CoMRhSi (M = Mn, Cr) alloys are presented by investigating the mechanical stability for the cubic configuration that is achieved by fulfilling the Born–Huang criteria [50] as follows:

$$C_{44} > 0, (C_{11} - C_{12}) > 0, (C_{11} + 2C_{12}) > 0, \text{ and } C_{12} < B < C_{11}, \tag{17}$$

where the three independent elastic constants (C_{11}), (C_{12}), and (C_{44}) are referred to longitudinal compression, transverse expansion, and the share modulus predictor, and B refers to the Bulk modulus. The inequality conditions that are shown in Equation (17) are fulfilled for these alloys and thus the mechanical stability is confirmed. The results are summarized in Table 3. The B values were predicted to be 216.88 and 211.45 GPa for CoRhCrSi and CoRhMnSi alloys.

The three independent elastic constants were used to calculate the shear modulus (G), Young's modulus (E), Cauchy pressure (C_p), Poisson's ratio (ν), and the anisotropy factor (A). The results of these calculations are shown in Table 3. The above G, C_p, and A parameters are defined as follows [50–52]:

$$G = (G_V + G_R)/2 \tag{18}$$

$$C_p = C_{12} - C_{44} \tag{19}$$

$$A = 2C_{44}/(C_{11} - C_{12}) \tag{20}$$

G_R and G_V in Equation (18) are the shear moduli of Reuss, and Voigt. The E (G) values were calculated as 185.23 GPa (68.35 GPa) and 216.13 GPa (81.29 GPa) for CoRhCrSi and CoRhMnSi alloys. The large positive E and G values indicate that these alloys are rigid. These findings are consistent with prior predictions for CoCrScAl, CoCrScSi, CoCrScGe, and CoCrScGa alloys [53]. The Pugh's ratio (B/G) values of CoCrRhSi and CoMnRhSi alloys are 3.17 and 2.60, which are more than the standard value ($B/G > 1.75$ [54]). These results indicate that the alloys have a ductile nature behavior, and are in good agreement with prior calculations [28,50]. Moreover, the Cauchy pressure is calculated to explain the bonding of the alloys. If the Cauchy pressure is negative, the material is classified to have a covalent bonding. For metallic bonding, the Cauchy pressure is positive [50]. According to the present results, the C_p values of the CoCrRhSi and CoMnRhSi alloys were calculated as 86.31 and 69.05 GPa, which indicates a metallic bonding behavior for these alloys. In addition, the anisotropy factor, A, is calculated to test the anisotropy of the materials. If A values are different than unity, the materials are anisotropic [55]. The present calculation for A was found to be less than unity, which indicates the alloys are anisotropic materials. These results agree with previously reported calculations for ZrTiRhGe and ZrTiRhSn [56].

In order to provide an understanding of the heat resistance of the material, the melting point (T_{melt}) is calculated using the following formula [29,54,57]:

$$T_{melt} = \left[553 \text{ K} + \left(\frac{5.91 \text{ K}}{\text{GPa}}\right) C_{11}\right] \pm 300 \text{ K}. \tag{21}$$

This formula shows that the melting temperature depends on the longitudinal compression (C_{11}) of the alloys. As a result, the alloy with the larger longitudinal compression has a higher melting temperature. In this case, the CoCrRhSi alloy possesses a higher melting temperature as indicated in Table 3. The values of T_{melt} are found to be 2624 and 2547 K for CoCrRhSi and CoMnRhSi alloys, which are in good agreement with those obtained previously for CoFeCrGe (2584 K) and CoFeTiGe (2484 K) [28].

3.4. Electronic and Magnetic Properties

The band structure, total density of states (TDOS), and magnetic properties of CoMRhSi (M = Mn, Cr) QHAs are investigated as shown in Figure 3 for the CoMRhSi (M = Mn, Cr) alloys. The results presented in this figure show that the electronic bands of the majority spin channel overlap with the Fermi level, which indicates a metallic behavior. However, the minority spin channel shows a semiconducting behavior with an indirect band gap between the Γ and X high-symmetry points at the valence band maximum (VBM) and the conduction band minimum (CBM), respectively. The band gap values in the minority spin channels were found to be 0.54 eV and 0.57 eV for CoCrRhSi and CoMnRhSi. These results are in agreement with other calculations [48,56–61]. The spin polarization of QHAs can be determined using the following relationship [62,63]:

$$P = \frac{\rho_\uparrow(E_f) - \rho_\downarrow(E_f)}{\rho_\uparrow(E_f) + \rho_\downarrow(E_f)} \times 100, \qquad (22)$$

where $\rho_\uparrow(E_f)$ and $\rho_\downarrow(E_f)$ are the majority and minority spin densities of states at the Fermi level (E_f). Thus, both the CoCrRhSi and CoMnRhSi alloys were found to have a 100% spin polarization due to the absence of the states at the minority spin channel. This is not the case for the majority spin channel, as indicated in Figure 3. The present results for the above alloys are in good agreement the results obtained for CoRhMnSi by Ghosh et al. [63]. This indicates that these alloys may demonstrate potential applications in spintronics.

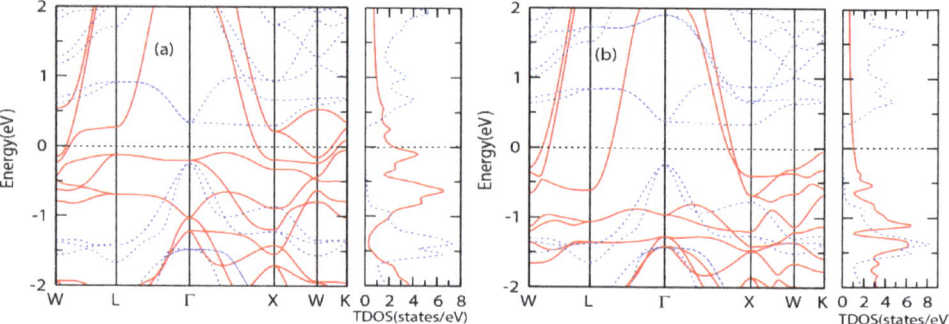

Figure 3. The electronic band structures and total density of states (TDOS) of (**a**) CoCrRhSi and (**b**) CoMnRhSi QHAs. The solid and dotted lines represent the majority and minority spin channels, respectively.

The magnetic structure of the QHAs was tested for three states: ferromagnetic, antiferromagnetic, and paramagnetic. The results show that the ferromagnetic state was found to be the most stable for these alloys. Table 4 presents the total and local magnetic moments of the CoMRhSi QHAs. Both the CoCrRhSi and CoMnRhSi alloys exhibit integer values of 4 μ_B and 5 μ_B. A similar value of the net magnetic moment of 5 μ_B for CoRhMnSi was obtained by Ghosh et al. [64]. These values are found to follow the Slater–Pauling rule for half-metallicity that is described as: $M_{tot} = (Z_{tot} - 24)\ \mu_B$, where M_{tot} and Z_{tot} refer to the total magnetic moment and the number of the total valence electrons. The significant contribution of the total magnetic moment of 2.41 and 3.31 μ_B for the CoCrRhSi and CoMnRhSi alloys comes from the Cr and Mn atoms. The Co, Rh, and M (M = Cr, and Mn) atoms have a ferromagnetic coupling between their local magnetic moments in the two QHAs.

Table 4. The calculated band gap values E_g (eV), spin polarization P (%), total magnetic moment m_{total} (μ_B), local magnetic moments per atom m_i (μ_B) (i = Co, Cr, Mn Rh, Z) for the CoRhMSi (M = Cr, Mn) alloys.

Compound	E_g (eV)	P (%)	m_{Co} [μ_B]	m_{Rh} [μ_B]	m_Y [μ_B]	m_{Si} [μ_B]	m_{total} [μ_B]
CoCrRhSi	0.54 (minority)	100	1.15	0.36	2.41	−0.02	4.00 / 4.00 [a]
CoMnRhSi	0.57 (minority)	100	1.25	0.44	3.31	−0.02	5.00

[a] Ref [43].

The linear relationship between the Curie temperature (T_C) and the total magnetic moment is one of the methods that can be used to calculate T_C as follows [65–67]:

$$T_C = 23 + 181 M_{tot}, \tag{23}$$

where, M_{tot} is the total magnetic moment. Based on this equation, the T_C values of the CoCrRhSi and CoMnRhSi alloys are found to be 747 and 928 K, which means the ferromagnetic structure is retained for temperatures much higher than room temperature. However, it is important to note that this is just one of the methods to estimate the T_C and, very often, the results deviate from the experimental output. Hence, these values just serve as an estimate of the actual values of the transition temperature.

3.5. Transport Properties and ZT

The transport properties were calculated for the CoMRhSi (M = Mn, Cr) QHAs' stable structures. The calculations were performed using Boltzmann transport theory with a constant relaxation time approximation [38]. The total S and σ of the majority and minority spin channels are calculated by using the two-current model as [27]:

$$S = \frac{S_\uparrow \sigma_\uparrow + S_\downarrow \sigma_\downarrow}{\sigma_\uparrow + \sigma_\downarrow}. \tag{24}$$

Here $S_\uparrow$ ($S_\downarrow$) and $\sigma_\uparrow$ ($\sigma_\downarrow$) represent the Seebeck coefficient and electrical conductivity for the majority (minority) spin channels.

The results of the total Seebeck coefficient are shown in Figure 4a,b as a function of the chemical potential (E-E_f) at 300 and 800 K. The figure shows that the total S values of these alloys increase by increasing the temperature. The CoMnRhSi alloy has higher values of the total S than those of the CoCrRhSi alloy. The values of the total electrical conductivity (σ) as a function of (E-E_f) at 300 and 800 K are depicted in Figure 4c,d. As seen in the figures, the n-type doping level shows higher σ values than those of the p-type. Moreover, the temperature effect on σ values appear to be minimal. Figure 4e,f exhibit the power factor (PF) as a function of (E-E_f) at 300 and 800 K. The PF values are found to be higher as the temperature increases with maximum values of 20.2×10^{11} and 31.1×10^{11} (W/m K^2 s) for the CoCrRhSi and CoMnRhSi alloys at 800 K.

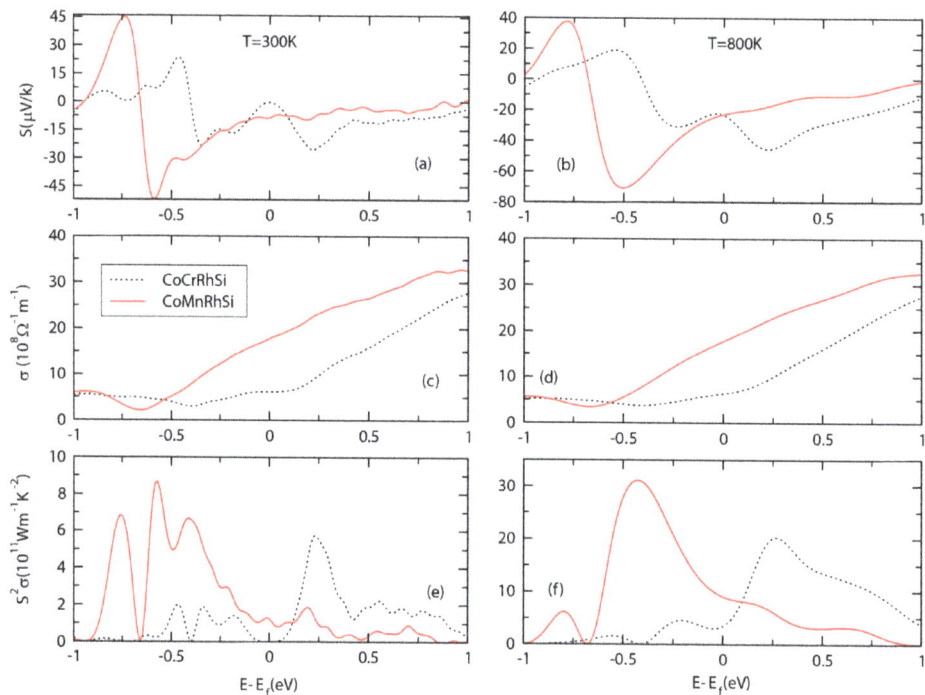

Figure 4. (**a**,**b**) The Seebeck coefficient (S), (**c**,**d**) electrical conductivity (σ), and (**e**,**f**) power factor ($S^2\sigma$) as a function of the chemical potential at temperatures of 300 K, 800 K for the CoRhMSi (M = Mn, Cr) QHAs.

Figure 5a,b present the electronic thermal conductivity (κ_e) of CoMRhSi (M = Mn, Cr) QHAs as a function of (E-E_f) at 300 and 800 K. From these figures, one can notice that the κ_e values of the CoMnRhSi alloy are higher than those of the CoCrRhSi alloy. The values of the lattice thermal conductivity (κ_l) of CoMRhSi (M = Mn, Cr) QHAs are calculated based on the Slack equation. The computed parameters Θ_D, γ, v_m, v_t, and v_l are presented in Table 5. The Θ_D (γ) values of the CoCrRhSi and CoMnRhSi alloys are found to be 420.78 (2.19) and 454.28 K (1.97). These results are in good agreement with the previous calculation [46]. Figure 5c presents κ_l as a function of temperature. The κ_l values are found to be 1.10 and 0.69 W/m·K at 800 K for the CoCrRhSi and CoMnRhSi alloys, which are significantly low as compared to those of other QHAs such as CoFeCrGe (11.01 W/m·K and CoFeTiGe (12.26 W/m·K) [28]. Figure 6 show the figure of merit ZT values as a function of (E-E_f) at 300 K and 800 K. The CoRhMSi (M = Mn, Cr) alloys have higher ZT values at 800 than those at 300 K. The highest ZT values at 800 K are the n-type 0.84 and the p-type 2.04 for CoCrRhSi and CoMnRhSi, respectively.

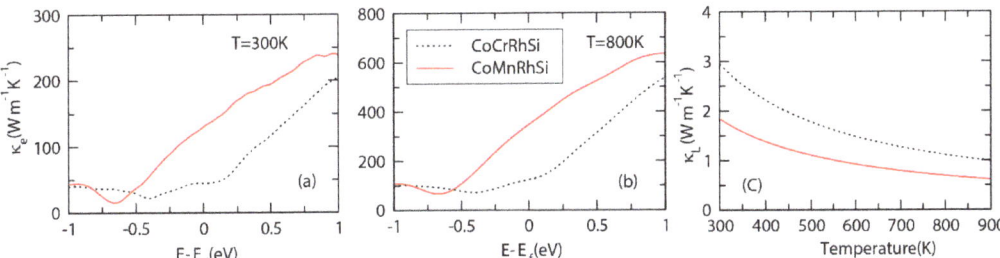

Figure 5. (**a**,**b**) The electronic thermal conductivity (κ_e) as a function of the temperature for CoRhMSi (M = Mn, Cr) as a function of the chemical potential at 300 K and 800 K, respectively. (**c**) The lattice thermal conductivity (κ_l) as a function of temperature for the CoRhMSi (M = Mn, Cr) alloys.

Table 5. The Debye temperature Θ_D (K), average sound velocity v_m (m/s), transverse sound velocity v_t (m/s), longitudinal sound velocity v_l (m/s), density ρ (kg/m^3), and Grüneisen parameter γ for CoRhMSi (M = Mn, Cr) QHAs.

Alloys	Θ_D	v_m	v_t	v_l	ρ	γ
CoCrRhSi	420.78	3274.84	2909.84	6176.90	8186.08	2.19
CoMnRhSi	454.28	3533.46	3151.32	6250.75	8073.10	1.97

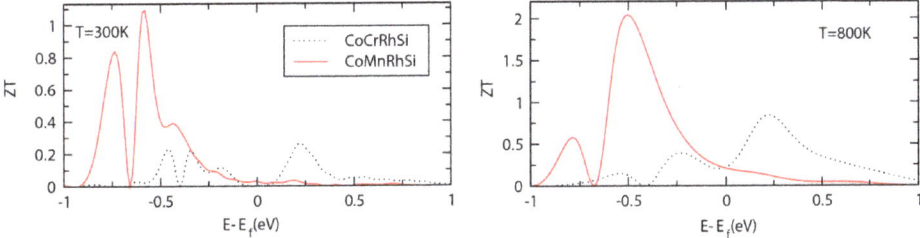

Figure 6. Calculated figure of merit (ZT) as a function of the chemical potential at (300 K, 800 K) for the CoRhMSi (M = Mn, Cr) alloys.

4. Conclusions

Density functional theory calculations were performed to investigate the structural, thermodynamic, dynamical, mechanical, electronic, magnetic, and thermoelectric properties of CoRhMSi (M = Mn, Cr) QHAs. Based on the total energy, phonon and elastic constant calculations, the Y-Type-II structure of CoRhMSi (M = Mn, Cr) QHAs was found to be the most stable configuration. Both CoCrRhSi and CoMnRhSi QHAs are predicted to be half-metallic with indirect band gaps of 0.542 and 0.576 eV in the minority spin channel, which yields a 100% spin polarization. The total magnetic moment of the CoCrRhSi and CoMnRhSi alloys was found to be 4.00 and 5.00 μ_B. In addition, the Curie temperatures of the CoCrRhSi and CoMnRhSi alloys were calculated to be 747 and 928 K. The half metallicity and the ferromagnetic structure with Curie temperatures higher than the room temperatures of CoRhMSi (M = Mn, Cr) QHAs appear to be promising for spintronic applications.

The thermoelectric properties for CoMRhSi (M = Mn, Cr) QHAs were obtained using the semi-classical Boltzmann transport theory. The maximum power factor values for CoCrRhSi and CoMnRhSi QHAs at 800 K are 20.2×10^{11} and 31.1×10^{11} (W/m K^2 s). The highest ZT values of the CoMnRhSi and CoCrRhSi alloys were also found to be 2.04 and 0.84 at 800 K. As a result, these alloys have the potential to be utilized in high-temperature thermoelectric applications.

Author Contributions: A.H.: validation, investigation, writing—original draft. H.A.: conceptualization, methodology, investigation, software, formal analysis, writing—review and editing. E.A.: investigation, writing—review and editing. B.H.: supervision, project administration, writing—review and editing. M.O.M.: writing—review and editing. All authors have read and agreed to the published version of the manuscript.

Funding: This research received no external funding.

Data Availability Statement: Data are contained within the article.

Acknowledgments: Abdullah Hzzazi and Hind Alqurashi were financially supported by the Saudi Arabian Cultural Mission. The calculations were performed at the high-performance computing center, University of Arkansas. This work is partially supported by the Open Access Publishing Fund administered through the University of Arkansas Libraries.

Conflicts of Interest: The authors declare that they have no conflicts of interest.

References

1. Zoui, M.A.; Bentouba, S.; Stocholm, J.G.; Bourouis, M. A Review on Thermoelectric Generators: Progress and Applications. *Energies* **2020**, *13*, 3606. [CrossRef]
2. Quinn, R.J.; Bos, J.-W.G. Advances in half-Heusler alloys for thermoelectric power generation. *Mater. Adv.* **2021**, *2*, 6246–6266. [CrossRef]
3. Peng, S.Z.; Zhang, Y.; Wang, M.X.; Zhang, Y.G.; Zhao, W. Magnetic Tunnel Junctions for Spintronics: Principles and Applications. In *Wiley Encyclopedia of Electrical and Electronics Engineering*; Wiley: Hoboken, NJ, USA, 2014; pp. 1–16. [CrossRef]
4. Oogane, M.; Sakuraba, Y.; Nakata, J.; Kubota, H.; Ando, Y.; Sakuma, A.; Miyazaki, T. Large tunnel magnetoresistance in magnetic tunnel junctions using Co 2MnX (X $\leq$ Al, Si) Heusler alloys. *J. Phys. D Appl. Phys.* **2006**, *39*, 834–841. [CrossRef]
5. Xie, W.-H.; Liu, B.-G.; Pettifor, D.G. Half-metallic ferromagnetism in transition metal pnictides and chalcogenides with wurtzite structure. *Phys. Rev. B* **2003**, *68*, 134407. [CrossRef]
6. Khandy, S.A.; Gupta, D.C. Investigation of the transport, structural and mechanical properties of half-metallic REMnO$_3$ (RE = Ce and Pr) ferromagnets. *RSC Adv.* **2016**, *6*, 97641–97649. [CrossRef]
7. Mir, S.A.; Gupta, D.C. Understanding the origin of half-metallicity and thermophysical properties of ductile La$_2$CuMnO$_6$ double perovskite. *Int. J. Energy Res.* **2019**, *43*, 4783–4796. [CrossRef]
8. Şaşioğlu, E.; Galanakis, I.; Sandratskii, L.M.; Bruno, P. Stability of ferromagnetism in the half-metallic pnictides and similar compounds: A first-principles study. *J. Phys. Condens. Matter* **2005**, *17*, 3915–3930. [CrossRef] [PubMed]
9. Graf, T.; Felser, C.; Parkin, S.S. Simple rules for the understanding of Heusler compounds. *Prog. Solid State Chem.* **2011**, *39*, 1–50. [CrossRef]
10. Wurmehl, S.; Fecher, G.H.; Kandpal, H.C.; Ksenofontov, V.; Felser, C.; Lin, H.J.; Morais, J. Geometric, electronic, and magnetic structure of Co$_2$FeSi: Curie temperature and magnetic moment measurements and calculations. *Phys. Rev. B Condens. Matter Mater. Phys.* **2005**, *72*, 184434. [CrossRef]
11. Alsayegh, S.; Alqurashi, H.; Andharih, E.; Hamad, B.; Manasreh, M. First-principal investigations of the electronic, magnetic, and thermoelectric properties of CrTiRhAl quaternary Heusler alloy. *J. Magn. Magn. Mater.* **2023**, *568*, 170421. [CrossRef]
12. Ishida, S.; Masaki, T.; Fujii, S.; Asano, S. Theoretical search for half-metallic films of Co$_2$MnZ (Z = Si, Ge). *Phys. B Condens. Matter* **1998**, *245*, 1–8. [CrossRef]
13. Sakuraba, Y.; Nakata, J.; Oogane, M.; Kubota, H.; Ando, Y.; Sakuma, A.; Miyazaki, T. Huge spin-polarization of L21-ordered Co2MnSi epitaxial heusler alloy film. *Jpn. J. Appl. Phys. Part 2 Lett.* **2005**, *44*, L1100. [CrossRef]
14. Seh, A.Q.; Gupta, D.C. Exploration of highly correlated Co-based quaternary Heusler alloys for spintronics and thermoelectric applications. *Int. J. Energy Res.* **2019**, *43*, 8864–8877. [CrossRef]
15. Haleoot, R.; Hamad, B. Ab Initio Investigations of the Structural, Electronic, Magnetic, and Thermoelectric Properties of CoFeCuZ (Z = Al, As, Ga, In, Pb, Sb, Si, Sn) Quaternary Heusler Alloys. *J. Electron. Mater.* **2019**, *48*, 1164–1173. [CrossRef]
16. Bainsla, L.; Suresh, K.G. Equiatomic quaternary Heusler alloys: A material perspective for spintronic applications. *Appl. Phys. Rev.* **2016**, *3*, 031101. [CrossRef]
17. Andharia, E.; Alqurashi, H.; Hamad, B. Lattice Dynamics, Mechanical Properties, Electronic Structure and Magnetic Properties of Equiatomic Quaternary Heusler Alloys CrTiCoZ (Z = Al, Si) Using First Principles Calculations. *Materials* **2022**, *15*, 3128. [CrossRef] [PubMed]
18. Yousuf, S.; Gupta, D.C. Thermoelectric and mechanical properties of gapless Zr$_2$MnAl compound. *Indian J. Phys.* **2017**, *91*, 33–41. [CrossRef]
19. Takahashi, Y.K.; Kasai, S.; Hirayama, S.; Mitani, S.; Hono, K. All-metallic lateral spin valves using Co$_2$Fe(Ge$_{0.5}$Ga$_{0.5}$) Heusler alloy with a large spin signal. *Appl. Phys. Lett.* **2012**, *100*, 052405. [CrossRef]
20. Zhang, L.; Cheng, Z.X.; Wang, X.T.; Khenata, R.; Rozale, H. First-Principles Investigation of Equiatomic Quaternary Heusler Alloys NbVMnAl and NbFeCrAl and a Discussion of the Generalized Electron-Filling Rule. *J. Supercond. Nov. Magn.* **2018**, *31*, 189–196. [CrossRef]

21. Eliassen, S.N.H.; Katre, A.; Madsen, G.K.H.; Persson, C.; Løvvik, O.M.; Berland, K. Lattice thermal conductivity of Ti$_x$Zr$_y$Hf$_{1-x-y}$NiSn half-Heusler alloys calculated from first principles: Key role of nature of phonon modes. *Phys. Rev. B* **2017**, *95*, 045202. [CrossRef]
22. Lue, C.S.; Chen, C.F.; Lin, J.Y.; Yu, Y.T.; Kuo, Y.K. Thermoelectric properties of quaternary Heusler alloys Fe$_2$VAl$_{1-x}$Si$_x$. *Phys. Rev. B Condens. Matter Mater. Phys.* **2007**, *75*, 064204. [CrossRef]
23. Yabuuchi, S.; Okamoto, M.; Nishide, A.; Kurosaki, Y.; Hayakawa, J. Large Seebeck Coefficients of Fe$_2$TiSn and Fe$_2$TiSi: First-Principles Study. *Appl. Phys. Express* **2013**, *6*, 025504. [CrossRef]
24. Alqurashi, H.; Hamad, B. Magnetic structure, mechanical stability and thermoelectric properties of VTiRhZ (Z = Si, Ge, Sn) quaternary Heusler alloys: First-principles calculations. *Appl. Phys. A Mater. Sci. Process.* **2021**, *127*, 1–11. [CrossRef]
25. Alqurashi, H.; Haleoot, R.; Pandit, A.; Hamad, B. Investigations of the electronic, dynamical, and thermoelectric properties of Cd$_{1-x}$Zn$_x$O alloys: First-principles calculations. *Mater. Today Commun.* **2021**, *28*, 102511. [CrossRef]
26. Kraemer, D.; Poudel, B.; Feng, H.-P.; Caylor, J.C.; Yu, B.; Yan, X.; Ma, Y.; Wang, X.; Wang, D.; Muto, A.; et al. High-performance flat-panel solar thermoelectric generators with high thermal concentration. *Nat. Mater.* **2011**, *10*, 532–538. [CrossRef] [PubMed]
27. Singh, S.; Gupta, D.C. Lanthanum based quaternary Heusler alloys LaCoCrX (X = Al, Ga): Hunt for half-metallicity and high thermoelectric efficiency. *Results Phys.* **2019**, *13*, 102300. [CrossRef]
28. Haleoot, R.; Hamad, B. Thermodynamic and thermoelectric properties of CoFeYGe (Y = Ti, Cr) quaternary Heusler alloys: First principle calculations. *J. Phys. Condens. Matter* **2019**, *32*, 075402. [CrossRef] [PubMed]
29. Ilkhani, M.; Boochani, A.; Amiri, M.; Asshabi, M.; Rai, D.P. Mechanical stability and thermoelectric properties of the PdZrTiAl quaternary Heusler: A DFT study. *Solid State Commun.* **2020**, *308*, 113838. [CrossRef]
30. Alqurashi, H.; Haleoot, R.; Hamad, B. First-principles investigations of the electronic, magnetic and thermoelectric properties of VTiRhZ (Z = Al, Ga, In) Quaternary Heusler alloys. *Mater. Chem. Phys.* **2021**, *278*, 125685. [CrossRef]
31. Kresse, G.; Joubert, D. From ultrasoft pseudopotentials to the projector augmented-wave method. *Phys. Rev. B* **1999**, *59*, 1758–1775. [CrossRef]
32. Kresse, G.; Hafner, J. Ab initio molecular dynamics for liquid metals. *Phys. Rev. B* **1993**, *47*, 558–561. [CrossRef] [PubMed]
33. Blaha, P.; Schwarz, K.; Sorantin, P.; Trickey, S. Full-potential, linearized augmented plane wave programs for crystalline systems. *Comput. Phys. Commun.* **1990**, *59*, 399–415. [CrossRef]
34. Perdew, J.P.; Burke, K.; Ernzerhof, M. Generalized gradient approximation made simple. *Phys. Rev. Lett.* **1996**, *77*, 3865–3868. [CrossRef] [PubMed]
35. Probert, M.I.J.; Hasnip, P.J.; Lejaeghere, K.; Bihlmayer, G.; Bjorkman, T.; Blaha, P.; Blum, V.; Caliste, D.; Castelli, I.E.; Dal Corso, A.; et al. Reproducibility in density functional theory calculations of solids. *Science* **2016**, *351*, 1415. [CrossRef]
36. Kishore, M.R.A.; Okamoto, H.; Patra, L.; Vidya, R.; Sjåstad, A.O.; Fjellvåg, H.; Ravindran, P. Theoretical and experimental investigation on structural, electronic and magnetic properties of layered Mn$_5$O$_8$. *Phys. Chem. Chem. Phys.* **2016**, *18*, 27885–27896. [CrossRef] [PubMed]
37. Madsen, G.K.; Carrete, J.; Verstraete, M.J. BoltzTraP2, a program for interpolating band structures and calculating semi-classical transport coefficients. *Comput. Phys. Commun.* **2018**, *231*, 140–145. [CrossRef]
38. Madsen, G.K.; Singh, D.J. BoltzTraP. A code for calculating band-structure dependent quantities. *Comput. Phys. Commun.* **2006**, *175*, 67–71. [CrossRef]
39. Lin, T.; Gao, Q.; Liu, G.; Dai, X.; Zhang, X.; Zhang, H. Dynamical stability, electronic and thermoelectric properties of quaternary ZnFeTiSi Heusler compound. *Curr. Appl. Phys.* **2019**, *19*, 721–727. [CrossRef]
40. Khandy, S.A.; Chai, J.-D. Thermoelectric properties, phonon, and mechanical stability of new half-metallic quaternary Heusler alloys: FeRhCrZ (Z = Si and Ge). *J. Appl. Phys.* **2020**, *127*, 165102. [CrossRef]
41. Hong, A.J.; Li, L.; He, R.; Gong, J.J.; Yan, Z.B.; Wang, K.F.; Liu, J.M.; Ren, Z.F. Full-scale computation for all the thermoelectric property parameters of half-Heusler compounds. *Sci. Rep.* **2016**, *6*, 22778. [CrossRef]
42. Ma, H.; Yang, C.-L.; Wang, M.-S.; Ma, X.-G.; Yi, Y.-G. Effect of M elements (M = Ti, Zr, and Hf) on thermoelectric performance of the half-Heusler compounds MCoBi. *J. Phys. D Appl. Phys.* **2019**, *52*, 25550. [CrossRef]
43. Slack, G. Nonmetallic crystals with high thermal conductivity. *J. Phys. Chem. Solids* **1973**, *34*, 321–335. [CrossRef]
44. Chen, Q.; Sundman, B. Calculation of debye temperature for crystalline structures—A case study on Ti, Zr, and Hf. *Acta Mater.* **2001**, *49*, 947–961. [CrossRef]
45. Hill, R. The Elastic Behaviour of a Crystalline Aggregate. *Proc. Phys. Soc. Sect. A* **1952**, *65*, 349. [CrossRef]
46. Horner, H. Lattice dynamics of quantum crystals. *Z. Phys.* **1967**, *205*, 72–89. [CrossRef]
47. Anderson, O.L. A simplified method for calculating the debye temperature from elastic constants. *J. Phys. Chem. Solids* **1963**, *24*, 909–917. [CrossRef]
48. Xiong, X.; Wan, R.; Zhang, Z.; Lei, Y.; Tian, G. First-principle investigation on the thermoelectric properties of XCoGe (X = V, Nb, and Ta) half-Heusler compounds. *Mater. Sci. Semicond. Process.* **2022**, *140*, 106387. [CrossRef]
49. Hoat, D.; Hoang, D.-Q.; Binh, N.T.; Naseri, M.; Rivas-Silva, J.; Kartamyshev, A.; Cocoletzi, G.H. First principles analysis of the half-metallic ferromagnetism, elastic and thermodynamic properties of equiatomic quaternary Heusler compound CoCrRhSi. *Mater. Chem. Phys.* **2021**, *257*, 123695. [CrossRef]
50. Kundu, A.; Ghosh, S.; Banerjee, R.; Ghosh, S.; Sanyal, B. New quaternary half-metallic ferromagnets with large Curie temperatures. *Sci. Rep.* **2017**, *7*, 1803. [CrossRef]

51. Zhao, J.-S.; Gao, Q.; Li, L.; Xie, H.-H.; Hu, X.-R.; Xu, C.-L.; Deng, J.-B. First-principles study of the structure, electronic, magnetic and elastic properties of half-Heusler compounds LiXGe (X = Ca, Sr and Ba). *Intermetallics* **2017**, *89*, 65–73. [CrossRef]
52. Wang, X.; Cheng, Z.; Wang, J.; Liu, G. A full spectrum of spintronic properties demonstrated by a $C1_b$-type Heusler compound Mn_2Sn subjected to strain engineering. *J. Mater. Chem. C* **2016**, *4*, 8535–8544. [CrossRef]
53. Berri, S. Computational Study of Structural, Electronic, Elastic, Half-Metallic and Thermoelectric Properties of CoCrScZ (Z = Al, Si, Ge, and Ga) Quaternary Heusler Alloys. *J. Supercond. Nov. Magn.* **2020**, *33*, 3809–3818. [CrossRef]
54. Chen, X.-R.; Zhong, M.-M.; Feng, Y.; Zhou, Y.; Yuan, H.-K.; Chen, H. Structural, electronic, elastic, and thermodynamic properties of the spin-gapless semiconducting Mn_2 CoAl inverse Heusler alloy under pressure. *Phys. Status Solidi (b)* **2015**, *252*, 2830–2839. [CrossRef]
55. Semari, F.; Boulechfar, R.; Dahmane, F.; Abdiche, A.; Ahmed, R.; Naqib, S.; Bouhemadou, A.; Khenata, R.; Wang, X. Phase stability, mechanical, electronic and thermodynamic properties of the Ga3Sc compound: An ab-initio study. *Inorg. Chem. Commun.* **2020**, *122*, 108304. [CrossRef]
56. Alqurashi, H.; Haleoot, R.; Hamad, B. First-principles investigations of Zr-based quaternary Heusler alloys for spintronic and thermoelectric applications. *Comput. Mater. Sci.* **2022**, *210*, 111477. [CrossRef]
57. Rached, H. Prediction of a new quaternary Heusler alloy within a good electrical response at high temperature for spintronics applications: DFT calculations. *Int. J. Quantum Chem.* **2021**, *121*, e26647. [CrossRef]
58. Benkabou, M.; Rached, H.; Abdellaoui, A.; Rached, D.; Khenata, R.; Elahmar, M.; Abidri, B.; Benkhettou, N.; Bin-Omran, S. Electronic structure and magnetic properties of quaternary Heusler alloys CoRhMnZ (Z = Al, Ga, Ge and Si) via first-principle calculations. *J. Alloys Compd.* **2015**, *647*, 276–286. [CrossRef]
59. Dag, T.S.; Gencer, A.; Ciftci, Y.; Surucu, G. Equiatomic quaternary CoXCrAl (X = V, Nb, and Ta) Heusler compounds: Insights from DFT calculations. *J. Magn. Magn. Mater.* **2022**, *560*, 169620. [CrossRef]
60. Elahmar, M.; Rached, H.; Rached, D.; Benalia, S.; Khenata, R.; Biskri, Z.; Bin Omran, S. Structural stability, electronic structure and magnetic properties of the new hypothetical half-metallic ferromagnetic full-Heusler alloy CoNiMnSi. *Mater. Sci.* **2016**, *34*, 85–93. [CrossRef]
61. Gencer, A.; Surucu, O.; Usanmaz, D.; Khenata, R.; Candan, A.; Surucu, G. Equiatomic quaternary Heusler compounds TiVFeZ (Z = Al, Si, Ge): Half-metallic ferromagnetic materials. *J. Alloys Compd.* **2021**, *883*, 160869. [CrossRef]
62. Rached, H.; Rached, D.; Khenata, R.; Abidri, B.; Rabah, M.; Benkhettou, N.; Omran, S.B. A first principle study of phase stability, electronic structure and magnetic properties for $Co_{2-x}Cr_xMnAl$ Heusler alloys. *J. Mag. Mag. Mater.* **2015**, *379*, 84. [CrossRef]
63. Bourachid, I.; Rached, D.; Rached, H.; Bentouaf, A.; Rached, Y.; Caid, M.; Abidri, B. Magneto-electronic and thermoelectric properties of V-based Heusler in ferrimagnetic phase. *Appl. Phys. A* **2022**, *128*, 493. [CrossRef]
64. Ghosh, S.; Ghosh, S. Site dependent substitution and half-metallic behaviour in Heusler compounds: A case study for Mn_2RhSi, Co_2RhSi and CoRhMnSi. *Comput. Condens. Matter.* **2019**, *21*, e00423. [CrossRef]
65. Candan, A.; Uğur, G.; Charifi, Z.; Baaziz, H.; Ellialtıoğlu, M. Electronic structure and vibrational properties in cobalt-based full-Heusler compounds: A first principle study of Co_2MnX (X = Si, Ge, Al, Ga). *J. Alloys Compd.* **2013**, *560*, 215–222. [CrossRef]
66. Elahmar, M.; Rached, H.; Rached, D.; Khenata, R.; Murtaza, G.; Bin Omran, S.; Ahmed, W. Structural, mechanical, electronic and magnetic properties of a new series of quaternary Heusler alloys CoFeMnZ (Z = Si, As, Sb): A first-principle study. *J. Magn. Magn. Mater.* **2015**, *393*, 165–174. [CrossRef]
67. Jain, R.; Jain, V.K.; Chandra, A.R.; Jain, V.; Lakshmi, N. Study of the Electronic Structure, Magnetic and Elastic Properties and Half-Metallic Stability on Variation of Lattice Constants for CoFeCrZ (Z = P, As, Sb) Heusler Alloys. *J. Supercond. Nov. Magn.* **2018**, *31*, 2399–2409. [CrossRef]

Disclaimer/Publisher's Note: The statements, opinions and data contained in all publications are solely those of the individual author(s) and contributor(s) and not of MDPI and/or the editor(s). MDPI and/or the editor(s) disclaim responsibility for any injury to people or property resulting from any ideas, methods, instructions or products referred to in the content.

MDPI AG
Grosspeteranlage 5
4052 Basel
Switzerland
Tel.: +41 61 683 77 34

Crystals Editorial Office
E-mail: crystals@mdpi.com
www.mdpi.com/journal/crystals

Disclaimer/Publisher's Note: The statements, opinions and data contained in all publications are solely those of the individual author(s) and contributor(s) and not of MDPI and/or the editor(s). MDPI and/or the editor(s) disclaim responsibility for any injury to people or property resulting from any ideas, methods, instructions or products referred to in the content.

www.ingramcontent.com/pod-product-compliance
Lightning Source LLC
LaVergne TN
LVHW070402100526
838202LV00014B/1371